Perilous Waters

ANTHONY E. CARLSON

Perilous Waters
Settlers, Swamps, and the State, 1775–1920

The University of North Carolina Press *Chapel Hill*

© 2026 The University of North Carolina Press
All rights reserved
Set in Arno Pro by Westchester Publishing Services
Manufactured in the United States of America

Library of Congress Cataloging-in-Publication Data
Names: Carlson, Anthony E. author
Title: Perilous waters : settlers, swamps, and the state, 1775–1920 /
  Anthony E. Carlson.
Description: Chapel Hill : The University of North Carolina Press, 2026. |
  Includes bibliographical references and index.
Identifiers: LCCN 2025045127 | ISBN 9781469694795 cloth | ISBN 9781469694801 paperback |
  ISBN 9781469687285 epub | ISBN 9781469694818 pdf
Subjects: LCSH: Reclamation of land—Government policy—United
  States—History | Reclamation of land—West (U.S.)—History |
  Agriculture and state | Swamps—Political aspects—United States |
  Settler colonialism—United States—History | BISAC: HISTORY / United
  States / 19th Century | TECHNOLOGY & ENGINEERING / Agriculture / Irrigation
Classification: LCC HD1683.U4 C37 2026
LC record available at https://lccn.loc.gov/2025045127

Cover art: *The Land of Evangeline* by Joseph Rusling Meeker, 1874, courtesy of Wikimedia Commons. Map of Humboldt County, Iowa, from *A. T. Andreas's Illustrated Historical Atlas of the State of Iowa, 1875*, by Alfred Theodore Andreas (Andreas Atlas Co., 1875), courtesy of Iowa Digital Library.

For product safety concerns under the European Union's General Product Safety Regulation (EU GPSR), please contact gpsr@mare-nostrum.co.uk or write to the University of North Carolina Press and Mare Nostrum Group B.V., Doelen 72, 4831 GR Breda, The Netherlands.

For Tera

Contents

List of Illustrations ix

Acknowledgments xi

INTRODUCTION
Ditches of Empire 1

CHAPTER ONE
Swamps, Nature, and Disease in America, 1775–1820 17

CHAPTER TWO
The Foundation of All Correct Tillage: Drainage, Farm Improvement, and the State, 1830–1865 35

CHAPTER THREE
An Empire of Ditches: Settlers, Railroads, and Wet Prairies in the Valley of the Red River of the North, 1877–1886 54

CHAPTER FOUR
Settler Activism and State Intervention: Draining the Red River Valley, 1886–1900 76

CHAPTER FIVE
Wasted Lands, Wasted People: Eugenics, Drainage, and the Legacy of Nathaniel Southgate Shaler 94

CHAPTER SIX
A Construction Agency: The Reclamation Service and Settler Home Creation, 1902–1910 122

CHAPTER SEVEN
Consulting Advisors: Federal Drainage Engineers and Settlers, 1902–1913 151

CONCLUSION
We Opposed the Draining of Our So-Called Swamp Lands 182

Notes 191
Bibliography 223
Index 255

Illustrations

FIGURES

"Impassable Marsh" in Section 33 of Iowa's Lake Township, Humboldt County (1875) 7

Image of a farmer laying clay drainage tiles at the bottom of a trench, gifted to John Johnson in 1859 45

James J. Hill, the architect of Red River Valley drainage 63

Tiling machine on a Minnesota farm, early 1900s 91

Nathaniel Southgate Shaler 102

Congressman Halvor Steenerson 124

Minnesota farmer laying drainage tile at the base of a trench, 1910 147

Hovland tile ditcher crew at work in Minnesota, 1910 148

The staff of the Office of Drainage Investigations, circa 1909 168

An Office of Drainage Investigations encampment and field party in Arkansas, 1909 169

Unidentified man laying clay tile in North Carolina 173

MAPS

The 1881 drainage-related flooding in Clay County 68

The distribution of wetland spaces available for drainage, 1907 175

The geographic distribution of federal drainage investigations 177

Acknowledgments

Perilous Waters originated during a visit to the National Archives and Records Administration (NARA) in Denver, Colorado, more than a decade ago. At the time, I intended to dig into the Bureau of Reclamation's archival records to analyze the relationship between the bureau and railroad corporations in developing the trans-Mississippi West's water resources during the late nineteenth and early twentieth centuries. Shortly after arriving in Denver, however, I stumbled upon dozens of boxes crammed full of correspondence, reports, and memoranda related to wetlands drainage. This proved baffling. I wondered, Why would a Western irrigation agency produce so many detailed records about draining swamps, marshes, wetland prairies, and other watery tracts across the Midwest, South, and Atlantic Seaboard?

Upon returning home, I shared the archival findings with my graduate advisor Don Pisani, and this project was born. Don remains one of the country's top irrigation and conservation scholars, but he heartily encouraged me to shift focus and begin untangling land drainage's outsized yet overlooked role in national water politics, conservation, state formation, and the building of a settler nation. The more I dug into drainage's complex and contested history, the more I discovered its weighty influence on US environmental and political history, especially its intersection with settler colonialism, federalism, scientific racism, eugenics, and land modification and water engineering. Drainage, it quickly became clear, touched on the lives of thousands of Americans across every generation and revolutionized governance at the local, state, and federal levels.

This book is the culmination of a long intellectual journey that began in Denver, and I am indebted to dozens of generous colleagues, institutions, and family members who have joined me on it. My faculty colleagues past and present at the School of Advanced Military Studies (SAMS)—Rick Herrera, John Curatola, Barry Stentiford, Matt Muehlbauer, Jacob Stoil, Amanda Nagel, Don Wright, Matt Yandura, and Nick Prime—have been a welcome source of encouragement and goodwill. Rick is responsible for introducing me to the wizardly Deb Gershenowitz and for encouraging me to submit the manuscript to the University of North Carolina Press, a wise piece of counsel. Deb, thanks for encouraging me to "let it go!" and for eagerly backing this project from the start. The administration and support staff at SAMS over the last decade, including Scott Gorman, Kirk Dorr, Candi Hamm, Andy Morgado, Bruce Stanley, and Rich Dixon, has likewise supported my academic endeavors. Dave Cotter, the Command and General Staff College's dean of academics, has gone to bat for me more than once during this project. All their support and encouragement have buoyed my spirit along the way.

Professional scholarship is a collaborative endeavor, and numerous historians have critiqued manuscript drafts, provided valuable feedback at conferences and over Zoom, and entertained my unending queries. Josh Nygren and Brian Frehner, my fellow Kansas City environmental historians and craft beer enthusiasts, sharpened my arguments and prose by reading and rereading multiple chapters. Josh especially endured countless phone calls and nagging emails to help flesh out ideas. Few colleagues have been more helpful and kind than Bob Morrissey. Bob read the entire manuscript and delivered sage advice about improving it during video calls, conference panels, and road trips together during conferences. He also generously assisted me with navigating the University of Illinois' online databases, pointing out valuable primary sources that I had overlooked. Finally, Patty Limerick has been a vocal and effusive supporter. Since meeting Patty in 2017, she has prodded me to unravel drainage's role in the construction of a settler order that erased the rich, varied historical experiences of Native and Black peoples in wetlands. Her friendship also generated numerous opportunities to share my research with promising young graduate students from the University of Colorado's Applied History Initiative. Sometimes we get lucky and meet splendid colleagues later in life, and Patty is no exception.

A small network of historians and geographers whose research touches on aspects of drainage deepened my appreciation of swampland reclamation's towering role in US history. John William Nelson, Jon Coleman, Kristin Greteman, Morgan Vickers, John Baeten, Joe Otto, Kevin Mason, Aaron Purcell, and Anya Kaplan-Seem have proven invaluable collaborators, sharing their research and probing my conclusions and arguments. Tim Hemmis, Gustave Lester, and my "line editor" Kevin Hymel also read chapters and provided stylistic and conceptual suggestions. Bobby Wright patiently produced this volume's maps, persevering through numerous emails requesting changes to the maps. Additionally, the monthly Works in Progress seminars at the University of Missouri–Kansas City's Linda Hall Library have offered a forum to test new ideas with rising historians of science and medicine from around the world.

I took a brief hiatus from this manuscript to finish a military history book, *The Other Face of Battle*, with three of today's leading military historians. Wayne Lee, David Silbey, and David Preston were superb coauthors, showing me much about the ins and outs of bringing a final book to press.

No one has exerted a more positive influence on me than Don Pisani. From the moment I read Don's *To Reclaim a Divided West* in graduate school, I fell in love with environmental history and was inspired to study water resources history. Don is a true gentleman, and his academic track record speaks for itself. Perhaps less known are his tireless dedication and bottomless generosity for students. Even in retirement, he read two early drafts of the manuscript, provided meticulous suggestions about improving it, and implored me to explore the fascinating and obscure conservation career of Nathaniel Southgate Shaler. He also never lets me buy the first beer. Many readers will detect evidence of Don's unmistakable influence in *Perilous Waters*, particularly localism's and federalism's enduring legacies in US water resources management and his conviction that much remains to be written about the conservation movement.

Research for this book took me to twenty repositories across thirteen states and Washington, DC. I am grateful for the archivists and staff at the Antioch College Special Collections; the Bancroft Library; the James J. Hill Reference Library; the Library of Congress; the Louisiana State Museum Historical Society; the Minnesota Historical Society; Minnesota State University–Moorhead's Archives; NARA in College Park, Denver, Fort Worth, Kansas City, San Francisco, and Washington, DC; the North Dakota Institute for Regional Studies; the Nathan Marsh Pusey Library (Harvard University); the George A. Smathers Library (University of Florida); the Smithsonian Institution Archives; the University of Illinois Archives; and the Western Reserve Historical Society Library. Archivists and librarians who deserve special praise for going above and beyond include Bruce Kirby, Steve Nielsen, Nick Duncan, Scott Sanders, Lori Cox-Paul, Mark Peihl, Pat Maus, Molly MacGregor, W. Thomas White, Eileen McCormack, Korella Selzler, and J. Bonnie Rehder. Finally, several institutions provided fellowships and grants that made this book possible: the Foundation for the National Archives, the Forest History Society, the American Heritage Center, the Charles Redd Center for Western Studies, the James J. Hill Reference Library, and the University of Oklahoma History Department.

My parents, Dennis and Charlotte, and my sister Andrea have been joyful cheerleaders of my passion for history since the first day of graduate school. I also appreciate the luxury of having two amazing in-laws, Larry and Nancy

Sands. I took my first research trip for this book with my grandparents, Dee and Opal Goracke. While I was working in the Minnesota Historical Society's archives, they held down the hotel room and eagerly awaited the arrival of each evening so we could try out St. Paul's downtown restaurants. I deeply regret that they did not live long enough to see this book in print. I miss them every day and am eternally grateful for their unconditional love, good cheer, and sacrifices on my behalf.

On the home front, Eden, Mia, and Quinn lost interest in dad's fascination with drainage long ago. Their teenage years have arrived far too quickly and suddenly, but parenthood has instilled an overwhelming joy that I never anticipated. Thanks to each of you for letting me tap away at the keyboard on many late evenings and weekends. I love you.

This book is dedicated to my best friend and life partner, Tera, who has shouldered the heaviest load during the book's completion. In addition to working full time while I attended graduate school, Tera has managed household affairs during my many extended research trips and evenings and weekends spent writing. Fate smiled at me when I found such a superb, funny, and beautiful partner, and meeting her remains the best day of my life.

As a federal government employee, I am obliged to state that the opinions expressed herein do not reflect the views of the US Government, the Department of Defense, or the US Army. All errors of fact, interpretation, or omission are mine alone.

Tony E. Carlson
Kansas City, Missouri

Perilous Waters

Introduction
Ditches of Empire

On July 2, 1886, the excitement of the crowd at the Crookston, Minnesota, opera house reached a fever pitch. Summoned to Crookston by local political and business leaders, the attendees spent two days learning about the promise of land drainage. Speeches by C. G. Elliott, James J. Hill, and Halvor Steenerson—three leaders of the emerging national drainage movement—particularly enthused the crowd, persuading them that the Red River of the North's Valley would never become a farmer's paradise until settlers transformed its wetland prairies and swamps into dry agricultural fields.

C. G. Elliott kicked off the Crookston convention with a stirring endorsement of drainage. As the country's most accomplished agricultural drainage engineer, he marveled at the Red River Valley's bottomless agricultural potential, especially the "exceeding richness and depth" of its soils. After sketching out the valley's topographical and hydrological features on a blackboard, he implored delegates to overhaul the watery landscape. All that was necessary to achieve a "high state of productiveness," he prophesied, was for the government to build a series of drainage ditches and other structures that separated soil from water. By diverting surface waters into rivers and creeks, settlers could convert the valley's marshy tracts into a global breadbasket and secure a future of prosperity.[1]

James J. Hill proved no less fiery and confident. One of the Gilded Age's most renowned captains of industry, he later served as the president of the Great Northern Railway, the final American transcontinental built to the Pacific. In late June, he arrived in Crookston eager to rally support for a general drainage program. As the owner of over 1 million acres of wet, uncultivable valley lands, Hill's railroad stood to profit handsomely from a system that dried out the soggy region. He offered to pay for half of the costs of a drainage survey if county governments covered the remainder, which would determine the feasibility of draining the valley. "The question of drainage was one of the greatest importance to every material interest in the whole valley," he crowed. In his closing remarks, he urged the audience to join him in organizing and bankrolling a preliminary survey.[2]

Finally, Halvor Steenerson strolled to the podium and briefly addressed the audience. An attorney, state representative, and future ten-term congressman, he also helped organize the Crookston drainage convention, leaning heavily on his local political celebrity and popularity. Like many valley settlers, Steenerson's two brothers resided on waterlogged farms that struggled to turn an annual profit. He celebrated drainage as a tool to boost annual crop yields, stimulate population growth, and replace uncultivable wet spaces with dry settlements. Seeking to reassure skeptical audience members, he promised that the diversion of water from wet prairies into the Red River of the North's basin would not drown out downstream towns and farms. Drainage carried no adverse repercussions. Settlers should immediately launch a thorough survey that evaluated the valley's drainage prospects and demand state intervention on their behalf.[3]

Dazzled by these glowing portrays of drainage, the excited audience, representing six northwestern Minnesota counties, heartily agreed to split the costs of a topographical survey that Elliott's engineering firm would lead. These efforts would produce a drained, developed, and domesticated landscape filled with flourishing settler homes and farms.[4]

After the convention adjourned and the delegates returned home, everyone awaited the results of Elliott's survey. For the remainder of the summer and into the autumn, the engineer and his team of assistants established a headquarters at Crookston, carried out a wide-ranging topographical survey, and formulated a drainage plan. On December 8, 1886, he unveiled his proposal to the reassembled delegates at Crookston. The plan envisioned a mosaic of institutions—local, state, and federal—cooperating to dig drainage ditches, channelize rivers and streams, and bury underdrainage clay tiles to completely reengineer the landscape by removing its water. Government intervention, Elliott evangelized, would prepare the valley for dense settlement and herald its agricultural ascendancy.[5]

The delegates immediately approved Elliott's plan, pledged additional resources, and stood up committees to lobby the state and federal governments for drainage aid. In 1887, these efforts bore fruit when state policy-makers passed legislation empowering county governments to organize as drainage districts. Six years later the hesitant state legislature finally created the Red River Board of Audit, a state drainage agency responsible for straightening and channelizing the Red River Valley's watercourses. By 1903 the Board of Audit and its successor commission had spent $200,000 constructing 117 miles of state ditches and extensions that connected to 620 miles of county-built drainage ditches. Overall, the maze of infrastructure, constructed in less than a decade, reportedly drained over a million acres of marshy land.[6]

Although the Red River Valley drainage movement occurred in a remote, sparsely populated corner of Minnesota, it was embedded in a much larger, more momentous, and largely overlooked story of American history: the conversion of the continent's abundant swamps and wetlands into dry settlements.

Indeed, in its scale, intensity, and disruptive impact on the human and nonhuman worlds, wetlands drainage from the late colonial era to the end of World War I constituted one of the most consequential and transformative episodes of landscape modification and water engineering in US history. By 1920—the year the federal government published its first drainage census—settlers in the Red River Valley and communities scattered across the United States had organized more than 53 million acres of land into drainage enterprises, a staggering quantity that dwarfed the country's 19 million acres of irrigated land. The enormous scale and dizzying pace of drainage, which has been marginalized as a minor footnote in American environmental and political history, triggered waves of ecological, social, and political consequences that reverberate to today. Drainage penetrated state authority into the rural periphery via new political structures. It revolutionized the relationship between settlers and the state, generated new fields of specialized knowledge, ignited successive bursts of technological experimentation and innovation, and much more.[7]

Perilous Waters probes and unravels drainage's complex and outsized role in US history, highlighting its intersection with Indigenous dispossession, settler colonialism, eugenics and scientific racism, federalism, and many other prominent themes. It argues that white settler communities, like those at Crookston, demonized swamps as one of the gravest environmental impediments to agricultural expansion and the establishment of prosperous communities. In doing so, they enlisted the state's knowledge, resources, and authority to create new political institutions that facilitated drainage.

As early as the mid-nineteenth century, the state elevated drainage into a paramount public policy objective. Counties, states, and the federal government fashioned new drainage organizations and mobilized their personnel and resources to assist settlers with reclaiming the land. Ultimately, settlers extolled drainage as one of the most decisive acts in solidifying their claims to Indigenous homelands, justifying and rationalizing Native people's removal. In the end, drainage recalibrated settlers' relationships with one another, as well as with the state, and contributed to the historical erasure of continental wetlands, a legacy that persists to this day.

DRAINAGE'S RISE TO PROMINENCE owed itself largely to the confounding nature of swamps. Neither fully land nor fully water, swamps and other forms of *wetlands*—marshes, bogs, sloughs, wet prairies, pocosins, prairie potholes, marshy river bottomlands, coastal estuaries, and so forth—were liminal landscapes that defied simple or binary categorizations. Ecologists estimate that before colonization the conterminous United States encompassed 221 million acres of swamps and other wetlands (11 percent of its surface area)—with the vast majority concentrated east of the Great Plains. Unquestionably, a large portion of European colonists and later settlers traveled near or lived in the vicinity of seasonal or annual wet tracts.[8]

Long before Red River Valley farmers hatched their drainage scheme, settlers designated most landscapes characterized by an abundance of surface water as *swamps*. As one congressional report elaborated in 1849, "Swamp lands, in common parlance, conveys the idea merely of wet or soggy lands, where the water issuing from the soil renders it too wet for cultivation." As the report hinted, "swamp lands" originated when some hydrological oddity or other environmental misfortune—perhaps even the vestiges of Noah's flood—left them submerged and uncultivable. The implication was clear: The presence of swamps or watery tracts signaled that nature was out of balance, and it behooved settlers to restore that balance by converting them into dry homes and crop fields.[9]

The murky origins of swamps from the start raised nagging doubts about their habitability. To a large degree, settlers' aversion stemmed from fears about the incompatibility of European bodies with soggy spaces. Flummoxed by repeated cycles of urban yellow fever outbreaks and rural malaria, physicians in the late 1700s believed that the rotting animal and vegetable matter in swamps spewed miasmatic poisons into the atmosphere, hastening a whole range of febrile illnesses. Virginia planter William Byrd II's 1728 petition to King George II for permission to drain the Great Dismal Swamp neatly captured this sentiment. "By draining the Dismal, it will make all the adjacent country much more wholesome," he promised. "This will happen by correcting and purifying the air, which is now infected by the malignant vapours rising continually from that large tract of mire and filthiness." Settler societies, Byrd's argument ran, possessed an obligation to divert surface water into rivers or streams, remedying nature's worst blemishes and redeeming the continent's suitability for settler bodies.[10]

The imagined connection between swamps and disease spurred on the conviction that they were uninhabitable, but so too did their linkage with outcasts, undesirables, criminal vagrants, escaped enslaved people, and potentially hostile or rebellious actors. It was bad enough that Indigenous warriors from King Philip's War (1675–76) to the three Seminole Wars (1817–58) deftly navigated the dense, overgrown terrain of swamps to resist the encroachment of white intruders, but swamps also lured fugitive enslaved peoples and the instigators of slave rebellions who threatened social and political order.[11]

In 1805, *The Literary Magazine, and American Register* objected that swamps "afforded an asylum and subsistence to fugitive negroes for several years." While swamps' pestilential airs, thick and tangled vegetation, and oppressive humidity deterred the presence of settlers, their ample wildlife, plentiful fruit, and "throng[s]" of fish nourished absconding enslaved people. Moreover, the proliferation of free Black communities across the South's secluded swamps by 1800 fueled a persistent myth that Black bodies boasted racialized immunities to diseases associated with swamps and tropical climates, including malaria and yellow fever, which slavers callously exploited to justify condemning them to lifetime enslavement.[12]

So terrified were colonial authorities about swamps doubling as sanctuaries that they codified instructions about hunting down formerly enslaved laborers. In 1705, Virginia instituted "An act concerning Servants and Slaves" that acknowledged "many times, slaves run away and lie out, hid and lurking in swamps, woods, and other obscure places, killing hogs, and committing other injuries to the inhabitants of this here majesty's colony." The law's language, which North Carolina inserted verbatim into its own later statutes, acknowledged the government's inability to control its perilous waters and unruly occupants, especially in the 2,000-square mile Great Dismal Swamp on the Virginia–North Carolina border. Put simply, swamps posed a clear and present danger.[13]

Not only could swamps potentially act as staging grounds for slave rebellions, they also disrupted commerce and movement. The 1805 article in *The Literary Magazine, and American Register* condemned swamps' impenetrable and "tenacious" vegetation, "stinging insects," and spongy soils for making travel hazardous and onerous. Three and a half decades later, the stigma persisted. Describing his 1839 visit to the Great Dismal Swamp in Nansemond County, the renowned Virginia farm reformer Edmund Ruffin howled in protest that the menacing landscape engulfed and fatally swallowed up settlers.

Several persons though residents, and well acquainted with the ground, have lost their way and their lives in the swamp, and not far from the boundary of cultivation. Of two who are known thus to have perished, the skeleton of one was found many years later, and identified by the remains of the iron tools which he had carried with him. Of the other no trace has ever been found. These facts . . . will serve to convey some idea of the difficulties of getting through the thickly overgrown and treacherous bog, and of the lost guidance of the sun, or some other sure indication of direction.[14]

In proclaiming swamps as the "boundary of cultivation," Ruffin signaled that they stubbornly resisted absorption into settler land regimes. Indeed, the experiences of federal land surveyors reinforced swamps' intransigence. After the national government seized control of public lands and adopted the Land Ordinance of 1785, which mandated the surveying and division of Indigenous territory into linear townships and sections, government surveyors spent the next century measuring, chaining, and marking off the land grid's boundaries. Swamps and shallow surface ponds defied the state's efforts to impose an abstract grid system over them. In the early 1850s, for instance, the commissioner of the General Land Office ordered surveyors to "meander [establish artificial lines demarcating a body of water] . . . all *lakes* and deep ponds of the area of twenty-five acres and upward; also navigable bayous; *shallow* ponds, readily to be drained, or likely to dry up, are not to be meandered." In practical effect, his directive enabled surveyors to bypass swamps that consisted of twenty-five-plus acres by meandering their perimeters. Consequently, surveyors often depicted them on land plats as blank "impassable marshes"—inaccessible, empty, and resistant to settlement and subdivision at the fringes of empire.[15]

INCUBATORS OF MIASMATIC POISON, portals to freedom, untraversable spaces, and bulwarks of Indigenous resistance, swamps subverted settler colonialism from the outset. Over the course of the nineteenth century, settlers concluded that unless swamps were drained, other infrastructure and amenities—roads, homes, crop fields, fences, ditches, railroads, telegraph wires, electrical poles, and so forth—could not be grafted onto the land. These symbols of settler permanence would remain incomplete or absent. As civil engineer C. G. Elliott—who addressed the Crookston convention in 1886, surveyed the Red River Valley, and later served as the US Department of Agriculture's (USDA) chief drainage engineer—laconically explained,

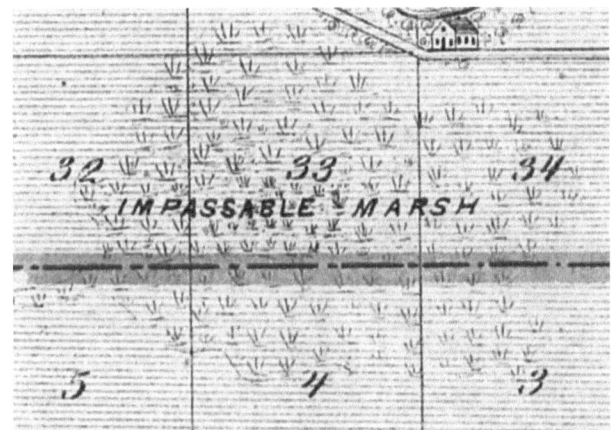

"Impassable Marsh" in Section 33 of Iowa's Lake Township, Humboldt County (1875). Reproduced from the *Illustrated Historical Atlas of the State of Iowa*, 1875, 50. https://digital.lib.uiowa.edu/node/814984.

"drainage is the multum in parvo of road making." In his view, drainage encompassed the opening prelude and foundational act of rural development that enabled all future infrastructure developments.[16]

Unsurprisingly, settlers broadly and imaginatively defined *drainage*. James J. Hill, who mobilized drainage support at Crookston and later encouraged Congress to nationalize drainage, imagined drainage as the art and science of correcting nature's hydrological imbalances through infrastructural and technological remedies. "Where there is too much [water] for profitable cultivation, [settlers] must draw off the surplus," he explained in 1910. "Upon such control of water supply depend the habitability of much of the earth's surface and its contribution to the total stock of wealth." He praised the drainage of England's fenlands and Holland's reclaiming of its "native" lands from the sea to carve out a wide-ranging coast. As a fluid category, drainage thus encompassed any activity that discarded surplus surface waters, including building drainage ditches, straightening and dredging watercourses, removing obstructions from flood-prone creeks and streams, and burying clay underdrainage tiles. The goal was one and the same: dry and densely cultivated communities.[17]

Despite the missionary zeal with which settlers and the state promoted drainage as the linchpin of rural society, it initially proceeded in a piecemeal and halting fashion. As long as the country's population density remained low and an abundance of dispossessed territory remained available for the taking, settlers had little incentive to choose a submerged tract and invest in draining it. As Missouri congressman John Scott attested in 1826, "Agriculturalists are not disposed to purchase lands covered with water, and incur the expense of draining it before it can be rendered useful and productive."[18]

By the middle of the century, however, cracks in Scott's theory surfaced. Urgent concerns about the security and stability of growing Western settlements on the vast wet prairies of the Old Northwest and in the Mississippi River Valley spurred Congress into action. In 1849, 1850, and 1860, it authorized three Swamp Land Acts, which took the unprecedented step of donating federal "swamp and overflowed" lands to fifteen states on the condition they sell the lands and invest the revenue in building drainage projects. The laws aimed at engineering a landscape completely devoid of surface water, free of disease, available for agricultural exploitation, and off limits to Indigenous combatants and self-liberated enslaved peoples.[19]

After Congress punted drainage to the states, legislatures began to empower drainage districts with land reclamation responsibilities. These quasi-independent governments unleashed market technologies—clay underdrainage tile technology and, later, steam dredges—in the fight against wetlands. Districts possessed the authority to "take, tax, and spend," which in one geographer's judgment rendered them the "indispensable engine of the American reclamation movement." Increasingly aided by specialized knowledge generated by engineers and scientists in the US Geological Survey (USGS) and the USDA, autonomous drainage districts assaulted the nation's swamps with ruthless effectiveness. As described earlier, the 1920 census recorded that over 53 million acres of land had been organized into drainage enterprises—an area roughly the size of Minnesota. Moreover, the census reported that more than one out of ten farms was drained and that settlers had dug 112,000 miles of open ditches and buried 45,000 miles of underground clay tiles to remove surface water.[20]

Settlers' voracious appetite for drainage sparked repeated surges of technological innovation and experimentation. The US Patent Office from 1849 to 1915 issued an astounding 1,250 patents for drainage tools and technologies. Inventors and farmers in every state and territory except Rhode Island and New Mexico secured patents for clay-tile making devices, tile cutters, ditchers, dredges, tile laying machines, road graders, subsoil plows, and many other drainage accoutrements. The imperative to drain triggered one of the most substantial and enduring transformations in US environmental history as settlers used these tools to overhaul swamps, marshes, wet prairies, bogs, and other watery tracts into dry, checkerboard fields with industrial efficiency and celerity.[21]

Yet therein lies a paradox. Despite the indispensability of drainage to the expansion of settler agriculture, historians have paid far less attention to swamps than to forests, rivers, grasslands, rangelands, national parks, wilderness, and especially deserts. In terms of water policy and land reclamation scholarship, drainage has wallowed in historiographical backwaters—even though drainage reengineered far more land than irrigation had done by the end of World War I.[22]

In large measure, drainage's emaciated historiography stemmed from irrigation's ascendancy in post–World War II water scholarship. For postwar political theorists and historians, the practice of irrigation dating to antiquity symbolized elites' (and later the state's) capacity to harness environmental knowledge, labor, and resources on a grand scale. The towering dams, the mammoth reservoirs, and the storage and diversion of entire watersheds subordinated individuals to collective aims. Irrigation structures—emblems of modernity, social cohesion, and technological prowess—were heralded over and over in photographs and artwork and hinged on the confiscation and redistribution of property; the designing of roads, communities, and electrical systems; and the advent of novel legal concepts for water appropriation. Quite simply, irrigation proved a far more alluring vehicle for analyzing the relationship between political power and water administration, as well as the application of technology, than drainage, whose scattered and inconspicuous infrastructure often remained literally buried underground and out of sight, locally administered and requiring little centralized control or coordination.[23]

Studies of the late-nineteenth- and early-twentieth-century US conservation movement followed this script. In his seminal *Conservation and the Gospel of Efficiency*, Samuel P. Hays touted the American federal irrigation program as a harbinger of the embryonic administrative and regulatory state. Hays argued that during the 1890s an emergent bureaucratic and scientific elite in Washington seized control over natural resources management. Perhaps the boldest, far-reaching achievement of Progressive Era conservationists, he maintained, came in the policy area of irrigation. In 1902, Congress created the Reclamation Service in the Department of the Interior to build reclamation (e.g., irrigation) projects in the arid American West. As an independent institution with its own funding mechanism, the Reclamation Service fulfilled Hays's prototype of centralized planning and administration by disinterested experts. Federalized irrigation leapt from the pages of *Conservation and the Gospel of Efficiency* as the purest embodiment of the incipient environmental-management state, etching out a historiographical course that shaped the trajectory of US conservation and water scholarship.[24]

Hays set in motion a wide-ranging conversation about the social and political implications of state-sponsored water management. Following his lead, a generation of scholars unpacked irrigation's impact on appropriative law, federalism, diplomacy, public-private relationships, culture, wealth concentration, and nature itself. The term *reclamation* became synonymous with irrigation in desert climates, anchoring water and conservation history to the American West and cementing the 100th meridian—the north-south axis running from Manitoba and North Dakota through Texas that roughly divides North America into arid and humid halves—as an impregnable historiographic boundary.[25]

Only at century's end did historians and geographers of the United States begin to pay meaningful attention to wetlands, America's "unknown landscape." This growing body of scholarship explores wetlands from a range of environmental, social, legal, institutional, and political lenses, demonstrating, for instance, how the later twentieth century softened cultural attitudes toward swamps in ways "far more radical than changes in action" or policy. The decoupling of water and conservation studies from irrigation unlocks fresh opportunities for framing new narratives about drainage's long-neglected role in national history.[26]

PERILOUS WATERS BUILDS on this scholarship by emphasizing the state's primacy in converting swamps and other liquid landscapes into dry settlements. Drawing on a broad and diverse collection of rarely cited archival materials, it integrates the state's drainage contributions into the broader conservation movement in the Gilded Age and Progressive Era. The book argues that settlers' contempt for swamps galvanized national sentiment in favor of state action to assist communities in draining and domesticating wet rural spaces. As one of the most organized and complex rural settler activities, agricultural drainage obligated all levels of government to join in implementing cooperative social institutions while also systemizing environmental and technological knowledge.

Beginning with the Swamp Land Acts, Congress championed drainage not by directly financing and building drainage projects (as it did with Western irrigation after the 1902 Reclamation Act). Rather, it did so by marshaling conservation science to amass specialized knowledge and disseminate that knowledge to local institutions (drainage districts and state drainage commissions) via new federal natural resource bureaucracies and publications. In the 1880s, for instance, the USGS commissioned a series of investigations under the direction of Harvard geologist Nathaniel Southgate Shaler that inventoried, quantified, and categorized national wetlands with the goal of transforming them into settler farms. Shortly after the turn of the century,

Congress created the USDA's Office of Drainage Investigations (ODI), which dispensed expert knowledge to drainage districts and other local drainage bodies while organizing hydrological and soil examinations in tandem with communities and universities.[27]

The state's approach to drainage snugly conformed with patterns of associational governance. Far from being "stateless" or beholden to laissez-faire values, the nineteenth-century federal government served, according to the historian Brian Balogh, as a "coordinating agent" and "developmental field marshall," working through local and state governments to deliver desirable political, social, and environmental results. "In many instances, the central government served as a coordinator," Balogh observes, "providing fiscal aid and a national perspective to state and local partners." Drainage typified this governance style; in hyping and aiding the drainage and settlement of wetlands, the national state preferred to enable rather than command, to incentivize and subsidize rather than directly control. Institutionally, the American drainage movement unfolded in a piecemeal, halting, and nonlinear fashion as independent local communities, such as those who gathered in Crookston, mustered support for drainage and then designed and implemented water engineering plans according to their own timing, contingent contexts, and environmental requirements.[28]

The state's sponsorship of drainage represented one of its most profound environmental interventions during the emergent conservation movement. Scholars once described the rise of conservationism in the late nineteenth century and the Progressive Era as an enlightened response to cultural anxieties about an imminent timber famine, the closing of the frontier, the wanton destruction of animal species, the breakdown of the myth of inexhaustible natural resources, and other real and perceived environmental crises. In the last quarter century, however, new studies have chipped away at this triumphal narrative, weaving the history of conservation into the contexts of scientific racism, class, imperialism, eugenics, and the building of a settler nation. In repositioning race, class, gender, and Indigenous dispossession at the vanguard of conservation studies, they have highlighted the complex and contradictory motivations of environmental reformers who spearheaded federal-led natural resources conservation. This book explores how drainage intersected with domestic racism, imperialism, settler colonialism, and especially eugenics, as highlighted in USGS scientist Nathaniel Shaler's 1890s proposals to turn drained swamps into eugenic sanctuaries for native-born Anglo-Saxons.[29]

Federal leaders of conservation weaponized drainage to dismantle and erase Indigenous connections to wet tracts by forcefully asserting the state's role in *reclaiming* them. The high tide of these efforts occurred in 1906 when Minnesota congressman Halvor Steenerson, the architect of the Crookston drainage convention, hatched a scheme to drain Anishinaabe peatlands, wet prairies, and wild rice beds. Over the previous decade and a half, Congress had forcibly confiscated more than 3 million acres of Anishinaabe homelands in Minnesota, but half of it in 1906 remained unsettled largely due to its swampiness. According to Steenerson, "The state should go on and reclaim the swamp lands in co-operation with federal authority" in order to open "homes for thousands of settlers." The state looked to expedite settlement through swifter land expropriation. In this manner, drainage married the goals of federal conservation (settler home creation and efficient resource usage) and Indian policy (colonial dispossession). And in Steenerson's calculus, the drainage of Anishinaabe homelands would underwrite what the historian Paul Frymer recounts as federal land policy's preeminent purpose during the long nineteenth century: "to move as many settlers as possible onto contested lands in order to overwhelm and dominate the preexisting population." On wet tracts, this was not possible until the state surveyed, divided, and commodified them—all of which drainage accelerated.[30]

Not coincidentally, the terms "reclamation" and "reclaim" permeated the correspondence, official documents, and speeches of Progressive Era conservationists like Steenerson. In 1911, Frederick H. Newell, one of President Theodore Roosevelt's closest natural resource advisors and the leader of the Reclamation Service, defined reclamation as the "regulation of water supply, putting it on where there is a deficiency, and taking it away where there is excess; substituting the will of man for the unregulated natural forces." For Newell and Steenerson, land reclamation simply entailed removing water from swamps (drainage) or applying it to deserts (irrigation) for the purpose of creating prosperous settler homes on divested territory.[31]

Newell's definition, however, concealed as much as it revealed. What precisely was being reclaimed? And for what end? The term *reclamation* presumed rightful ownership. Through the intervention of drainage (and irrigation), the state aimed to RE-claim environmentally marginal spaces that it had seized from Indigenous communities by developing and then *conserving* them for future farmers. The state operated as if the dispossessed tracts, due to their alleged "unoccupied" status and unrealized agricultural potential, had always somehow belonged to settlers.

In 1907, Gifford Pinchot, head of the US Forest Service, similarly touted settler home creation as the preeminent objective of Roosevelt's conservation policy. Speaking to the National Irrigation Congress in Sacramento, Pinchot declared that "the single object of the public land system of the United States ... is the making and maintaining of prosperous homes. That object cannot be achieved unless such ... lands as are suitable for settlement are conserved for the actual home maker. Such lands should pass from the possession of the Government directly and only into the settler who lives on the land." For Pinchot and like-minded conservationists settlers would by definition be white freeholders, and the act of reclamation permanently extinguished and erased Indigenous claims. In this manner, land reclamation served as *both* the opening act of settler rural improvement and dispossession's culminating event, consolidating the combined results of violence, military conquest, duplicitous diplomacy, and removal.[32]

Reclamation's very logic and rhetoric thus envisioned the deliberate disruption of Native relationships with land, soil, and water by systematically replacing them with agroecological regimes (drained monoculture fields) tailored to settler agriculture. Hence the incubation and maturation of federal conservation science and knowledge related to swamps and drainage, which reached their apogee in the Progressive Era, completely ignored Black and Indigenous knowledge and experiences. Consequently, it underwrote what the Indigenous philosopher and scholar Kyle Whyte asserts as one of settler colonialism's overriding objectives: "the establishment of its own ecology, at the expense of Indigenous ecologies ... US settler colonialism, in terms of collective continuance, is a complex process because settlement inscribes the settler ecology." Conservation and colonization were flip sides of the same coin.[33]

The inscription of settler ecologies across the United States helps dissolve the seeming contradiction between the conservation movement's twin impulses for development (e.g., land reclamation and settler home creation) and preservation (e.g., the establishment of national parks and forest reserves). The simultaneous populating and depopulating of Indigenous homelands proved mutually supporting because they both engraved settler ecologies that blotted out previous occupation and usage by normalizing colonial land regimes. Whether reclaiming swamps and deserts for dense, prolonged white settlement or violently emptying *wilderness* areas to create settler "pleasuring-grounds," conservation undergirded settler colonialism's modus operandi: "destroy[ing] to replace."[34]

PERILOUS WATERS TELLS the story of settlers, swamps, and the state from the early republic to 1920. Settlers judged swamps as one of the greatest environmental impediments to expansion and enlisted state authority to drain them, seeking to permanently stake their claims to the land. After showing how early Americans vilified swamps as pathogenic spaces linked to epidemics and unhealthy climates and briefly surveying contrasting Indigenous relationships with wetlands (chapter 1), it narrates how antebellum farm reformers cast drainage as a cornerstone of agricultural improvement and progressive farming. Congress and state legislatures responded to this groundswell of enthusiasm by enacting policies that accommodated settler demands for a dry, healthy, and densely occupied continent (chapter 2). With the passage of the Swamp Land Acts (1849, 1850, and 1860), the federal government subsidized drainage with public land donations while state legislatures propped up intermediary drainage institutions. Drainage districts—one of the state's primary mechanisms for fashioning dry settler agroecologies—altered settlers' relationships with each other and nature, pitting communities and new technologies against wet environments and their nonhuman populations.

Federal subsidies and local drainage institutions, however, seldom proved sufficient. Chapters 3 and 4 zoom in to the valley of the Red River of the North, which separates present-day Minnesota and North Dakota. In the 1880s, Minnesota settlers and the Great Northern Railway embarked on what contemporaries hailed as the largest US agricultural drainage project ever undertaken. It failed spectacularly. The railroad's faulty and haphazard drainage system flooded dozens of farms and triggered a decade of contentious litigation. The fracas repudiated settler mythmaking about the uncomplicated nature of drainage. At Crookston, the railway hired C. G. Elliott, the nation's leading agricultural drainage engineer, to conduct a detailed topographical survey and design a comprehensive drainage plan. In the end, the state of Minnesota joined the railway and local counties in organizing a state drainage commission that carried out the railway's ambitious plans to renovate the valley into a farmer's paradise, a reputation it has not relinquished.

The Red River Valley case study lays bare the dearth of settler knowledge about wetlands. And since settlers always painted themselves as passive victims of a harsh and unforgiving nature, they welcomed federal assistance. In 1884, the USGS stepped in and hired Nathaniel Shaler to execute the government's first investigation of wetland resources (chapter 5). Although the Harvard professor and USGS scientist limited his first study to New England, he published a comprehensive report six years later covering the entire country. Shaler rejected the idea that swamps were the remnants of Noah's flood or some other mythic event. He argued that they were integrated into nature's complex, interdependent webs. Nevertheless, he trumpeted that their agricultural value exceeded that of all other uses. Although he commandeered federal conservation science to address fears about land scarcity and race mixing, his publications divulged how government investigations and specialized knowledge were necessary to convert wetlands into farms.

The final two chapters bring the story into the twentieth century. During President Theodore Roosevelt's administration, a coterie of federal engineers, hydrologists, and bureaucrats officially launched the American conservation movement. Equating efficiency with top-down control, federal conservationists sought to shift sovereignty over natural resources planning and administration from states and localities to expert-staffed federal bureaucracies.

In one of the bitterest (and most overlooked) episodes of the conservation movement, Frederick Newell's Reclamation Service sought to nationalize all drainage responsibilities in aid of Representative Halvor Steenerson's efforts to reclaim traditional Anishinaabe territory and other "ceded Indian lands" (chapter 6). The service's pursuit of institutional aggrandizement quickly failed due to federalism, partisanship, and the USDA's intense opposition. The latter rejected nationalization because its tiny new Office of Drainage Investigations (ODI) was experimenting with a form of environmental governance that fused federal expertise with local autonomy (chapter 7). Dismissing Newell's attacks on local governance as elitist and out of touch, the ODI trumpeted that settlers guided by the technical and apolitical wisdom of federal engineers could control their own political destinies. Under the leadership of Elwood Mead and then C. G. Elliott, the ODI dispatched engineers across the country to support drainage districts in overcoming intractable engineering problems and shepherding them through the process of organizing under state laws. These disciples of drainage extended the tentacles of the state into remote rural spaces by conducting topographical surveys, publishing technical bulletins, studying soils, mediating landowner disputes, and nurturing drainage districts to maturity.

Drainage became indispensable to the perpetuation of settler agriculture on a watery continent. The processes, technologies, resources, and administrative machinery necessary to drain a sprawling, densely wooded swamp in Louisiana's Atchafalaya basin, however, differed drastically from what was needed for a homestead in northern Iowa's wet prairies. In striving for a smooth narrative, *Perilous Waters* sometimes generalizes rather than overly particularizing, except in its case study of the Red River Valley, where the innumerable political, legal, and ecological complexities of drainage are laid bare.

Finally, the pages that follow broadly differentiate between two forms of drainage: farm drainage and swamp drainage. *Farm drainage* describes efforts by settlers, engineers, agricultural reformers, railroad corporations, and drainage districts to furnish existing farms (or groups of contiguous farms) with drainage systems. The material and public health improvements of already established farms were its intended outcomes. On the other hand, *swamp drainage* refers to attempts by politicians, government scientists, capitalists, and other elites to drain, develop, and then subdivide large watery landscapes that were not yet occupied by settlers. Motivated by anxieties about the imminent exhaustion of arable land, the decline of the family farm, the perils of unfettered urbanization, and the assimilation of "new" immigrants, its promoters sought to resurrect a vanishing rural past for nostalgic and romantic purposes. In many instances, the inspirations behind farm drainage and swamp drainage overlapped and reinforced one another, as did their desired end state: a landscape devoid of water and crowded with dry homes and crop fields. The following pages explain how this landscape came to be.

CHAPTER ONE

Swamps, Nature, and Disease in America, 1775–1820

Few landscapes have inspired more fear, abhorrence, and mistrust than wetlands and especially swamps. Historically, wetlands of all forms and varieties have been villainized for creating unhealthy climates, poisoning the atmosphere with "miasma," and interfering with commerce and sedentary agriculture. Europeans as far back as antiquity despised wetlands for impeding travel, depressing property values, blocking otherwise productive land from cultivation, facilitating the growth of strange and noxious weeds, and providing sanctuary for harmful predators, reptiles, and stinging microorganisms.

Medieval Christians popularized the myth that wetlands were sinister and forbidding. In their literature, the stagnant water, oozing mud, darkness, humidity, and putrid smells of swamps and bogs formed part of the iconography of hell. The seventeenth-century English poet John Milton went even further, imagining swamps as Satan's personal stomping ground. In contrast, the freshness, limpidity, and sweet aroma of flowing springs and gentle brooks symbolized God's goodness and undeserved grace. By demonizing swamps as the den of reprobate sinners and Lucifer himself, this literature implored righteous and lawful Christians to avoid the corruption associated with fetid, stagnant waters.[1]

The vilification of swamps reached a fever pitch during the Enlightenment. By the time Americans secured their independence, European physicians had revived the ancient Hippocratic medical tradition attributing certain fevers to marshes, poorly drained lowlands, and other unhealthy environs. According to this body of medical thought, the putrefying vegetable and animal matter in swamps and marshes discharged miasmas and other unwholesome airs into the atmosphere that, after being inhaled or absorbed by the skin, disrupted or weakened the body's normal functioning, prompting fevers and possibly even death. As a result, a growing number of European physicians and scientists probed the associations among air, weather, elevation, and disease using new apparatuses such as the barometer, hygrometer, thermometer, and eudiometer. Their discoveries legitimized the idea that eliminating miasmatic landforms improved and moderated unhealthy climates, which premodern Europeans conceptualized as microenvironments where the interaction of air, soil, and water influenced the health of nearby organisms. Inspired by the Enlightenment's unfettered optimism about humanity's ability to improve itself, these medical practitioners recommended draining marshes, filling lowlands, culling forests, and cleaning up urban filth. The result of these efforts, they preached, would be genial climates and healthy bodies.

The Hippocratic tradition's resurgence coincided with an international controversy about North America's climates. During the second half of the eighteenth century, the celebrated French naturalist Georges-Louis Leclerc, Comte de Buffon, theorized that miasmatic marshes dominated North America's topography. The continent's pervasive wetness, he argued, produced an assortment of biological and atmospheric oddities: North American quadrupeds were smaller than their Old World counterparts, imported animals degenerated or died, insects and unwanted reptiles proliferated, and unwholesome airs saturated local climates. In stigmatizing the marshy continent as a pathogenic wasteland, Buffon orchestrated a bitter dispute with American settlers about the size and vitality of the landmass's biological organisms. In response, Thomas Jefferson and others argued that the proliferation of European methods of land management, including land drainage and deforestation, had already dispersed miasmatic concentrations and moderated erratic climates. The only good wetland, everyone agreed, was one that had been drained and cultivated.

Historians have largely explained settlers' historical antipathy toward swamps from the perspectives of economics, commerce, and agrarianism. As the US population grew during the nineteenth and early twentieth centuries, they contend, drainage emerged as an important social tool to open up more farmland, perpetuate the nation's agrarian heritage, and offset problems related to industrialized urban life. Yet the collective impetus to drain wetlands initially took form in response to collective fears about the deleterious impact of watery landscapes on settler bodies. The eighteenth-century Hippocratic revival and pneumatic chemistry revolution ushered in the belief that draining wetlands comprised an essential ingredient of enlightened settler land management. On a continent afflicted by endemic malaria and recurrent yellow fever eruptions, drainage became one of the keys to achieving healthy, stable, and secure settlements.[2]

The New Hippocratism in Early Modern Europe

In the late 1600s, European physicians began resurrecting the ancient Hippocratic concept of an environment-disease nexus. By the 1750s, they had concluded that communities could liberate themselves from fevers by identifying and eliminating miasmatic sources. Civilized societies could master their climates because nature responded positively to specific forms of human intervention. Disease prevention through judicious environmental management emerged as the hallmark of this so-called new Hippocratism, which flourished in eighteenth-century European (and especially British) intellectual circles.[3]

The new Hippocratism's intellectual roots stretched back to ancient Greece. The Hippocratic Corpus, a series of works attributed to the Greek healer Hippocrates of Cos (circa 460–377 BCE) and his followers, correlated the onset of fevers and epidemics with specific landforms, personal habits, and physical constitutions. Many of the corpus's seminal works emphasized the association between air and good health, which required a proper balance of the body's four humors (blood, phlegm, black bile, and yellow bile). Early Hippocratic practitioners believed that an amalgam of the humors comprised the human body. Illness and disease resulted when impure airs, such as "miasmas," disrupted humoral equilibrium. The precise composition of miasma was difficult to define. From a health perspective, miasma referred to the contaminating, polluting, or impure substances released into the atmosphere through various sources: sluggish waters, putrefying vegetable and animal matter in swamps and marshes, excrement, decaying corpses, subterranean gases, and so forth.[4]

The Hippocratic tradition preached disease avoidance and treatment rather than prevention. Since diseases proved endogenous to particular geographic features, itinerant Greek healers schooled themselves in the topographical features and prevailing wind patterns of the communities they served. Elevation, offensive odors, fog, heat, humidity, moisture, soil composition, sunlight, winds, and *especially* the pestiferous waters of marshes and swamps determined a place's epidemiological qualities. As the scholar James C. Riley observes, the Hippocratic tradition suffered from a pessimistic "fatalism." Although Greek physicians treated fevers by administering barley water, hydromel (a water-and-honey mixture), or oxymel (vinegar and honey), they considered themselves powerless to eliminate the environmental sources of disease. Human bodies were the captives of their physical surroundings.[5]

As early as the second century BCE, Hippocratic writers and architects vilified marshes and fens as some of the biggest environmental threats to communities. The scholar Marcus Terentius Varro warned against erecting structures near marshes because they provided habitat for tiny, invisible creatures that, after being ingested, prompted fevers. To preserve the health and longevity of Roman citizens, Marcus Vitruvius Pollio encouraged their segregation from marshes. According to the agriculturalist Lucius Junius Moderatus Columella, marshy areas corrupted the atmosphere and encouraged the proliferation of dangerous snakes, reptiles, and stinging insects. Finally, the geographer and philosopher Strabo judged it foolish to build cities near lakes since high summer temperatures receded water levels, leaving a patchwork of miasmatic marshes at the water's edge.[6]

The Hippocratic tradition thus dichotomized circulating and stationary waters. Flowing waters in rivers and streams symbolized life, hope, longevity, and refreshment. Stagnant waters hastened death, despair, and decay. Watercourses signified health and purity; marshes denoted sickness and contamination. The limpid waters of brooks and streams were invigorating and therapeutic; turbid fen waters proved enervating and corrupt. As "liminal zones" where water and land intermingled and surrounding airs were defiled, watery landscapes pointed to nature's imperfections and flaws. Water supposedly belonged in circulating streams, creeks, and brooks where motion precluded the formation of miasma.[7]

Early modern Europeans appropriated ancient Greco-Roman medical ideas for their own geopolitical and social reasons. Soaring curiosity about the influence of air and weather on human health spurred on the Hippocratic renaissance just as Europeans embarked on their colonial adventure in North America. Beginning in the late 1600s, English physicians and other intellectuals, influenced by the New Science's experimental methodology and the Royal Society's patronage, encouraged the systematic observation, documentation, and publication of climatic phenomena for medical purposes. The growing availability of weather instruments—barometers, hygrometers, thermometers, and wind and rain gauges—enabled the precise measurement of meteorological events. Dedicated to the pursuit of natural philosophy, the Royal Society promoted the meticulous documentation of barometric pressure, humidity, temperatures, wind patterns, and precipitation levels. The Society disseminated measurements from domestic and overseas weather diarists in its journal, the *Philosophical Transactions*, thereby enlisting meteorology as medicine's patron.[8]

Pneumatic chemists also joined the project of cataloguing meteorological and climatic phenomena. Until the late eighteenth century, most scientists and naturalists believed that the atmosphere constituted a single substance rather than an assortment of gases. Cracks emerged in this theory with the 1766, 1772, and 1774 discoveries of nitrogen, hydrogen, and finally "dephlogisticated air" (oxygen). Knowledge that the atmosphere contained gases of varying wholesomenesses had clear medical and environmental implications. Specifically, the invention of the eudiometer (Greek for "measure of good air"), an apparatus designed to quantify the volume of various gases, equipped physicians with a scientific means of measuring atmospheric quality. As one historian of science puts it, "Medical and chemical traditions converge in eudiometry." In the 1770s, the Italians Marsilio Landriani and Felice Fontana developed a marketable eudiometer that administered a nitrous air test to measure an air sample's wholesomeness. Eudiometry reinforced physicians' belief that elevated phlogiston levels acted as a public health menace no less hazardous than miasma. In the 1770s, Italian and British physicians organized eudiometrical tours in the countryside, measuring and comparing the aerial goodness of different landforms and elevations.[9]

Eudiometry tied sick bodies to particular landforms. In the spirit of the Hippocratic tradition, eudiometer-wielding chemists identified congested structures and stagnating waters as pathogenic spaces. Unventilated homes, cellars, cemeteries, crowded prisons, narrow city streets, tanneries, butcher shops, hospitals, barracks, markets, slaughterhouses, and open sewers poisoned cities with effluvia. The unimproved countryside was an even more potent generator of sickly climates: Swamps and sluggish streams spewed miasmas. Dense forests allegedly magnified the problem by shielding damp soils from sunlight and blocking winds from diluting miasmatic concentrations. Neo-Hippocratic physicians, to some degree anticipating the later emergence of sanitary science, touted several ameliorative strategies: widening city streets, razing dilapidated buildings, properly disposing of human and animal waste, erecting spacious courtyards for air to circulate, providing cities with fresh water, thinning forests, and, especially, draining swamps and other wetlands.[10]

Two factors lent urgency to British efforts to eradicate unhygienic climates. First, during the 1700s, the "culture of sensibility" transformed British middle-class life as individuals developed polished manners, refined tastes, and a heightened awareness of their feelings. As the middle classes adorned themselves in lavish clothing, indulged in tea and coffee, enjoyed carriage rides, and played indoor card games, physicians worried that they might become too "soft" or effeminate, magnifying their susceptibility to harmful airs. Since modern conveniences had become too alluring for people to resist, managing local geographic spaces through proper land management practices became imperative.[11]

Second, overseas imperialism and foreign wars increasingly brought British bodies into contact with supposedly miasmatic and undeveloped landscapes and *climates*, which neo-Hippocratic practitioners cast as microenvironments rather than latitudinal zones. Colonial administrators, military physicians, and settlers feared that European bodies' vulnerability to diseases and fevers increased in proportion to the distance they strayed from their home climates. The new Hippocratism instructed colonial administrators, military planners, Creole physicians, and colonists that temperate lifestyles, abstemious diets, and the proper placement of military encampments would shield bodies from corruption until the implementation of European methods of land management—such as swamp drainage and deforestation—uncorrupted the colonial climates and mended the air. In overseas colonies, healthy climates followed the settler's axe and drainage ditches.[12]

Soggy Land, Erratic Climates:
Nature and Place in North America

North America provided a laboratory to test the new Hippocratism's premises. As early as the first decades of the 1600s, the continent's mysterious and fitful climates had perplexed European settlers. Unlike in Europe and tropical destinations, North America's weather phenomena, influenced by the Little Ice Age's climatic vicissitudes—including oppressive humidity, dramatic temperature fluctuations, intense thunderstorms, hurricanes, tornadoes, and waterspouts—proved unpredictable and violent. In addition, settlers found climate's historical latitudinal definition to be inadequate. When the cultivation of tropical crops failed in the Chesapeake and settlers discovered that Nova Scotia suffered severe winters despite its latitudinal proximity with London, the mystery deepened. Even worse, many colonial boosters' and promoters' sanguine claims about North America's invigorating air proved illusory, as disease and starvation became a fixture of everyday colonial life. The ongoing demographic collapse of Indigenous communities due to imported diseases also carried ominous overtones. Tragically, this "Great Dying," which likely led to the extensive abandonment of agricultural lands and the regrowth of vegetation across the hemisphere, may have triggered the Little Ice Age's coldest extremes as atmospheric CO_2 levels plummeted.[13]

Speculations about North America's fatal climates and their unsuitability for settler bodies fueled a bitter international controversy. During the third quarter of the eighteenth century, this well-known "dispute of the New World" erupted. In his verbose forty-four-volume *Histoire naturelle*, the renowned French naturalist Comte de Buffon insisted that a calamitous flood had submerged North America just prior to its European settlement. "When the waters on the surface of the earth cannot find vent to flow, they form marshes and fens," he lectured. "America may be said to be one continued marsh, throughout all its plains." The expansive remnants of the deluge—swamps, marshes, bogs, pocosins, Midwestern wet prairies, and other sodden tracts—degenerated the continent's flora and fauna and poisoned the atmosphere. Indeed, the alleged paucity of large quadrupeds, the prevalence of slothful small animals, and swarms of insects and reptiles highlighted the continent's inhospitable nature. Touting large quadrupeds as a yardstick for nature's vitality, Buffon sneered that camels, dromedaries, giraffes, hippopotamuses, and lions were nowhere to be found in North America. The chilly and soggy continent produced an indolent and unimpressive natural order: "In this state of abandon, everything languishes, decays, stifles. The air and the earth, weighed down by the moist and poisonous vapors, cannot purify themselves nor profit from the influence of the star of life. The sun vainly pours down its liveliest rays on this cold mass, which is incapable of responding to its warmth; it will never produce anything but humid creatures, plants, reptiles, and insects; and cold men and feeble animals are all that it will ever nurture."[14]

European intellectuals soaked up Buffon's grim assessment. In his *Recherches philosophiques sur les Américains*, the Dutchman Cornelius de Pauw ridiculed North America as a "fetid and boggy terrain." British natural philosophers juxtaposed their country's salubrious climates with those of North America. In 1767, Adam Ferguson explained that "the climates of America ... are observed to differ from those of Europe. There, extensive marshes, great lakes, aged, decayed, and crowded forests, with the other circumstances that mark an uncultivated country ... replenish the air with heavy and noxious vapours." Scottish historian William Robertson's enormously popular *History of America* endorsed the claim: "Prodigious marshes overspread the [North American] plains. . . . When any region lies neglected and destitute of cultivation, the air stagnates in the woods, putrid exhalations arise from the waters; the surface of the earth, loaded with rank vegetation, feels not the purifying influence of the sun; the malignity of the distempers natural to the climate increases, and new maladies no less noxious are engendered."[15]

These caricatures posed an existential dilemma. Following the logic of European naturalists and medical practitioners, American political leaders recognized the new Hippocratism as the fountainhead of early modern climatology. They did not dispute claims about the continent's wetness. Rather, they countered European attacks by arguing that deforestation, drainage, and cultivation were already moderating their climates. The only reason that swamp waters proved perilous, they asserted, was because Indigenous inhabitants for centuries had neglected their agricultural improvement and occupation. In 1814, the political economist Tench Coxe, a former assistant secretary of the treasury under Alexander Hamilton, articulated this assessment in his famous report on arts and manufacturers for the federal government. According to Coxe, the United States possessed "*unused*, a superabundant quantity of fenny, marshy, boggy or swamp land." In conveniently erasing centuries of Indigenous inhabitation and exploitation of wetlands, he championed Lockean notions of land ownership and clamored for the state to drain its "unoccupied swamps." Once that occurred, the dried-out swamps would support the raising of "heavy fleeced breeds of long woolled sheep," attract dense settlement, and emerge as a key source of national wealth.[16]

Cadwallader Colden became one of the first Americans to blame a local disease outbreak on the sluggish pace of intensive land exploitation. In 1743, the physician and natural scientist partly attributed a series of "epidemical distemper" outbreaks in New York City to "noxious vapors" arising from the "moist slimy ground" of marshes. "Stagnating waters," Colden maintained, "have been infamous from all antiquity for their noxious quality, and for that reason by the ancient poets described under the representation of the hydra, throwing out a poisonous deadly breath." To prevent future disease outbreaks, Colden recommended draining all marshes, swamps, and stagnant ponds within city limits and also cutting down dense forests, which would improve air circulation and disperse miasmatic concentrations. According to Colden, older settlers supported his theory that the settler's axe had moderated the climate and decontaminated the air. "The climate grows every day better as the country is cleared of the woods, and more healthy, as all the people that have lived long here testify. . . . I therefore doubt not but it will in time become one of the most agreeable and healthy climates on the face of the earth." Drainage and deforestation were the panaceas for unruly climates.[17]

Over the next few decades, physicians and many others followed Colden's lead. In 1769, New Jersey's Edward Antill informed the American Philosophical Society that land drainage and deforestation had already improved the air. "Whoever compares the present state of the air," Antill explained, "with what i[t] was formerly, before the country was opened, cleared and drained, will find that, we are every year fast advancing to that pure and perfect temperament of air." In a 1770 paper presented before the society, physician Hugh

Williamson identified deforestation and drainage as the principal catalysts of ongoing climatic moderation. According to Williamson, "Tall timber greatly impedes the circulation of air," making it difficult for "fresh" ocean breezes to invigorate the atmosphere of interior settlements. Before the beginning of European colonization, "the face of this country was clad with woods, and every valley afforded a swamp or stagnant marsh ... and [because of] a general exhalation from the surface of ponds and marshes, the air was constantly charged with a gross putrescent fluid." The adverse climatic impact of swamps could not be doubted.[18]

The most famous defense of American climates flowed from Thomas Jefferson's pen. In *Notes on the State of Virginia* (1785), the future president ridiculed Buffon's accusation that the continent's frigid temperatures and abundance of moisture diminished the size and diversity of its animal species. In a largely overlooked passage paraphrasing a portion of Buffon's argument, Jefferson admitted that North America had "more waters ... spread over its surface by nature, and fewer of these drained off by the hand of man." Yet the land's marshiness did not reduce overall biodiversity or birth dwarf species. Jefferson compared the animal species of the Old World and New World to underscore that the latter's fauna, as a whole, were superior in size, stature, and diversity. He also rebutted Buffon's contention that North America had cooler climates than European countries of similar latitudes. Parroting the findings of other settlers, Jefferson unoriginally contended that during the previous generation the climates had warmed, snow accumulation had decreased, and local floods had diminished in frequency.[19]

Clergymen and intellectuals eagerly contributed to these climatic discussions. Ezra Stiles, the president of Yale, compiled six volumes of daily temperature measurements from 1763 to 1795. Many other clergymen kept meticulous meteorological and demographic records. Literary journals, including the *North American Review* and *Columbia Magazine*, published weather charts and temperature measurements from colonists living in every part of the continent. The American Philosophical Society's *Transactions* routinely featured articles about the amelioration of ungenial climates through drainage and deforestation.[20]

There was a component of this climate discourse that touched on broader Biblical and cultural themes. The great flood in Genesis, Noah's flood, had occurred as a punishment for sin, rebellion, wickedness, and disobedience. The continent's vast swamps could easily be seen as the vestiges of the flood, or at least as a visible reminder of God's wrath toward unrepentant sinners. By making the earth healthy and productive, drainage restored God's prelapsarian creation and satisfied the biblical injunction to "fill the earth and subdue it."[21]

Culturally, however, Americans stigmatized swamps for an assortment of reasons beyond prevailing climatology, principally because they were seen as being *beyond civilization*. Not only were they allegedly disease-ridden and inhospitable, swamps served as magnets and sites of resistance for Indigenous combatants, runaway enslaved persons, the organizers of slave uprisings, and a vagabond and criminal class of poor whites. In the sixteenth and seventeenth centuries, upper-class authors and travelers mercilessly disparaged the itinerant white inhabitants of swamps as "bogtrotting Irish." Entering the English language in 1682, "bogtrotter" referred to impoverished, landless people living near swamps who managed their own illicit economies. As the historian Nancy Isenberg frames it, swamps held a prominent role in early modern climatology, but they also demonstrated how "class structure was tied to geography and rooted in the soil."[22]

Swamps attracted marginalized outcasts beyond the control of the state, but they also inspired local folklore about bestial humanoids and gigantic reptiles lurking amidst their perilous waters. Legends such as the Skunk Ape, Rougarou, La Llorona, Jack O'Lantern, and other Bigfoot-like creatures signaled swamps' otherworldliness. The Beast of 'Busco, a gargantuan alligator snapping turtle, was rumored to inhabit Indiana's Black Oak Swamp. In most instances, these folktales functioned to reduce the prospect of violence, whether to protect the social outcasts from settlers or to protect society from the menacing denizens of swamps by keeping both groups separated.[23]

The First American Climate Crisis

Optimism about the continent's improving climates coincidentally overlapped with an almost three-decade absence of yellow fever in the mainland United States. From 1766 to 1792, yellow fever, which had made its first appearance in 1649 in Spanish Florida, vanished from American cities. Since many medical practitioners attributed yellow fever outbreaks to miasmatic climates, the disease's prolonged hiatus seemed to validate colonists' sweeping declarations about the moderation of climates due to drainage, deforestation, and other forms of European land management.[24]

The optimism soon disappeared. In the summer of 1793, yellow fever struck Portsmouth (New Hampshire), New York City, and Philadelphia, the young nation's capital. In Philadelphia, 4,000 people—nearly 10 percent of the city's population—lay dead by the time the first frost ended the outbreaks.

During the next dozen years, the disease annually visited cities ranging as far north as Portland, Maine, to as far south as Savannah, Georgia. It decimated New Orleans four times between 1796 and 1804. From 1693 to 1799 yellow fever claimed at least 17,088 lives in North America and probably many more.[25]

Yellow fever's ghastly symptoms terrified settlers. Transmitted by the bite of the female *Aedes aegypti* mosquito, the acute viral disease—colloquially known as "yellow jack" and "black vomit"—was one of the most dreaded diseases in North America and the Caribbean. In addition to triggering jaundice as a result of liver damage, yellow fever's onset sparked multiple symptoms in nonfatal and fatal cases. In the former, fevers, muscle pains, headaches, nausea, and prostration resulted. In fatal incidences, victims secreted blood through their nose, ears, and mouth; experienced delirium; passed bloody stools; and vomited black coagulated blood brought on by gastric bleeding. Since *A. aegypti* eggs required clean, fresh water to develop into larvae and pupae, the mosquito preferred to breed in barrels, buckets, cisterns, clay pots, kegs, and other household and agricultural items that collected rainwater. With a limited flying range of a few hundred meters, *A. aegypti* thrived in dense human concentrations such as cities or sugar plantations. The historian J. R. McNeill argues that environmental changes triggered by the growth of the Caribbean sugar industry, including deforestation, the proliferation of ports, and the ubiquity of water storage containers used to refine sugar, spawned fertile conditions for *A. aegypti* beginning in the late seventeenth century.[26]

The yellow fever outbreaks ruptured Americans' climatic self-confidence. Just as importantly, it inspired a contentious medical discourse about the disease's etiological origins, pitting *localists* against *contagionists*. Localists such as Dr. Benjamin Rush and Noah Webster argued that diseases originated from the presence of locally based miasmatic sources. Conversely, contagionists asserted that the disease was imported into the United States and could be transmitted between individuals. The localists' narrative ultimately triumphed because they published more, were better organized, appealed to common sense, occupied university positions, and trained future generations of medical practitioners committed to local etiological explanations of disease. In any case, by reinforcing the new Hippocratism's landscape-disease nexus, the localists identified marshes and swamps as some of the biggest incubators of unhealthy and miasmatic climates that fueled the yellow fever scourge.[27]

Yellow fever's resurgence stoked renewed fears about swamps. In 1795, for instance, Philadelphia physician William Currie delivered a paper before the

American Philosophical Society arguing that the putrefaction of vegetable and animal matter in marshes sparked a chemical reaction, which diminished atmospheric oxygen levels. Curry cited the eudiometrical experiments of Dutch physician Jacob Van Breda to confirm that the 18:48 ratio of oxygen to nitrogen near marshes plunged beneath the normal atmospheric level of 1:3. "The causes of the unwholesomeness of low and moist situations," Currie opined, "is not owing to any invisible miasmata or noxious effluvia, which issue from the soil and lurk in the air, but to a very different cause, viz. to a deficiency of the oxygenous portion of the atmosphere in such situations." Over the previous two decades, advancements in the medical community's understanding of human anatomy had demonstrated oxygen's centrality to cardiovascular and pulmonary functions. Starved of oxygen, bodily functions performed "imperfectly and languidly." As Currie explained, oxygen deprivation rendered the "vessels on the surface of the body powerless, and atonic." The cardiovascular system's weakened state made the body vulnerable to miasmas. "It appears more than probable," Currie concluded, that "febrile contagion ... is rendered virulent and powerful in proportion to the absence or defect of the oxygen and the degree of heat to which the living body has been exposed."[28]

Currie proposed that supplementing oxygen levels around marshes could enhance people's resistance to fevers. By building a system of "drains, deep trenches, and wells" that conveyed surface water into flowing watercourses, settlers increased their oxygen levels. According to Currie, physicians had a duty to implore farmers to fill low, miry spots with clay, sand, or lime; set fire to dead and decomposing weeds, grass, and trees; and sow grasses, "plants of vigorous growth," and vegetables for the purpose of "replenish[ing] the atmosphere with oxygen." Concentrations of surface water deserved elimination.[29]

The era's defining study on the swamp-disease nexus was published by Charles Caldwell, a student of Benjamin Rush and founder of the University of Louisville medical school. In 1802, Caldwell argued that yellow fever and malaria were symptomatic of a broader problem related to uneven water distribution. Entitled *An Oration on the Causes of the Difference ... Between the Endemic Disease of the United States of America, and Those of the Countries of Europe*, Caldwell's book juxtaposed nature in Europe with that in North and South America. Taking a hemispheric perspective, he argued that the New World's "diversity of latitudes and climates ... aid a foundation for a state of things unknown in other quarters of the Globe." Given the towering size and majesty of New World forests, mammals, lakes, mountains, rivers, and waterfalls, which dwarfed their European counterparts, Caldwell theorized that it was entirely predictable that the Western Hemisphere's "bold" and "gigantic" topographical and hydrographic features, including vast stretches of swamps, engendered fatal climates.[30]

In Caldwell's estimation, North America's combination of excessive summer heat and limitless wetlands stirred up fertile conditions for endemic fevers. "Compared with most parts of Europe," he wrote, "the super-abundance of *marsh miasma* in the United States ... surpasses the super-abundance of our summer heats, and would seem to be more immediately and powerfully influential in the production of our diseases." The continent's flat topography, abundance of precipitation, suffocating heat, and flood-prone rivers produced long stretches of stagnant surface waters that rotted vegetable and animal matter, discharging miasma and unwholesome airs. From the coastal salt marshes along the Atlantic seaboard to the Mississippi River's alluvial floodplains, North America was home to "vast factories of this febrile poison." Little wonder that Caldwell identified wetness as one of the continent's defining topographical features.[31]

The new Hippocratism seduced Caldwell. Casting doubt on the hydrological purpose of swamps and flood-prone tracts, he argued that landscapes where water languished were "physical imperfections." They posed an omnipresent threat. "The most unlettered husbandman," Caldwell lectured, knew "that [marshes'] hurtful properties depend on its super-abundance of stagnating humidity, and that the only method of rendering it innoxious to health, and useful in agriculture, is to circumscribe and intersect it with a number of ditches sufficient to carry off [their] redundant waters." Settlers who ignored drainage were poor environmental stewards. As Caldwell put it, "A neglect of this rational, salutary, and lucrative practice, subjects thousands in the United States to the malignant action of mash miasma, who would otherwise escape this deleterious poison."[32]

In framing North America's disease-friendly climates from the perspective of swamps, Caldwell appealed for state intervention. He applauded communities that paid bounties for exterminating undesirable predators such as wolves, but the continental scale of unhealthy climates attested to the urgent need for drainage incentives. By spurring farmers to undo nature's "physical imperfections" by building artificial outlets where watercourses did not exist, local communities could secure a disease-free environment. Blissfully unaware that the clay tile technology, engineering expertise, and new forms of political organization necessary for agricultural drainage did not yet exist in America, Caldwell nevertheless became one of the first Americans to advocate that the state assume a more forceful role in draining swamps.[33]

After the turn of the century, a few medical societies heeded Caldwell's advice and took action. Organized in 1804, the Georgia Medical Society enlisted direct government aid to curb wet rice cultivation. "With a semitropical climate, such as ours," the society's charter explained, "there could be no worse or more malignant incidental cause of disease than the stagnant water, which remains on a rice field exposed to an ardent summer's sun, and the saturated soil which is next exposed, when the water is drained off." In 1810, the society implored the Savannah City Council to implement a program aimed at eliminating wet rice cultivation outside of city limits. In 1817, the city finally approved the plan, imposing taxes on white city residents to raise revenue to pay nearby rice planters to drain their land and convert to dry rice culture. The program enjoyed limited success since most cultivators, despite the subsidies, found it too costly to drain their land and convert to dry cultivation.[34]

Three years after the program's approval, a yellow fever epidemic struck Savannah, killing more than 700 people. The Georgia Medical Society blamed the epidemic on planters' leisureliness in draining old rice fields, the slow pace of swamp drainage in the hinterland, and the absence of tree barriers outside of Savannah. In neglecting drainage, planters endangered themselves and their families.[35]

Wetlands and America's Native and Black Peoples

America's Indigenous and Black populations looked at wetlands differently—very differently. While Jefferson, Coxe, Caldwell, and others disparaged swamps and marshes as uninhabitable and unhealthy wastelands, Indigenous and Black peoples eagerly harnessed their plentiful resources. Despite claims to the contrary, America's swamps were neither unoccupied nor unexploited during the colonial period and early republic.[36]

In America's productive and biodiverse swamps, Native peoples found a cornucopia of edible and medicinal plants, dyes, waterfowl, fish, and numerous other sources of raw materials. Through the centuries, they amassed a sophisticated knowledge base gained and refined through experiential and managerial activities. Instead of avoiding swamps due to their menacing and/or pathogenic characteristics, Native peoples across the United States prized their biological and botanical endowments. As the contemporary Abenaki educator Judy Dow aptly puts it, wetlands historically served as vital "supermarkets" and "grocery stores" that nourished and sustained Indigenous bodies.[37]

Three wet places showcase how Native peoples calibrated and adapted subsistence strategies around watery spaces: the Ohio River Valley, California's Great Central Valley, and Chicago's muddy portage. Native women in the expansive Ohio River drainage basin, especially the villages abutting the tributary rivers and streams flowing south from Lake Erie, incorporated wetland resources into an integrated system of dry farming and hunting. As men supplied household proteins by hunting in forests and fishing, women planted beans, corn, squash, melons, and other fruits and vegetables in fertile riparian soils. The women also diversified household diets by collecting aquatic plants from the valley's maze of wetlands. They particularly sought out tuckahoe, spatterdock, pickerelweed, wild rice, macopin, nodding onions, Solomon's seal, and other edible plants to feed themselves and their larger communities.[38]

Other highly valued plants thrived in the Ohio River Valley's wetlands. Native women gathered the common milkweed and broad-leaved waterleaf for medicinal purposes. In addition, they experimented with plants, roots, and tree barks to fashion dyes that adorned clothing, deer tails, porcupine quills, and body paint. The kaleidoscope of colors manufactured by the women spanned the rainbow's spectrum. Roots from the aquatic madder red plant created red dyes; the puccoon's root produced orange; American yellow root gifted yellow hues; the hickory tree's bark yielded a green dye; and sumac flowers and white walnut barks made black dyes. Far from being empty and devoid of human exploitation, the Ohio River Valley's wetlands anchored a vibrant and prosperous Indigenous culture that thrived for centuries until violent federal military campaigns shattered them in the 1790s.[39]

In California's Central Valley, Native peoples likewise gravitated toward wetlands. The Central Valley's medley of wetlands, freshwater lakes, grassland vernal pools, tule (e.g., large bulrush) marshes, and seasonally wet bottomland forests formed the basis of a rich, biodiverse region that attracted elk, grizzly bears, pronghorn antelope, large fish populations, and millions of wintering migratory waterfowl from the Pacific Flyway. Over long periods of time, California's Indigenous communities innovated creative methods and technologies to capture and kill waterfowl. For instance, hunters from the Sacramento River Valley's Konkow and Patwin communities used nets, nooses, and decoys to lure and seize geese and ducks. In the Delta region, the Plains Miwok trapped waterfowl in tule swamps by draping large nets over them. Another technique involved hoisting nets into the path of low-flying flocks, which knocked the waterfowl onto the ground. In addition to enhancing their diets, California's Indigenous peoples selected other wetland raw materials to fashion cordage (which they weaved into blankets) and towering tule bulrushes (which they constructed into flat rafts).[40]

Natives in the Central Valley also wielded fire to manage and enhance wetland ecosystems. In the San Joaquin Valley, the Wukchumni Yokuts deployed controlled burnings to improve botanical diversity and boost animal populations. Selectively burning away old growth in tule marshes, the Wukchumni Yokuts also set fire to reed-choked spaces, allowing sunlight to penetrate to the water's surface. The rays of sunlight stimulated the sprouting of new vegetation and opened passageways for waterfowl to move, feed, and nest. Such practices indicate a keen Indigenous familiarity with wetland environments, accumulated and transmitted across several generations, that clashed with settler narratives about wet spaces.[41]

Like the Ohio and Central Valleys' Indigenous populations, Anishinaabe people in Chicago's muddy portage between the Chicago and Des Plaines Rivers exploited a waterscape of botanical and biological abundance. In the summer months, the Anishinaabeg combined crop cultivation with fishing for bass, catfish, perch, sturgeon, and suckers in the portage's meandering inland watercourses, sloughs, reeds, and tall grass marshes. In the autumn, the Anishinaabeg dependence on wetlands spiked when immense flocks of ducks, geese, passenger pigeons, sandhill cranes, swans, and teals arrived en masse to feast on wild rice and other aquatic plants and creatures. Throughout the eighteenth century, men paddled rafts through the maze of water and mud to capture and kill large quantities of waterfowl. At the same time, women ricers navigated canoes through thick stands of ripening wild rice. From inside of their canoes, they thumped loose rice grains, which fell into their vessels' hulls for transport back to their villages. Thrashed grains that fell outside of the canoes, however, germinated in subsequent years, perpetuating wild rice stands that fed Native bodies and enticed millions of migratory waterfowl during their annual continental passage.[42]

While wetlands served as Indigenous supermarkets and grocery stores, they also became cherished sanctuaries for thousands of self-liberated Blacks in South Carolina, Alabama, North Carolina, and Virginia. The extensive Great Dismal Swamp, which straddled the Virginia–North Carolina boundary, hosted one of America's largest maroon communities. In 1784, a Scottish-born visitor marveled that "run-away Negroes have resided in these places for twelve, twenty, or thirty years and upwards, subsisting themselves in the swamp upon corn, hogs, and fowls, that they raised on some of the spots not perpetually under water." He deplored how the swamp's physical attributes "have always been perfectly impenetrable to any of the inhabitants of the country around, even to those nearest to and best acquainted with the swamps." As a result, fugitive enslaved people "in these horrible swamps are perfectly safe, and with the greatest facility elude the most diligent of their pursuers." Swamps were secluded sites of Black resistance beyond the control of the settler state.[43]

These snapshots lay bare how beliefs about swamps and other wetlands were contested and contingent. Rather than stigmatizing watery tracts as disease-ridden and uninhabitable, Indigenous and Black peoples across the United States esteemed them as supermarkets and portals to freedom. Over the course of centuries, Natives converted their botanical and biological wealth into protein, medicine, dyes, clothing, rafts, and other material items. Likewise, many self-liberated Blacks envisioned swamps as autonomous oases where they could defy the settler state's racial order. The ongoing contestation of wetlands in early America signaled that their ultimate historic erasure was neither preordained nor inexorable.

Conclusion

Early modern climatology cast a long shadow over settlers' environmental attitudes and values. By framing unhealthy climates and disease outbreaks from the perspective of America's swamps, the new Hippocratism tied the fates of sick settler bodies to specific geographic phenomena. The resurgence of yellow fever in 1793, as well as the presence of endemic malaria in the rural United States, lent a particular urgency to land drainage. Despite the prevailing climatological thought, however, very little land drainage had occurred by the start of the nineteenth century. Given the abundance of easily acquirable arable land, drainage still made little economic sense. Moreover, the emergence of clay tile and steam dredging technologies and forums to disseminate drainage techniques and best practices (such as agricultural journals and agricultural societies) awaited a later day. Small wonder that land drainage, despite the period's rich and varied climatological discourse, remained the principal concern of intellectuals and physicians rather than of ordinary settlers.[44]

Nevertheless, by establishing an urgent and compelling rationale for drainage, the neo-Hippocratic discourse by the 1820s had removed all doubt that America's wetlands must be drained to preserve settler bodies, enable the expansion of settler agriculture, and impose control over unruly waters and peoples. Beginning in the antebellum period, a new generation of agricultural reformers drew on and expanded this climate-based rhetoric to galvanize support for drainage. Concerned about the stagnation of US agriculture and the fate of settlements on the sprawling wet prairies of the Midwest, they hailed drainage as a decisive act in staking settler claims to Indigenous homelands by forging stable and secure settlements. By the middle of the nineteenth century, they would recruit Congress and legislatures to authorize policies aimed at converting all of the nation's swamps into dry, verdant farms.

CHAPTER TWO

The Foundation of All Correct Tillage
Drainage, Farm Improvement, and the State, 1830–1865

Edmund Ruffin scoffed at settlers' inaction. Writing in 1839 for *The Farmers' Register*, the influential Virginia agricultural reformer condemned farmers and landowners for ignoring land drainage. "It is not only the poor and ignorant, but the rich and better informed, who neglect this incalculable source of value," he lectured readers. "For with the exception of a very few such improvements ... the drainage of the swamp lands is either not thought of by proprietors, or considered as among the many visionary schemes of bookfarmers." Ruffin bemoaned settlers' inattention to drainage, which carried hefty consequences. Left in a soggy condition, Virginia's swamps and marshes languished as "unproductive of everything but malaria, sickness, and death."[1]

In attributing the backwardness of American farming practices and nature's disorderliness to the presence of swamps, Ruffin insisted that the stability and security of rural settlements depended on drainage. In the 1830s, his writings marked the high tide of a lifelong struggle to promote drainage and secure its place in settler agriculture and land management. As one of the leading and most respected voices of the antebellum agricultural reform movement—perhaps the most notable eruption of farming innovation in American history—he rose to prominence following his discovery that adding marl to soil offset acidity. At the same time, he launched the Virginia-based *Farmers' Register*, an agricultural journal, and used it to promote farm improvement methods such as marling, fertilization, crop rotation, and drainage.[2]

From his estate in Virginia, Ruffin inspired a rising generation of agricultural reformers to take drainage seriously. As far back as the 1820s, a coterie of patrician agriculturalists, alarmed by the dwindling supply of arable land and poor farming practices, organized hundreds of new state and local agricultural and horticultural societies. These organizations, mostly clustered in the Northeast and Midwest, published farm journals, organized county fairs, held statewide agricultural meetings, and promoted progressive agricultural techniques—crop rotation, diversification, fertilization, manuring, and drainage. Among these improvement practices, reformers consistently hyped drainage as "the foundation of all correct tillage" and the "first great lesson of agricultural improvement." They propagandized swamp reclamation as the fountainhead of progressive settler agriculture because other forms of improved farming proved futile if the soil remained submerged and uncultivable. Moreover, they leveraged the rapidly expanding rural press, whose circulation surpassed 1 million by 1880, to celebrate the benefits of drainage and provide a forum for farmers to discuss drainage technologies, techniques, and best practices.[3]

The premise that water belonged inside of the banks of watercourses, where it could not impede agriculture, produce unalluring vegetation, stir up miasmas, or impose other hardships on settlers, underpinned reformers' support for drainage. For his part, Ruffin consistently denigrated swamps as "the boundary of cultivation," that is, the space where improved nature (e.g., farms, dry cultivated fields, roads, fences, cleared lots, and other settler infrastructure) faded into a disordered maze of soggy soils and tangled, unseemly flora and fauna. Beyond this "boundary," nature resisted the encroachment of settler agriculture and swarmed with potentially hostile actors such as runaway enslaved persons, Indigenous warriors, white criminals, and vagrants. Reformers touted drainage as one of the most decisive acts in correcting nature's disorderliness and preparing it for settlement. Indeed, proponents of pre–Civil War scientific farming insisted that settlers fulfilled a social and moral obligation by diverting surface waters back into rivers, streams, or creeks—where water allegedly belonged and could not jeopardize local climates.[4]

This chapter takes a fresh look at how antebellum farm reformers packaged drainage as the preeminent form of agricultural improvement. In doing so, it evaluates how reformers deployed the agricultural press, which served as the unofficial mouthpiece of the farm reform movement, to spread the gospel of drainage. While many political leaders and settlers continued to stigmatize swamps for providing sanctuary for society's dangerous and unwanted, reformers recalibrated support for drainage around three broad themes: human and livestock health, profit generation, and the bringing of order to disordered nature. To some degree, this recalibration reflected shifting geographic realities. In the colonial era and early republic, settlers who lacked access to drainage technologies and capital could easily bypass wetlands in search of drier, arable plots. Once the floodgates of Western emigration burst open and settlers fanned out across the vast wet prairies of the Old Northwest and the fertile river valleys of the upper South and the Mississippi River watershed, drainage became a prerequisite for agricultural expansion.

Despite the drainage euphoria, Ruffin and other reformers identified a few key obstacles. Draining swamps and other wetlands defied individual efforts.

Swamps, sloughs, bogs, marshes, and wet prairies routinely infringed abstract property boundaries and demanded collective action. The neglect of one or more settlers (such as an absentee owner) to drain their property could render a broader reclamation scheme unsuccessful. Moreover, landowners without access to a natural outlet—a stream, river, or other watercourse—found themselves powerless to dispose of surface water. To rid themselves of water, settlers had to secure permission from a neighbor (or neighbors) to dig a ditch across their properties to connect with an outlet. Drainage thus obliged individual initiative *and* responsibility from all settlers on a given plot. Early in the 1830s, Ruffin publicized the significance of individual responsibility, deploring how the indifference of a minority of landowners could scuttle drainage. "If the owners of low grounds would act according to their true interest, this plan would be extended as far as the nature of the land required it, without regard to whom might be the owner of any particular spot," he explained. "But that is not now to be counted on, and each person must expect his drains to end with the termination of his land."[5]

The unruliness of surface water eventually prompted antebellum farm reformers to enlist the state's assistance. Local and national legislators responded to their demands by assembling an associational form of environmental governance that blended local and federal action. At the state level, legislators empowered settlers to form drainage districts and other local institutions that collectively planned, financed, and managed cross-boundary drainage ventures. At the federal level, Congress subsidized drainage through lavish land grants. The 1849, 1850, and 1860 Swamp Land Acts ceded federal "swamp and overflowed" lands to fifteen Midwestern and Southern states on the condition that they sold the lands and used the revenue to build drainage projects. Envisioning the drainage and domestication of an entire biome, the laws' intended outcome was exceptional and unprecedented in environmental policymaking. By the eve of the Civil War, the state had constructed an associational framework that, in time, enabled Americans to assault national wetlands with devastating efficiency and ruthlessness.[6]

Draining for Health

Antebellum farm journals disdained the baneful impact of waterlogged soils on biological systems. Miasmas produced by decaying vegetable and animal matter in swamps allegedly triggered febrile illnesses, invited prolonged periods of torpor, and sometimes hastened death. In livestock and domestic animals, the inhalation of miasmas and the ingestion of hydrophytic plants precipitated biological degeneration. As a result, midcentury contributors to farm journals championed the elimination of every source of marsh miasma—or "malaria" (literally translated "bad air")—as Americans increasingly referred to atmospheric poisons discharged by soggy tracts.[7]

Accelerating Western migration put a premium on drainage. The 1810 US Census counted 7.23 million people, but only 1 million of them lived in the trans-Appalachian West. By 1850, however, the overall number had jumped to 23.2 million; 10 million settlers, or 44 percent of the population, lived west of the Appalachians. The trans-Appalachian West by that time had entered the public imagination as a marshy death trap rather than as a laboratory of republican agrarianism. Endemic malaria greeted settlers who fanned out across the upper Midwest's wet prairies and flood-prone valleys. Westerners fashioned a broad lexicon to describe malarial ailments: Arkansas chills, autumnal fever, bilious fever, black swamp fever, intermittent fever, malaria, remittent fever, seasoning, and swamp fever.[8]

Epidemiologists now recognize that widespread malaria spread by the female anopheles mosquito—not miasmas—triggered the "shakes," "fevers," and "chills and ague" that afflicted settlers near swamps, bogs, sluggish streams, sloughs, and shallow lakes, which were the anopheles' summertime breeding habitats. Despite this etiological confusion, by midcentury malaria was killing more settlers in Illinois than any other factor. Eruptions of fevers and shakes became a dreaded everyday reality. In 1821, an Illinois settler quipped, "An illness native in the prairie country was fever and ague. There was burning fever following chills which left the patient so weak he could not work. It came with perfect regularity." Lorin Blodget, the era's leading climatological authority, similarly compared the South's lethal climates to notorious tropical destinations: "India itself has not been more certain to break the health of the emigrant than the Mississippi valley."[9]

The backcountry's unhealthfulness led farm reformers to puzzle over miasma. In 1837, "R. B. J****" elaborated on malaria's "modus operandi" for Ruffin's *Farmers' Register*. The author speculated that "the effluvium arising from marshes, is a subtle, highly attenuated and undefined substance, the nature of which is unknown." The only thing certain about miasma's physical nature was its uncertainty: "Theory succeeding theory has been exploded without arriving at anything like certainty about the substance of miasm itself."[10]

While malaria's chemical properties defied explanation, its spatial characteristics were indisputable. In 1856, *The Illinois Farmer* instructed farmers in uncultivated regions to inhabit their homes' upper levels since malaria clung to the earth's surface. In 1870, A. J. Murray, a Detroit veterinary surgeon and former professor at England's Royal Agricultural College, informed the *Western Rural* that "miasma is a product of marshes and of excessive moisture in the ground" and could be carried for up to six miles by the wind. Without rendering a firm judgment, he referenced Italian physicians who located its origins in the "myriads of spores from an alga peculiar to the marshes."[11]

Most contributors agreed that miasma owed itself to the decomposition of submerged vegetable and animal matter. In 1823, "Rusticus" told the *American Farmer* that inland swamps emitted "nothing but the most pestilential miasma, thereby contaminating the otherwise wholesome atmosphere, and spreading disease and death through a whole region of a fine fertile country." Little changed over the next half century as letters, speeches, and medical treatises reprinted in farm journals reinforced the wetland-malaria nexus. In an 1870 issue of *The New England Farmer*, for instance, "A Medical Man" opined about how miasmas materialized: "By [miasma] is meant the effluvia, exhalations, [etc.], which emanate from vegetable and animal matter while undergoing decomposition" in swamps and marshes.[12]

Malaria and swampy tracts had a particularly deleterious effect on livestock. Soggy tracts endangered livestock in three ways. First, they promoted aquatic vegetation and "watery succulent herbage" deficient in fortifying nutrients. *The Cultivator* linked livestock's occasional "loss of flesh" to grazing on hydrophytic flora, which confirmed those plants' "want of nutrient." Second, poor soil drainage stunted plant, hay, and grass growth. As Ohio farmer E. Woolverton warned *The Genesee Farmer* in 1856, "The growth of the plants [on wet soils] is retarded—the health of that plant which is to be used for man or beast, is materially injured, and the health of the consumer is injured accordingly. Thus by [the] neglect [of drainage], the health, strength, vigor, and even life of plants may be extinguished."[13]

Finally, malaria heightened livestock's vulnerability to a mysterious class of airborne bovine diseases. According to *The Cultivator*, cattle that breathed in malaria or foraged on nutrient-deficient aquatic plants degenerated into weak, sickly, and scrawny creatures. "Several diseases of domestic animals, such as 'liver-complaint' in cattle, and 'rot' in sheep, are known to be connected with the same causes which produce the diseases in man," the journal editorialized in 1849. "The effects of malaria and watery succulent herbage, in producing the rot, have long been known." In 1819, *The American Farmer* explained that livestock foraging on dry plots were "superior in size and quality, and less subject to disease" than those with a diet consisting of hydrophytic plants. In a St. Louis-based journal in 1860, D. A. A. Nichols doubted the wisdom of raising sheep near Mississippi Valley swamps since they "are subject to miasma—my own opinion would be, that it would not be very healthy for sheep or the shepherd."[14]

Drainage vanquished malaria. In 1849, Alabaman N. T. Sorsby explained that conveying stagnant water off the land "destroys the noxious miasmata that wet soils and decomposing vegetable matter so rapidly generates during the summer and fall, to [settlers'] great annoyance and danger." Replacing overgrown watery tracts with dry crop fields corrected pathogenic climates. As Dr. T. Kerr laconically put it in an 1856 issue of *The Prairie Farmer*, "cultivation frees a country from malaria."[15]

The reduction of human and livestock fatalities in England's formerly fenny and swampy sections highlighted the dichotomy between wholesome streams and pestilential surface waters. In 1849, *The Cultivator* reminded settlers that "the rural population of drained districts in England have often remarked [on] the favorable effects of drainage on the health and improvement of animals, by which losses of stock have been prevented to a great extent. . . . As might be expected, the health of sheep and cattle has been benefitted by drainage to an equal or greater degree than that of the human race." In 1847, *The Ohio Cultivator* editorialized that "tracts of land in England, which were liable to fevers and agues . . . by a complete drainage have become salubrious, and are now upon an average standard of longevity with other parts of the country." As the only landscapes themselves capable of sickening or killing settlers, swamps and other wetlands deserved immediate drainage.[16]

Draining for Wealth

In addition to mending the air and ensuring healthy and stable settlements, drainage reinforced the antebellum reform movement's major themes of productivity, profitability, and improvement. As the historian Emily Pawley argues, Northern farm reformers "had at their root the belief that profit was itself an underlying feature of the nonhuman world, that hidden wealth had been divinely laid up for those willing to seek it." Seen this way, swamps and marshes did not function solely as incubators of miasmatic poison or menacing spaces teeming with undesirables at the fringes of empire. Rather, they were storehouses of untapped wealth gifted to humanity by a benevolent Creator or, to again quote Edmund Ruffin, the "obstructed and dormant but mighty resources of the state."[17]

Wealth abounded for drainage-minded settlers. Antebellum reformers advertised a number of ways drainage promoted larger annual harvests. First and foremost, drainage aerated heavy soils, making them "friable."

Prior to the Civil War, *friable* leapt from the pages of farm journals as a ubiquitous buzzword capturing how surface water removal broke down heavy, impervious soils. Once farmers diverted surface water into natural outlets via drainage ditches or underground clay tiles, "a contraction of the soil soon follows, and cracks are formed." According to *The Cultivator*, the "contraction" transformed the previously dense soil "into a state which allows the water readily to pass through it, the former difficulties of [soil particles] running together and baking, are obviated; the soil remains open and friable." Writing to *The Cultivator* in 1854, New Yorker John Johnston argued that his farm's drained soils were "as mellow as any loam; whereas had it not been drained it would have broke[n] up in lumps as large as the heads of horses or oxen." In 1858, another writer in the same periodical proclaimed that drainage energized the soil, making it "alive" and "ready to reward the labor of the farmer."[18]

Friable soils were ventilated soils. Since agricultural improvers theorized that plant roots could not reproduce without ample air access, healthy crops required subterranean air circulation. In an 1849 "Report on Drainage" delivered to the Oberlin (Ohio) Agricultural and Horticultural Society (and reprinted in *The Ohio Cultivator*), Henry Cowles and two coauthors illustrated how drainage encouraged ventilation: "A soil covered with surface-water cannot *breathe*. The pores of the soil are its nostrils and lungs. Cover them with surface-water and no air can enter. Atmospheric air is the vehicle of light and heat and of various fertilizing gases—all of which are essential to vegetation." In liberating the earth of redundant waters, drainage made sodden soils dry out, shrink, and crumble into finer pieces, facilitating subsurface air movement. Heftier yields were the farmer's reward.[19]

Drainage's second contribution to soil fertility involved fertilizer absorption. Farm reformers, including Edmund Ruffin, ridiculed settlers who applied manure to poorly drained fields. In 1858, "A. D. G." explained to *The Cultivator* that "draining also facilitates the work of enriching land. Manure applied to the surface, instead of being washed off by the rains and lost, is carried downward, and its juices incorporated with the soil." In an 1859 issue of the same periodical, "R" compared spreading manure onto undrained fields to "putting it into a pond, so far as any visible effect upon the crop is concerned." The lesson was indisputable: All forms of progressive farming—marling, manuring, and diversification—were ineffective without drainage.[20]

As a foundational farm improvement practice, drainage enabled precipitation to enrich soils. According to the majority of reformers, rainfall contained an assortment of nutrients. "Rain water, as it falls from the clouds, contains a small portion of ammonia," the *Ohio Cultivator* elaborated in 1852, so "the importance of securing this valuable substance in the soil, instead of allowing it to run off the surface, is one of the strong arguments in favor of underdraining." The next year "J. H." estimated to *The Genesee Farmer* that "rain water which falls on an acre of land in a year, is estimated to contain over 100 lbs. of ammonia, or sufficient for the growth of 17 bushels of wheat." In carving out microscopic passageways in the soil for ammonia to trickle downward and enrich crop roots, drainage gifted settlers with immense profits.[21]

Third, drainage counteracted drought. In 1849, *The Cultivator* editorialized that viscid clay soils "[ran] together" and resembled "mortar, which, when the water has evaporated, becomes like sun burnt bricks—unworkable, and totally unfit for the growth of plants." During periods of minimal precipitation, crops cultivated on such soils withered and died because their roots could not penetrate the mortar-like substratum. By making stiff and impenetrable soils friable, the author concluded, drainage "increase[d] the depth of the soil, to render it more permeable to the roots of plants, and less liable to be affected by drouth." Another author trotted out condensation as an explanatory metaphor: "During a hot day of summer you fill a tumbler with ice water, and after standing it in your room for a short time, you notice that the outer surface of the tumbler is covered with drops of water. The tumbler 'sweats.' How is this?" The author answered his own question by opining that the tumbler's sweat was "the condensed vapor of the atmosphere coming in contact with a substance colder than itself." Like the tumbler, drained soils possessed a lower mean temperature than the summertime air, inviting an infusion of moisture that invigorated parched crops.[22]

Finally, farm journals lauded the fecundity of marshy soils. As described earlier, antebellum settlers peddled the idea that the rotting animal and vegetable matter beneath swamp waters created a soil base exceptionally laden with nutrients and organic matter. Farmers who drained and cultivated a watery tract thus reaped high annual yields, averted soil exhaustion, and tapped into nature's hidden bounty. With a touch of hyperbole, an 1843 *Boston Cultivator* author explained that "for ages and ages the vegetable matter has been washing down from the high lands to the low lands, and the vegetable growth on the low lands has been decaying there, and from these two sources there has been an accumulation of rich vegetable matter of more importance to the country than would be the mines of Golconda." An 1823 correspondent with *The American Farmer* boasted about swamps' "inexhaustible fertility." Cultivation on drained plots required minimal manure or fertilizer applications, padding settlers' bottom line.[23]

For all these reasons, drainage made good economic sense. In 1850, Guernsey County (Ohio) farmer John Foster bragged in *The Ohio Cultivator* about draining his 320-acre farm. As late as November 1846, half of Foster's property comprised uncultivable "wet bottom[land]" because of two meandering creeks that routinely overflowed. Determined to dry out the bottomlands, Foster hired a ditcher to straighten the creeks. After five years, the ditcher shortened the distance the creeks traveled across his farm from 953 rods to 250 rods. "The water now passes off so fast in the new, short, and straight channels, that the creeks do not now overflow my land," he boasted. For an investment of $750, Foster estimated that his land, which he had purchased for $20 per acre, could fetch $30 per acre at auction. Furthermore, a previously uncultivable lowland section of his property measuring "13 acres and 29 rods" now supported cultivation. All in all, Foster calculated that the increased value of the land—$3,200—brought a return of $2,450 in addition to larger annual harvests.[24]

Foster's success story illustrated how extracting wealth from swamps increasingly hinged on technological innovation. Unsurprisingly, new technologies mesmerized agricultural reformers, and one of the most significant technological transfers in American agricultural history occurred during the midst of the antebellum farm improvement movement: the diffusion of Scottish clay underdrainage tiles.

The breakthrough occurred on the New York farm of John Johnston. In 1821, Johnston had emigrated from Scotland to North America, eventually settling on a 112-acre tract near Geneva on the shores of Seneca Lake. The fine-textured clay subsoil on Johnston's new farm invited seasonal waterlogging that annually drowned his winter wheat. All through the 1820s, Johnston's harvests seldom exceeded twenty bushels per acre despite liberal lime and manure applications. In 1833, however, his fortunes briefly reversed when he harvested 2,700 bushels of wheat from sixty-four acres, a bounty of forty-two bushels per acre. The bumper crop secured Johnston's credit with local lenders and he borrowed heavily to expand his farm to more than 700 acres.[25]

A tireless innovator and voracious reader, Johnston relished the era's jubilant attitude toward reform. He subscribed to Scottish agricultural journals and handcrafted his own tools. While growing up in Scotland, Johnston also had heard about clay tiles that Scottish farmers used to drain their farms. Once the farmers placed the U-shaped tiles at the base of shallow trenches and covered them with straw or dirt, the tiles acted as subterranean gutters, conveying away surface water and lowering the water table. In the mid-1830s,

Johnston asked his Scottish grandfather to ship clay tile samples to his Geneva farm.

Once the tiles arrived, Johnston hired a local manufacturer to produce new variants. In 1838, he received the first batch and paid Irish laborers to bury them at the base of trenches two and one-half feet deep and thirteen inches wide. Incredulous neighbors sneered that only "a most consummate ass [would] put crockery under ground," but the effort spectacularly succeeded. The tiles diverted away seasonal surface waters, dried out Johnston's wheat fields, and boosted his annual yields. A drainage mania quickly erupted across upper New York, spurred on by the Seneca County Agricultural Society's purchase of a Scragg's Patent Tile Machine from England. In one day, the machine churned out 12,000 uniform tiles (enough tile to cover 3,600 feet) at half the cost to farmers. In 1848, the craze continued as a dozen manufacturers within a ten-mile radius of Johnston's farm fabricated their own tiles. New York emerged as the nation's first drainage epicenter. Before the end of the Civil War, the state's farmers had buried 19,170 miles of underdrains and dug 7,460 miles of open drains, a combined distance that could encircle the earth. Johnston himself remained at the movement's vanguard. By the time of his 1880 death, America's "father of tile drainage" had buried seventy-two miles of tiles, bumping his annual yields to over sixty bushels per acre and improving the climate that, in his words, had once "bred pestilence."[26]

The clay tile revolution dazzled the rural press. Johnston became a household name among farm reformers and he authored a steady stream of articles describing the costs, availability, varieties, best practices, appropriate depth, and installation of drainage tiles. The tile craze soon spread to the Midwest and became no less indispensable to the expansion of settler agriculture than Cyrus McCormick's reaper, John Deere's steel plow, and the growth of agricultural societies. Indeed, between the 1850s and 1879, clay tiles followed the plow and tile factories sprung up across the Midwest, soaring in number from 66 to 840. Although the clay tile industry remained in its infancy before the Civil War, the rural press knitted together a network of improvement-minded farmers and reformers who exchanged knowledge about tile prototypes, manufacturing techniques, and installation methods. And it provided a forum for Johnston to showcase underdrainage's effectiveness. In 1860, he was perfectly justified in gloating that "twenty-one years ago, I was the only man using tiles on this continent, and there was but one person manufacturing them; now, the demand for tiles wherever a machine has been erected cannot be met." Soil and water must be separated.[27]

Farmer laying clay tiles at the bottom of a trench. Reproduced from a goblet set gifted to John Johnston by the New York State Agricultural Society in 1859. Courtesy of Historic Geneva.

Draining for Order

Drainage conferred health and material rewards, but it also served as a yardstick for measuring how well communities were imposing order on disordered nature. In the pages of farm journals, reformers consistently tracked the progress of settler agroecology through visual markers, namely replacing the chaotic and fluid with an ordered agricultural countryside. Until settlers drained swamps and other watery tracts, they could not create aesthetically pleasing farms where the presence of dried crop fields, fences, roads, and straightened creeks and streams collectively testified to the arrival of a new agroecology where nature was rationalized and bereft of destabilizing liminal landscapes.

As part of this transition, the rural press seized on how the conversion of watery landscapes into farms enlivened the dispositions of settlers. Writing to *The Cultivator* in 1864, a Long Island farmer marveled that drainage had renovated his murky, low-lying farm into crop fields "pleasing to the eye." "Those who were familiar with this swamp in bygone years would now scarcely recognize the spot. A more forbidding spectacle could scarcely be imagined; the whole being densely covered with sumach, alders . . . [and] the whole landscape heretofore marred and unsightly." By eradicating the native vegetation, channeling away murky waters, and planting crops, the farmer transformed the plot into "an object of pleasurable contemplation to the admirers of the beautiful." Another author disparaged swamp vegetation as an "eye sore." In 1861, New Englander H. W. Lester similarly promised that wide-scale drainage would bring the region's "richest land . . . into cultivation, and the [former] place of the bullfrog, water snakes, bulrushes, cat-tails and wild grass, hillocks, miasmas and pestilence, will excel the western prairies in productiveness."[28]

Wet soils supported alien, unappealing, and exotic hydrophytic vegetation that was both aesthetically displeasing and unproductive. Contributing to *The Prairie Farmer* in 1864, Edgar Sanders described a tract south of Lake Michigan in which the owner drained half and left the remainder "wet the same as nature left it." He surmised that the visually appealing drained portion yielded two tons of high-quality grass per acre while the unimproved section supported a crop that was "scarcely sufficient to pay for cutting."[29] During an 1850 speech to the Toledo agricultural fair reprinted by *The Ohio Cultivator*, Lewis Lambert chided audience members for paying less attention to drainage than to other forms of progressive farming. As a result of their inaction, "the stagnant pool or deadly marsh remains undrained," he thundered, "preventing the growth of useful vegetation, sending forth their poisonous vapors, their death bearing miasma; spreading disease and destroying human life, decimating our population, and frightening the emigrant to other lands, a stench and a curse to the neighborhood, and a scene of sickening disgust to the traveller."[30]

In addition to bringing order to nature and heralding the ascendancy of settler land management, drainage kindled the growth of flavorful and succulent fruits. In 1855, "Sub-Soil" warned *The Genesee Farmer* that poor surface drainage diminished fruit quality. Raspberries, cherries, currants, and strawberries grown on dry soils boasted a richer taste, brighter color, and size uniformity. In contrast, fruits grown on a "cold or damp piece of land . . . are found almost without fine flavor." Growing luscious fruits required "select[ing] a site, if possible, where water never lies upon or near the surface." Farmers who purchased soggy tracts should position "well-laid pipes [under] every sour spot" to convey away surface water.[31]

Drainage eased the burdens of farm life in other ways. As *The Cultivator* put it in 1837, bogs and marshes were "dangerous for a person to walk across," as the scattered presence of mired wagons, abandoned farm implements, and livestock skeletons attested. On a few occasions, settlers claimed to have discovered "entangled and mired" livestock carcasses in marshy areas "without any effort having been made for their recovery." Furthermore, surface accumulations shortened the lifespan of fence posts, one of sedentary agriculture's fixtures. Erecting fences constituted one of the most expensive, grueling, and time consuming of all antebellum farm tasks. In 1857, *The Cultivator* editorialized that prolonging fences' lifespans alone justified the investment: "The importance of a good drain under every post fence, is not generally understood.... Wherever post holes retain water, they are sure to be heaved by frost, and the fence thrown out of shape; and the post cannot last so long." Durable and sturdy fences were the hallmarks of dry, domesticated landscapes.[32]

Farm reformers thus contended that water belonged only in flowing watercourses. In 1863, for instance, *The Cultivator* reprinted a lecture by A. B. Conger at the New York State Fair in which he implored farmers to prevent water from escaping from creeks, streams, or rivers and forming topographical "defects." Wherever "watercourses are deficient in number, imperfect in flow, or obstructed in their outlet," he sermonized, "the first essay of the drainer is to remedy these defects." As an expansive category, *drainage* referred to any settler activity—digging ditches, installing underground clay tiles, channelizing watercourses, removing debris from creeks and streams, and building levees and other flood-control structures—that removed surplus water. Lewis Lambert similarly juxtaposed urban sanitary reform and agricultural drainage, implying that stagnant surface waters were the rural equivalent of pestiferous urban sewage. "If the citizens of Toledo thus guard against what threatens disease, why may we of the country not do the same," he asked. "Is life to them more sweet? Is health to them a greater blessing? Are deadly stagnant pools to city eyes more unsightly and revolting than to ours?" In an 1856 letter to *The Cultivator* crowing about underdrainage's many advantages, John Johnston argued that farm drainage offered a "radical cure for all the *ills* that land or its products are *heir* to."[33]

Since contributors to the rural press applauded farm drainage as one of the noblest forms of progressive farming, they demonized settlers who ignored it as lazy, recalcitrant, unenlightened, and irresponsible. In 1858, *The Cultivator* chided the "majority" of settlers who ignored drainage. "On the one hand we see or hear of farmers who eagerly avail themselves of every opportunity of extending the drainage of their farms ... [but] on the other hand, we behold the spectacle of hundreds and thousands—the great majority, indeed—shaking their heads and turning away, seemingly unconvinced or determined not even to try, when the most satisfactory and irrefutable proofs and demonstrations are placed before them ... that draining is always a paying and highly advantageous operation."[34]

Similarly, *The Ohio Cultivator* lambasted such farmers as "sleeping residents, who never knew or dreamt of such a thing as enterprise or public spirit."³⁵ Writing to *The Cultivator* in 1853, Iowa farmer W. G. Edmundson reproached his Mississippi Valley counterparts: "Little or no attention is given to the drainage of the land.... no pains are taken to drain the soil by the use of the plow, and underdraining, even on the most retentive soils, is never practiced." As a result, crop yields plummeted, climates deteriorated, and unsightly aquatic flora proliferated.³⁶

Reformers believed that drainage, by bestowing a medley of aesthetic and material benefits and unlocking nature's hidden wealth, heralded settlers' progress in correcting nature's blemishes and imperfections. In just a few short decades, reformers and contributors to the rural press had elevated drainage into a cornerstone of antebellum environmental thought.

Mobile Nature and the Beginnings of State Intervention

W. G. Edmundson's admonishments reflected the fact that farm drainage remained in its infancy at midcentury. Despite the rural press's celebration of drainage and the diffusion of new technologies, reformers were only beginning to awaken to how separating soil and water required the steadily increasing involvement of the state. Indeed, draining wetlands entailed complex social, economic, and legal negotiations that impinged on settlers across multiple property boundaries. In sum, the cultural mandate to drain laid bare an incongruity at the heart of American agricultural development: the incompatibility of the country's grid-based land division system with dynamic nature.

Boundaries have defined American land tenure since the war for independence. When the federal government adopted the 1785 Land Ordinance, it institutionalized the partition of land into rectilinear townships, sections, and quarter sections. In transforming expropriated lands into real estate, the grid-based system made it easier for settlers to record transactions and buy and sell property. Yet the abstract divisions that commodified the land clashed with the mobility of many natural resources, including soil, water, weeds, and wildlife.³⁷

This was especially true with wetlands. In the decades preceding the Civil War, settlers experienced how fluid nature—surface and subterranean waters—intersected with the static grid and stymied drainage. As Edmund Ruffin himself acknowledged in 1833, "Cheap and profitable [drainage projects] ... are rendered impossible under our existing laws, because the concurrence of every individual owner of the swamp is necessary for the execution of the work." Moreover, "If by possibility, only a single proprietor opposed the scheme, while all the others were in favor of it, he alone might obstruct the execution." Widespread drainage awaited state intervention.[38]

Ruffin's observation dovetails with the historian Mark Fiege's concept of a *hydrological commons*—that is, a geographic space where the superimposition of private farms atop watery landforms turned water control into a social and environmental problem. Flowing across boundaries above and beneath the surface, water aggregated landowners' interests because controlling it (e.g., disposing it into outlets) required cooperation, shared investment, and structures built across multiple properties. Unsurprisingly, Ruffin and many other reformers indicted individualism and the "sacred rights of property" as drainage's most implacable antagonists.[39]

Increasingly mindful that agricultural drainage demanded new forms of social and political organization, farm reformers implored the state to act. In the pages of farm journals and at county and state fairs, they exhorted policymakers to create institutions that harmonized drainage efforts across multiple boundaries. As *The Ohio Cultivator* editorialized in 1851, drainage mandated "the necessity of co-operation among farmers for the performance of [drainage] work; and in many cases the aid of the law is necessary to compel unwilling ones to perform." In a similar vein, a speaker at an Ohio fair reminded his audience that "the energy of an individual is sufficient to secure this improvement for his own farm; yet through the district are many and extensive [marshy] regions which require the co-operation of owners of adjacent lands, and this co-operation is not easily secured." He concluded that "to ensure the necessary action in such cases, a better law is needed, and legislative aid should be invoked." Likewise, in 1854 the New Orleans–based *DeBow's Review* cast Southern swamps as ideal candidates for "being controlled by wise legislation and agricultural energy."[40]

But what represented "wise legislation?" And what was the appropriate level of government to plan and coordinate drainage? How could the state facilitate collective action and promote the expansion of settler agriculture without assembling a hierarchical bureaucracy at odds with US political culture? Or more simply, to borrow from South Carolina Governor Whitemarsh B. Seabrook's 1849 annual message, how could government "enable farmers to drain their lands through their neighbor's possessions?"[41]

As antebellum farm reformers recast drainage as a pillar of progressive farming and settler land management, the obstacles involved in managing a transient resource across boundaries came into sharp focus. On the state level, beginning in the 1810s legislatures created levee boards, ditch laws, commissions of sewers, and other institutions. Loosely modeled after colonial "commissions of sewers," these quasi governments coordinated drainage across a patchwork of private boundaries. In 1857, the evolution of these institutions reached an apogee when Michigan authorized semiautonomous drainage districts.[42]

Districts served as subgovernments once a specific percentage of landowners in a given area petitioned a governing body—usually the county or a local court—to create one. If the petitioners gathered enough signatures and offered evidence that the district benefited public health, utility, or welfare, the body approved a district, which had the authority to impose taxes, issue bonds, condemn land, and compel the participation of opposing property owners. By 1929, thirty-five of the forty-eight states had sanctioned drainage districts. The district model signified legislatures' preference for piling overlapping jurisdictions on top of one another rather than replacing the grid's arbitrary lines with a land tenure system that conformed to hydrological realities. The proliferation of drainage districts, especially in the first two decades of the twentieth century, ushered in a silent public policy revolution that forever changed how settlers interacted with their neighbors, government, and nature while dealing with surface water.[43]

Seen from one perspective, the establishment of ditch laws, levee boards, commissions of sewers, and drainage districts created a chaotic and atomized structure. The local, ad hoc approach capped a decades-long period of experimentation that occurred with little unity of purpose. Despite the improvisation, the structure did provide settlers with the institutional and financial means of managing a mobile resource (surface and subsurface waters) across property boundaries. Moreover, the power to condemn lands and coerce the participation of uncooperative landowners later became a key fixture of drainage districts.

Through the creation of intermediary drainage institutions (e.g., drainage districts and other local drainage institutions), the state prioritized the expansion of settler agriculture. However, state lawmakers did not anticipate how federal landownership would negate the effectiveness of intermediary institutions in the Midwest and South. Unlike the original thirteen colonies, states carved out of the territory west of the Appalachian Mountains entered the union with large quantities of federally owned land inside of their borders. In those new states, farmers and plantation owners bristled at the intermingling of federal and private land, which impeded drainage because it was illegal to tax US property, condemn it, or forcibly absorb it into local drainage institutions. As a result, settlers living next to federally owned wetlands, mainly in the Midwest and the Mississippi River watershed, had no incentive to organize water control entities because their drains and levees had to cease at federal property lines.

During the late 1830s and 1840s, a coalition of Southern and Midwestern congressmen cried foul about how federal land ownership imperiled the associational arrangement. Nearly four dozen drainage-related petitions and resolutions from legislatures, counties, local drainage conventions, and private citizens poured into Congress from Arkansas, Florida, Illinois, Indiana, Iowa, Louisiana, Mississippi, Missouri, and Wisconsin asking it to subsidize drainage in public-land states by ceding "worthless" and "refuse" federal wetlands so that the states could sell them to pay for drainage projects. Other petitions favored the advent of a federal drainage program, steep reductions in the price of unsold swamplands, or congressional authorization to build drainage ditches and levees on federal lands.[44]

Just prior to midcentury, the loosely orchestrated campaign finally paid off. In 1849, Congress passed the first of three Swamp Land Acts. This act ceded the federal government's "swamp and overflowed" lands in Louisiana to the state government on the condition that it sell those lands and invest the revenue in drainage. In 1850, Congress extended the program's provisions to twelve Western, Midwestern, and Southern states: Alabama, Arkansas, California, Florida, Illinois, Indiana, Iowa, Michigan, Mississippi, Missouri, Ohio, and Wisconsin. Finally, in 1860 Congress instituted the third and final Swamp Land Act, which applied to Minnesota and Oregon shortly after they achieved statehood.[45]

During congressional floor debate, policymakers appropriated the rhetoric of the early republic's naturalists and physicians and antebellum farm reformers. They repeatedly denigrated swamps as "utterly worthless," "valueless," "unsalable," "injurious to human health," and "prolific . . . of disease," as well as being "a standing nuisance." Nevertheless, the swamp and overflowed lands were less wastelands than wasted lands. As one congressman explained, "These lands were of very little value in their present condition, but their reclamation would benefit the general treasury, as it would bring lands into the market which were now unsalable." Indeed, the floor debates signified how policymakers viewed wetlands as a formidable obstacle to expansion because they could not be sold, settled, and drained until they were surveyed and subdivided. For policymakers and agricultural reformers alike, watery tracts possessed no intrinsic value. Their only redeeming quality lay in their potential for being drained, commodified, and then incorporated into a settler agroecology that featured dried and domesticated soils.[46]

In the long run, a legacy of graft, fraud, and corruption marred the federal "swamp and overflowed" cessions to the states, but a less appreciated consequence emerged in environmental federalism. The cession of federal "swamp and overflowed" lands cast a long pall over environmental policymaking by devolving wetlands management into an exclusively local role. Yet the laws also achieved their aim—converting federal wetlands into settler farms—with astonishing alacrity and efficiency. In total, they transferred 65 million acres of wetlands to the states, much of which was privatized, incorporated into drainage districts, and plowed over by the end of World War I.[47]

Congress signed off on this gigantic land transfer while clinging to unscientific stereotypes about wetlands. Moreover, not until the mid-1880s would the USGS finally launch the government's first investigation into the quantity and geographic distribution of wetlands. That particular study would also be the first to categorize different classes of wetlands based on their botanical, geographic, hydrological, and pedogenic characteristics. In the meantime, lawmakers condemned every "swamp and overflowed" tract as a homogenous entity that manifested nature's disorderliness, be it in the form of excessive precipitation, an obstructed or meandering river, or some other factor. The belief that wetlands lay outside of nature's complex, interrelated webs made it easy for Congress to give them away with the expectation of their swift drainage and conversion into settler homes.

Conclusion

Antebellum farm reformers popularized drainage as one of the wisest forms of agricultural improvement. Along with crop rotation, diversification, marling, and manuring, drainage boosted land values and annual yields on existing farms in the East while also facilitating the creation of new farms in the Midwest. In repeatedly extolling drainage's health, wealth, and aesthetic benefits, the rural press hardened drainage into a social and cultural imperative. The next several generations of settlers would organize drainage districts or mount efforts to involve local and state governments to drain the land with the ultimate goal of forging a settler agroecology—crop fields, homes, barns, fences, roads, and so forth.

Farm reformers relied on the rural press to spread their message. The fecund and inexhaustible quality of wetland soils made them an untapped source of wealth. As "W. B." argued in an 1858 edition of *The New England*

Farmer, "A very great proportion of the future agricultural fertility of New England lies in her now profitless swamps and quagmires." Endowed with latent but recoverable wealth, swamps should not be avoided but rather occupied, reclaimed, and domesticated. Their settlement staked settler claims and took aim at nature's alleged hydrological imbalances. In now arguing that submerged soils were less perilous than profitable, the rural press encouraged settlers to seek out wet tracts and complete their agricultural conversion.[48]

Reformers and contributors to the rural press invited the state to intervene. Government—at the state and federal levels—intervened powerfully in the mid-nineteenth century to assist drainage-minded settlers in managing mobile nature across boundaries. By authorizing ditch laws, levee boards, and drainage districts, state governments propped up intermediary institutions that coordinated and administered surface water removal. Congress also chipped in. While the national state lacked the administrative capacity (or political will) to oversee drainage, it did provide liberal public land grants so that the states could create their own drainage programs.

State governments, however, were seldom up to the task. As a group of Norwegian immigrant farmers in northwestern Minnesota soon discovered, the control of surface water, despite the advent of the clay underdrainage tiles and new forms of legal and political organization, was still subject to contingencies, ecological realities, and corporate greed. It is to their story that we now turn.

CHAPTER THREE

An Empire of Ditches
Settlers, Railroads, and Wet Prairies in the Valley
of the Red River of the North, 1877–1886

The 1881 spring planting season initially went according to script for Andrew Lommeland. Born in Norway in 1823, Lommeland and his family had immigrated to the United States seeking prosperity and cheap land. After notching his fifty-fourth birthday in 1877, Lommeland settled in Clay County in far western Minnesota, filed for a homestead, plowed under the native prairie grasses, and planted his first small crop the next year. Within three years, the widower and father of two daughters owned three mules, a cow and calf, and three lambs. The lambs were a constant thorn in his side. Late in 1879, Lommeland slaughtered one for food and then watched the others succumb to a baffling illness—having produced a scant twelve pounds of fleece. Raising two daughters alone in a frontier region would have tested the mettle of even the hardiest settler, yet he persevered. Providing for his twelve- and eight-year-old daughters, Amalia and Sarra, offered plenty of motivation for him to plow up more of the prairie sod each year in the fertile valley of the Red River of the North, which at the time was eclipsing Nebraska and Kansas as the hub of Western settlement.[1]

Fate soon dealt a cruel blow. In the spring of 1881, after planting forty acres of wheat and oats, Lommeland eagerly awaited his best crop. A few weeks later, however, a shallow sea of water suddenly submerged his fields. Stunned, he described the sheet of water as "coming just a rolling on the prairie." The eighteen-inch-deep sea puzzled Lommeland and his Scandinavian immigrant neighbors. The nearby Buffalo River was not cresting. No nearby dams or levees had breached. Rainfall had been minimal. Yet the shallow lake stubbornly lingered for a month, drowning Lommeland's recently sprouted wheat and oats.[2]

The next year brought no relief. In the spring of 1882, another flood inundated eighteen square miles of Clay County's Moland and Morken Townships. Peter Boen, who lived a mile northeast of Lommeland, was incensed that the water loitered on his fields for over a month. Determined to unravel the mystery, Boen drove a team of animals directly into the current of the sheet of water. After several hours, he arrived at a large perpendicular ditch connected to the train track embankment of mogul James J. Hill's St. Paul, Minneapolis and Manitoba Railway (SPM&M), the predecessor of the transcontinental Great Northern Railway (GN). Railroad engineers had constructed the ditch to protect the embankment from washing out after rainstorms and to drain the company's unsold land grant. Rather than connecting the ditch with the Buffalo River, the engineers emptied it into a shallow prairie "swail," where it discharged significant volumes of water toward Moland and Morken. As Boen explored the ditch, a torrent rippled above the ground in the direction of his farm. Lommeland soon made his own trip to the ditch, where he discovered a "swift" current that "was over my boots." Ultimately, the water destroyed $2,000 of his wheat and oats for a second consecutive season. Envious of neighbors who had escaped the deluge, Lommeland protested that "I had just as good wheat as the rest of them, and they got from 20 to 25 bushels [per acre]."[3]

Despite the staggering negligence of SPM&M engineers, the Clay County floods symbolized far more than a narrative of victimized farmers rising up against corporate recklessness and rapacity. Indeed, the SPM&M's drainage program, as well as its impending legal battle with Lommeland and dozens of settlers, connected to several broader themes in the history of drainage. When the Swamp Land Acts devolved drainage responsibilities to the states, they did so without authorizing surveys of the location and extent of wetlands or dispensing any specialized knowledge. Congressmen dismissed drainage as an easy task that could be accomplished locally and without any additional federal assistance.

In many instances, communities succeeded. State legislatures empowered drainage districts with the authority to plan and administer farm drainage across private boundaries. During the last two decades of the nineteenth century, the quantity of farmland improved by drainage—concentrated mostly in Ohio, Indiana, Illinois, northern Iowa, southern Wisconsin, and southern Michigan—rapidly accelerated as large landowners and farmers organized drainage districts and availed themselves of clay tiles manufactured by dozens of regional start-up factories. The onslaught against Midwestern wet prairies and other wetlands maximized agricultural productivity, boosted land values, and served as a prelude for the Progressive Era's explosion of agricultural drainage. Indeed, prior to 1880 unimproved wetlands fetched an average of $7 an acre. By the end of the decade, drained and improved land boasted a price of $25 per acre; in the 1890s, it sold for $60 to $75 per acre; and at the end of World War I an acre of drained land commanded $400 on the open market.[4]

The conquest and transformation of Midwestern wetlands into an uninterrupted expanse of crop fields never followed a uniform playbook. Subject to a range of contingencies and contexts, the progress of drainage remained tied to local geography, hydrology, soils, politics, and landownership patterns. In the Red River Valley, no one doubted that wet prairies must be drained to accommodate the vanguard of white settlement, fortify settler health, and satisfy the ravenous appetite of Minneapolis flour mills. In the early 1880s, however, settlers and railroad executives discovered that the valley's absence of deep and defined watercourses left farmers with almost nowhere to dispose of the water from their farms. They asked the SPM&M, the valley's largest landowner, to step in and fill the void, but its ill-conceived drainage program flooded more farms than it drained, leading to a bitter and lengthy legal contest. These lawsuits, numbering in the dozens, forced the SPM&M to abandon any responsibility for water resources development. The valley's agricultural drainage awaited a later day when settlers could muster the social and political will to persuade the legislature to act.

The story of the initial phase of drainage in the Red River Valley unmasks the limits of environmental associational governance. Intermediary institutions always formed the bedrock of the associational model, but when they failed, so too did federal environmental policy (e.g., the Swamp Land Acts). Ironically, the breakdown of associational governance invited a broader process of state formation at the local and state levels as settlers mobilized to demand that Minnesota live up to its responsibilities under the Swamp Land Acts, enthusiastically welcoming a dramatic expansion of state government into their lives. In 1882, as Lommeland and his neighbors united to confront James J. Hill's company in court, these themes came into sharp focus, illustrating how farm drainage formalized new relationships between settlers, corporations, and the state.

The Red River Valley: The Remnant of a Glacial Lake

The origins of Lommeland's tragic story can be traced to the conclusion of the Wisconsin Glaciation. The valley of the Red River of the North, which constitutes the present-day boundary between Minnesota and North Dakota, is the remnant of an ancient glacial lake. When the Wisconsin Glaciation ended 12,000 years ago, retreating glaciers in Canada blocked the northerly flowing Red River from emptying into Hudson's Bay. The obstruction created a mammoth 400-foot-deep glacial lake—later named Lake Agassiz, in honor of the famed geologist Louis Agassiz—spanning 123,500 square miles.

As North America's largest glacial lake, Agassiz persisted in various forms for 4,500 years, submerging much of present-day Manitoba, Ontario, west-central Saskatchewan, and the Red River Valley. Before draining 9,000 years ago, Agassiz shaped the Red River Valley's topography. Whipped up by winds, waves smoothed the lake's floor, creating a flat, uniform surface bereft of natural outlets. Rivulets flowing into Agassiz transferred mineral matter from the glacial till into the lake, causing heavier clays to settle in the lake's center, silts in shallower water, and sand deposits along the lake's outer shores. As the lake waters slowly retreated, tall grass prairie colonized the former lakebed, and an exceptionally rich soil accumulated to a depth of five feet.[5]

Immature drainage systems are a hallmark of former glacial lakebeds. Indeed, Lake Agassiz's recession left in its wake a paucity of well-defined watercourses. On the Minnesota side of the Red River, the river's few tributaries flowing from the east to the west followed meandering courses and possessed shallow and undefined banks. They often overflowed their banks while traversing the flattest portions of the former glacial lake's floor, creating sprawling seasonal marshes. During the nineteenth century, American and Canadian settlers described this phenomenon as the watercourses "losing themselves on the prairie." In addition to the river-fed marshes, long stretches of wet prairie overlaid Agassiz's former floor. Wet prairie ecosystems, which dominated the predrainage tall grasslands on the flat upland till plains of Ohio, Indiana, Illinois, southeastern Wisconsin, southwestern Minnesota, northern Iowa, and the Red River Valley, sprang up where spring snowmelt and late summer thunderstorms engulfed the prairie with more water than the ill-defined drainage system could carry away or the impervious clay soils could absorb. In the Red River Valley, ecologists estimate that wet prairies and other wetlands comprised about 20 percent of the surface when Minnesota achieved statehood in 1858.[6]

Despite its seasonal wetness, the Red River Valley was emerging as a wheat-producing mecca when Lommeland arrived in western Minnesota. From the mid-1860s to the early 1880s, a complex chain of events set in motion a process that culminated in the "Red River Boom," a period of dense agricultural settlement when the valley overtook Kansas and Nebraska as the epicenter of Western settlement and the region acquired its esteemed agricultural reputation. It was also then that Lommeland and SPM&M engineers discovered firsthand the challenges of growing wheat and laying railroad tracks across the featureless, watercourse-starved bottom of a former glacial lake.[7]

The Red River Boom: Railroads, Twin City Millers, Wheat, and Bonanza Farms

In Minnesota as elsewhere, fierce competition over railroads characterized early territorial growth. After Illinois Senator Stephen Douglas persuaded Congress in 1850 to subsidize the Illinois Central Railroad with a public land grant, a political firestorm erupted as other communities vied for similar munificence. During the 1850s, Congress bowed to localism and granted some 22 million acres to eleven Southern and Midwestern states to pay for railroad construction. By the end of 1870, the amount of public land subsidies had jumped to 155 million acres to eighty railroad corporations.[8]

Minnesota benefitted from the giveaways. On March 3, 1857, Congress authorized public land subsidies for Minnesota railroads, granting the odd-numbered sections in six-mile-wide corridors to companies building in accordance with a predetermined network. For the Red River Valley, the two most important corridors extended from St. Paul to the west, and to the northwest toward the Red River's intersection with the international boundary. That same year, Minnesota's legislature incorporated the Minnesota and Pacific Railroad to construct a line from Stillwater to St. Paul and all the way west to Breckenridge, with a branch line connecting to Pembina on the forty-ninth parallel. The onset of the Panic of 1857, however, dried up capital, sapped investors' enthusiasm, and left speculators and banks hesitant to invest in a vast territory with a sparse population of 150,000. In April 1858, the legislature finally issued bonds, not exceeding $5 million, to fund construction. Within a year, recognizing that the young state could not redeem the bonds, the legislature terminated their issuance at $2.3 million and the railroad did not lay a single track.[9]

After the outbreak of the Civil War, legislators redoubled their efforts. In 1862, Minnesota still lacked a single mile of operational railroad, but the legislature incorporated the St. Paul and Pacific Railroad to assume control of the Minnesota and Pacific's franchises, possessions, and land grant. The company made torturously slow progress and, by summer's end, had barely completed a ten-mile line to St. Anthony. Once better economic circumstances returned following Appomattox, the St. Paul and Pacific quickened its construction pace and laid 283 miles of track by the early 1870s; in October 1871, the company finally reached Breckenridge on the Minnesota–Dakota Territory boundary. In spite of receiving much of its lavish 2.46-million-acre land grant, the company struggled to manage its high debt load and investors' expectations.[10]

The Saint Paul and Pacific was by no means the only Minnesota line to benefit from congressional largesse. In 1864, Congress chartered the Northern Pacific Railway (NP) to construct a transcontinental line from Lake Superior to Puget Sound. Though Congress decided against subsidizing the NP with generous loans—as it had the country's first transcontinental railroad, the Union Pacific—it approved a gargantuan land grant encompassing twenty sections per mile in the states it crossed and forty sections per mile through territories. In the summer of 1870, workers laid the first tracks near Duluth and the corporation hired Jay Cooke—a brilliant financier who had bankrolled the Union's Civil War efforts—to underwrite the immense undertaking. Cooke's support of the NP ultimately overextended his financial firm and threw the nation into the Panic of 1873, which plunged the company into receivership.[11]

As the Panic of 1873 ravaged the national economy, the city of Minneapolis, incorporated in 1867, was coming into its own as a global milling headquarters, churning out high-quality white flour. The city's strategic location near the falls of St. Anthony, as well as the discovery that Minnesota's abbreviated growing season proved well suited for hard spring wheat, spurred on its economic ascendancy. In spite of hard spring wheat's adaptability to Northern climates, it befuddled Minneapolis millers. The wheat's bran—the hard, brittle outer layer of the grain—disintegrated during the milling process, obstructing the millstones. Seeking a solution, the city's millers appropriated an eastern European technique that employed a series of rollers, separated by diminishing gaps, that detached the germ and bran from the wheat's kernel. After this first step, an air burst isolated the bran and germ. The innovation enabled Minneapolis millers, including the likes of William Washburn and C. A. Pillsbury, to refine the type of grain best suited for the city's agricultural hinterlands, including the extraordinarily fertile Red River Valley.[12]

Minneapolis's appetite for hard spring wheat afforded the NP with an opportunity. James B. Power, the NP's wizardly land commissioner, orchestrated a program in the 1870s that enabled NP bondholders to exchange their notes for immense tracts of the company's 47-million acre land grant, 10.7 million acres of which was located in present-day North Dakota. Power's exchange program instantly succeeded, particularly in the Red River Valley. From September 1875 to August 1878, bondholders' exchanges accounted for over 70 percent of NP land transactions. Over half of the 1.2 million acres disposed of during that period went to forty people. At the same time, a series of poor European harvests elevated American commodity prices, enticing additional bondholders to exchange their notes for large tracts christened as "bonanza farms." By 1880, the Red River Valley boasted at least ninety-one bonanzas on the Dakota side of the boundary, each of which consisted of at least 3,000 acres. The size of many of the operations, often owned by absentee landowners, staggered visitors, including President Rutherford B. Hayes in 1878: 76,000 acres, 65,000 acres, 37,000 acres, 36,000 acres, 34,000 acres, 28,000 acres, 22,000 acres, and so forth.[13]

Towering over the valley's opening phase of settler agricultural development, the bonanzas combined professional management, intense capitalization, the latest machinery, and cheap seasonal labor to turn the tall grass prairie into uniform factory farms. They also attracted national acclaim and publicity. Newspaper reporters converged on the bonanzas during the late 1870s, publishing glossy accounts of the easy wealth awaiting settlers. In the late 1870s, Power lavished praise on the bonanzas in stirring letters to *The Country Gentleman*, the *New-York Tribune*, and other periodicals. The bonanzas' dazzling productiveness likewise prompted Power's personal and public attacks against government scientist John Wesley Powell. A famous Western explorer and the future director of the USGS, Powell had invited Power's ire by casting aspersions on the agricultural potential of the arid western United States at an April 1877 meeting of the National Academy of Sciences. "Major Powell may be a first class 'rock sharp,'" Power confided to a friend in 1877, "but it is clear that as a judge of [the] agri[cultural] value of western lands he can be written 'an ass.'" The bonanzas' fortunes signaled that climate apparently imposed no restrictions on the onward march of settler colonialism. On both sides of the Red River, bonanzas devoured more and more land, but the valley also welcomed a rising number of midsized farms purchased from railroads and supplemented by government land sales. In 1882, the valley had 82 farms between 1,000 and 3,000 acres; by the end of the boom decade, the number climbed to 323. The number of farms ranging in size from 500 to 1,000 acres also exploded. On the Minnesota side of the valley in 1880, there were 70 farms comprising 500–1,000 acres; ten years later the number quadrupled to 258, with many of them purchased from the St. Paul and Pacific and its successors.[14]

Collectively, the bonanzas and railroad settler recruitment drives laid the foundation for the Red River Boom. Seeking to lure settlers and dispose of their land grants, the NP and the St. Paul and Pacific dispatched representatives to England, Germany, Holland, Iceland, and Switzerland. Hypnotized by tales of easy riches, land seekers from North America and northern Europe converged on the valley. By 1881, the amount of public land being occupied in the upper valley was the highest in the United States. As the boom approached its apogee in 1881, the federal land offices at Grand Forks, Fargo, and Crookston (Minnesota) respectively ranked third, fourth, and sixth for the entire country for total acreage in land sales and land entries. Population and cultivated land totals skyrockted as well. During the 1880s, the valley's inhabitants tripled from 56,000 to 166,000, the amount of land integrated into farms jumped from 1.6 million acres to 5.4 million acres, and the percentage of farmland in crop climbed from 22 percent to 65 percent. The breakneck pace of federal land claims and purchases in the decade by small farmers like Andrew Lommeland dwarfed the bonanzas' growth. At the federal land offices at Fargo, Grand Forks, and Crookston, settlers claimed a combined 6.4 million acres under the 1862 Homestead Act and 3.5 million acres under the 1873 Timber Culture Act between 1879 and 1895. Small farms had finally overtaken the bonanzas. In 1890, the valley counted 316 farmers who owned tracts of 1,000 acres or more, 890 farmers who owned tracts between 500 and 1,000 acres, and another 18,000 settlers who owned farms that averaged 263 acres.[15]

Whether big or small, valley farms shared a common feature: wheat cultivation. The region's flatness, connection to Minneapolis via railroads, and machinery innovations such as the McCormick twine binder bolstered the wheat economy. Railroad haulage statistics reflected wheat's dominance. In 1874, the NP and the St. Paul and Pacific carried a combined 2.4 million bushels of wheat; a decade later, they hauled 26.4 million bushels. In 1890, the six counties on the Minnesota side of the upper valley had 644,118 acres planted in wheat. Collectively, the number of acres planted in oats, barley, corn, potatoes, flax, hay, forage, and rye tallied 464,000 acres.[16]

Altogether, the bonanzas' legendary success, railroad propaganda, and the proliferation of small farms fueled a belief that easy riches awaited new settlers. The truth, however, proved more complex. As the best lands quickly passed into private ownership by 1885, many farmers discovered firsthand how seasonal wet prairies and undefined watercourses impeded the expansion of settler agriculture. It was left to an ambitious Canadian-born entrepreneur to furnish the valley with the drainage necessary to complete its transformation into the hemisphere's breadbasket.[17]

The Rise of James J. Hill and Railroad Drainage

That Canadian was James J. Hill, a domineering figure whose rise to prominence and prosperity was meteoric even by even Gilded Age standards. By the time improved milling techniques intensified Minneapolis's appetite for hard spring wheat, Hill had established himself as a successful St. Paul businessman. Born in 1838 in Ontario, Hill was the third child of James Hill, a Scots-Irishman who in 1829 had relocated his family to Ontario, and Ann Dunbar, a Scots-Irish Presbyterian. In 1856, the restless Hill moved to St. Paul and found work on the Mississippi River's levees, where he absorbed the ins and outs of the American-Canadian fur trade. On the heels of the Civil War, he founded the James J. Hill Company in St. Paul, positioned himself to become the St. Paul and Pacific's primary forwarding agent, and entered the steamboating business on the Red River of the North.[18]

In the late 1870s, Hill and a group of associates conspired to take over the St. Paul and Pacific. By 1878, Hill and four business partners had secured the line from its Dutch bondholders, rebranding it as the SPM&M to reflect its north-south orientation. From the moment Hill became the line's general manager, he hurried to meet construction deadlines to secure a Red River Valley land grant. When all was said and done, the SPM&M's grant encompassed the odd-numbered sections for ten miles on both sides of the track from Breckenridge to the Canadian boundary, making it the owner of 25 percent of the upper Red River Valley in Minnesota.[19]

As the Red River Boom entered its heyday, the valley figured prominently in Hill's economic strategy. First, the SPM&M stood to profit handsomely by hauling an ever-growing amount of wheat to mills in Minneapolis and Duluth and also by supplying settlers with equipment, household goods, livestock, lumber, machinery, and seed. Second, as the landowner of a quarter of the upper valley in Minnesota, the SPM&M could reap a fortune by marketing and selling its land grant. Small wonder that Hill tied the SPM&M's economic destiny to the valley's dense settlement. His efforts paid immediate dividends. In 1880 alone, the SPM&M sold almost 100,000 acres of valley land and its freight revenue exceeded $2 million.[20]

Despite this initial success, Hill despaired that poor drainage hindered the valley's agricultural development. From his correspondence with farmers and local leaders, he realized that wet prairies and immature rivers locked a large amount of land out of cultivation and delayed farmers from timely spring planting. Both of these factors jeopardized the SPM&M's growing carrying trade and marketing of its land grant, the twin pillars of Hill's business strategy. Moreover, the SPM&M suffered from repeated track washouts every spring when the prairie could not absorb snowmelt, driving up construction and maintenance costs.

Hill early on concluded that the valley required drainage to reach its full potential, but who bore responsibility for planning, building, and administering a drainage program? On this question, a growing number of Minnesotans

James J. Hill, the architect of Red River Valley drainage. Reprinted with the permission of the Minnesota Historical Society.

denounced the failure of federal drainage policy. As with almost every other state included in the federal Swamp Land Acts, Minnesota had brazenly flouted the law. The state received 4.7 million acres of "swamp and overflowed" public lands, the fifth largest donation to an individual state. Yet Minnesota's cash-strapped legislature used the donation for every purpose but drainage, including subsidizing road construction, an asylum, a "deaf and dumb" institute, schools, railroads, and a state prison.[21]

The breakdown of federal drainage policy created a power vacuum. Just as troublesome, by 1880 Minnesota had not authorized drainage districts. As a result, Hill's railroad became the only entity with the workforce and organizational capacity to plan and direct a drainage program. This phenomenon was not isolated to northwestern Minnesota. As historians Donald J. Pisani and Richard Orsi have argued, Western railroads often spearheaded the initial planning and management of forests, irrigation, rangelands, and wilderness parks, serving as de facto "surrogate governments" where local or federal authority was weak or entirely absent. As private institutions, railroads could fill the gap between local and national authority when state governments failed to intervene in environmental planning and administration. Although corporate profit obviously motivated railroad executives to act as government proxies, Hill sought to forge a community of interests with farmers that maximized corporate profits and conformed to prevailing environmental values, which enshrined drainage as a central ingredient of settler land management.[22]

Hill's interest in drainage thus blended pragmatism and idealism. Like most Gilded Age Americans, he subscribed to the belief that nature was infinitely malleable and should be made subservient to human needs, namely agriculture. "Nature," he later wrote, "holds out in one hand her horn of plenty and in the other her scourge." The fact that the environment "follow[ed] laws of its own" manifested itself most clearly in the uneven distribution of water over the earth's surface. He concluded that "man must adapt the distribution of water, by which the earth's productiveness is regulated, to suit his needs." By "man" Hill obviously meant white settlers, and he viewed the valley as an unoccupied region that was settlers' birthright. Proper management required the elimination of hydrological imbalances on landscapes that suffered from too much water (wet prairies and river-fed marshes) and too little (deserts). In his judgment, drainage and irrigation comprised the key activities for achieving settler permanence. Utility served as Hill's yardstick for evaluating nature, which had no intrinsic value beyond submitting itself to intensive agriculture.[23]

In the Red River Valley, Hill lamented that the lack of drainage disrupted wheat cultivation's delicate seasonal rhythms. If heavy winter snowfall inundated the prairie and postponed a settler's spring planting, the crop was at a heightened risk of not fully maturing before autumn frosts. Moreover, when late summer downpours waterlogged the valley's stiff soils, a new settler was incapable of "backsetting" his fields. Backsetting was the initial step of cultivation. After farmers tore up the prairie grasses in June and July, the land was backset—tilled in the opposite direction of the first breaking—before the ground froze to pulverize large clumps and clods, streamlining the spring seeding. If surface water lingered on the prairie in the autumn, settlers could not backset before the first frost, likely delaying planting in the spring. Finally, surface water jeopardized timely harvesting. Unlike corn, wheat and other small grains are unprotected by a husk and become especially vulnerable to natural elements as they mature. If loitering surface water delayed the harvest, the matured wheat, once pummeled by autumn rains, winds, or hailstorms, tumbled to the ground and spoiled, reducing annual yields.[24]

George C. Reis, a Polk County farmer, galvanized Hill's interest in drainage. In a September 1879 letter to Hill, Reis bemoaned that it had been four weeks since the last rainstorm, yet water still swamped his fields. Even worse, he was "not . . . able to put my Plows to backsetting." Strong willed and resourceful, Reis briefly considered draining the prairie on his own. He quickly discovered the vexing nature of drainage on a glacial lakebed. It required scientific precision in assessing proper levels and gradients. The ditches themselves required a uniformity of depth so that the water's movement to an outlet was not impeded. But there was more. Unless a farmer's property abutted a creek, river, stream, or pond, drainage ditches were useless and could not be emptied without jeopardizing someone else's property. A frustrated Reis conceded that it was impossible for him to "do his own Draining especially if he can't tell which direction to go, or where the lowest place is."[25]

Undeterred, Reis pressed Hill for assistance. In his mind, it made sense for the SPM&M to build a series of drainage ditches since much of the company's Polk County land grant, like his own farm, languished beneath water. "It will pay your Co. to look after this a little," Reis vowed, "as your lands can't be Sold if left in present shape." In late 1879, Hill dispatched engineers to dig ditches for Reis and other nearby settlers, but the effort failed. The ditches were too small, cheap, and rudimentary. The next May, Reis erupted in anger. He had 500 acres of land broken and ready for seeding, but all of it lay under water.[26]

Heavy precipitation from 1878 to 1885 magnified settlers' challenges. The Red River Boom coincided with a period of heavy precipitation. Early in April 1880, the Glyndon-based *Red River Valley News* sarcastically declared "Mud is king." The *News* reported that the abundance of surface water rendered county roads "impassable," meadows "too wet" for cutting hay, and wheat crops "stunted" and "yellow." The *Moorhead Weekly News* joked that "in some places fishing would be more profitable than farming and would look more reasonable."[27]

The abundance of water led Reis's exchanges with Hill to take on an urgent and desperate tone. Unless the SPM&M built additional parallel ditches alongside its roadbed that doubled as outlets for his own ditches, Reis contended that his crops "will all be drowned out ... and it will be a very serious loss to me and a great draw back to this whole country—we need a good sized Canal cut towards Red River, can your Co. help us some how[?]" He continued that only the SPM&M, and not local or state agencies, "have the force and men to do it." Reis's correspondence prodded Hill to visit the area, and he agreed to build more ditches to relieve Reis's land.[28]

Ineptitude and miscalculation marred the undertaking. No archival evidence indicates that the company's engineers conducted even baseline surveys or sought the assistance of agricultural drainage engineers, a profession still in its infancy. The ad hoc endeavor lacked an overarching strategy. In several locations in Clay, Polk, and Kittson Counties, Hill's engineers dug at least fifteen additional ditches, totaling some forty-five miles. The ditches, on average, measured two to four feet wide and four to six feet deep, and ranged from a half mile to three miles in length. Securing easements or rights-of-ways across multiple properties was one of the most daunting challenges of drainage, but SPM&M engineers encountered little resistance. In 1880, Clay County farmer Ole O. Brevik invited railroad engineers to excavate a ditch across a sixteen-foot strip near his property line after they promised that "the ditch would be a valuable improvement to [his] lands ... and that said lands would be effectually drained by said ditch."[29]

Hill boasted that his expanding empire of ditches would "drain that whole country away from the track" with a "comparatively small outlay." The *Red River Valley News* agreed. In the summer of 1880, it celebrated the SPM&M's initiative: "This enterprise, while serving well the company's interest, will also be a boon to settlers along the line, the lot of whom has this wet season been of special hardship." In a similar vein, the *Minneapolis Star Tribune* crowed that Hill's ditches had "created permanent streams, running down to the Red River, and have demonstrated the practicability of thoroughly draining the wet lands heretofore regarded as of little worth."[30]

The SPM&M's efforts persuaded newspaper editors that "ordinary" and "unfortunate" settlers lacked the capital, manpower, and expertise to drain their farms. How could they lay out or finance easements or rights-of-ways? What experience did they possess in establishing accurate levels and grades? Cursed by an absence of viable outlets, the valley afforded only a fortunate few—mostly large landowners—with access to watercourses. In Dakota Territory, the *Grand Forks Herald* encouraged the territorial government to open a series of "main" ditches through low, flat sections into which farmers could empty lateral ditches from their own fields.[31]

In Minnesota, state legislator Bernard Sampson spearheaded the first political attempt to secure drainage funding. In 1881, he introduced legislation appropriating $1,000 for the governor to hire a professional engineer to carry out surveys and submit a detailed report. Sampson's proposal mirrored the *Grand Forks Herald*'s proposal: after the survey, the state should dig main outlets into which settlers could convey their fields' surface water. The *Fisher Bulletin* endorsed the proposal, envisioning it as "dig[ging] a huge ditch—almost a canal in fact—from the swamps on State land to the Red River, and then let[ting] the owners of adjoining property run laterals, at their own expense, into the main conduit." Nothing came of Sampson's proposition, but a growing chorus touted surface water removal as a legitimate state function and the linchpin of securing settler homes. As the *Moorhead Weekly News* explained in late 1882, "The soil must be drained and the sooner farmers combine their efforts and establish a general system of drainage the better." Uncoordinated and haphazard economic development must yield to a centrally planned and managed program.[32]

The Failure of Hill's Drainage Empire

One of the SPM&M's first drainage ditches extended east to west for three miles across Moland Township in Clay County. Completed during the winter of 1880–81, the Moland ditch drained an enormous quantity of water from the company's parallel ditches and adjacent farmers' fields, but its most glaring structural flaw was that it emptied into a prairie "swail" or "depression," rather than a watercourse.[33]

In April 1881, as Andrew Lommeland tended to his small crop of wheat and oats, melting snow and precipitation filled the Moland ditch to capacity and it released large volumes of water into the prairie swale. The water rushed across the surface, accumulating to a depth of eighteen inches and blanketing an area three miles wide (east to west) and nine miles in length (north to south). The shallow flood occurred just after farmers' crops had sprouted. Andrew Lommeland was but one of dozens of settlers whose farms lay in the torrent's path. The water stubbornly lingered for a month.[34]

Hogen M. Hogenson lived two miles from Lommeland and six miles from the ditch's mouth. Born in 1857 in Rock County, Wisconsin, Hogenson and his Norwegian parents, Peter and Sonva Hogenson, moved to Olmstead County in southeastern Minnesota when he was six years old. After growing up and helping his parents establish a profitable homestead, Hogenson longed for a farm of his own. The opportunity presented itself in 1880, when he left home and filed for a homestead in Moland Township. Arriving in the summer, he worked hard to plow up the native tall grasses and backset his land before winter. The twenty-three-year-old settler was somewhat of an outsider: He was one of the few native-born Americans among the Norwegian and Danish immigrants.[35]

The 1881 flood destroyed Hogenson's first crop, and he never doubted its source. Outfitting his draw animals one afternoon, Hogenson drove them in the direction of the current and reached the SPM&M's ditch. He also circulated a petition among his neighbors addressed to Hill. Though the petition is lost to history, it likely asked Hill to dam the ditch or provide some other form of relief. In a perfunctory response, Hill pledged that the matter would

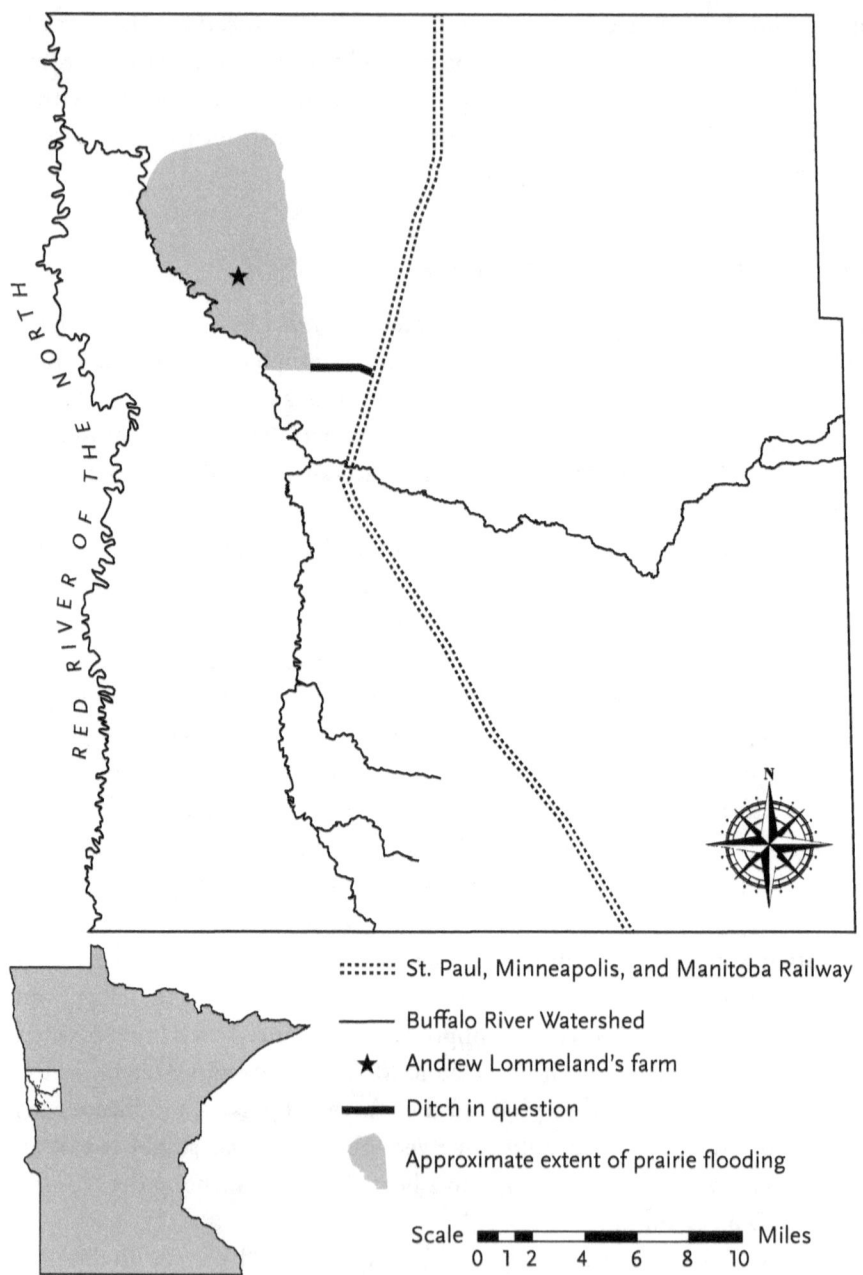

1881 drainage-related flooding in Clay County. Map created by Bobby Wright.

"receive attention at the earliest possible moment." Focused on extending his lines and addressing a bevy of similar flooding problems north of Fargo, he paid little attention to Hogenson's complaints.[36]

The assortment of problems revealed the limitations of corporate-led water management. It also laid bare the need for more comprehensive, integrated approaches. Early in June 1882, Hill tasked James B. Power—the SPM&M's land commissioner, former architect of the NP's bonanza program, and owner of the Helendale bonanza—with investigating drainage and preparing a general reclamation plan.

In a June 1882 letter to Hill, Power outlined his vision. After conferring with prominent landowners, he identified two basic conditions that inhibited drainage. First, the valley's absence of well-defined outlets prevented settlers from conveying their fields' water into natural watercourses via either open ditches or buried clay tiles. Second, rivers on the Minnesota side of the valley followed torturous and winding courses, possessed few tributaries, were choked with natural debris, and regularly overflowed and created marshes on the flattest portion of their journey into the Red River. Water conveyed into these rivers would enlarge the marshes. Power also found that a handful of bonanza farmers had drained their land by putting in "small and inefficient drains" at the expense of nearby settlers "that are so unfortunate as to be in the line of the extra overflow from these reclaimed lands. This I fear will be the result in every instance when individual owners attempt to reclaim their own lands and the drainage of these few large farms will in the end damage an area much larger than will be benefitted." Andrew Lommeland would have agreed.[37]

Power floated a proposal that followed the general contours of Bernard Sampson's plan. The valley's seasonal wetness, he argued, would not improve until some entity "open[ed] a series of main drains" into which settlers could empty lateral ditches or underground tiles. Once constructed, the main drains would empty into watercourses improved by railroad personnel. "Effectual reclamation of the valley land," Power concluded, "can only be accomplished by cleaning out the natural streams ... of the obstructions now preventing free flowage and adding to the natural drains a number of artificial water ways." The two-phase scheme promised substantial returns. In one section of Polk County, there was 312,000 acres—62,700 of which was owned by the SPM&M—of "practically useless" wet prairies. Power predicted that a valley-wide drainage program would furnish the company with an "additional carrying business of not less than $250,000 a year" and enhance the value of its unsold land grant by $100,000.[38]

Power tested his vision at the Keystone bonanza. Cooperatively owned by St. Paul's Springer Harbaugh and Charles Lockhart of Pittsburgh, the mammoth Polk County operation encompassed 9,000 acres. As the first bonanza operation established along the SPM&M's St. Vincent Extension to the Canadian border (completed in 1877), the Keystone resembled a bustling village. By May 1885, it included thirty-six buildings, ten granaries exceeding 80,000 bushels in capacity, nine barns, and a dairy equipped with running cold water to refrigerate bottled milk. The Keystone's proprietors owned 140 horses and mules and eighty-five cattle, and cultivated 5,550 acres of wheat, oats, rye, barley, and timothy.[39]

The Keystone's agricultural success had not been certain. In the wet seasons of 1880 and 1881, vast stretches of the bonanza lay underneath water. The situation grew especially dire in the summer of 1881 when hundreds of seasonal laborers stood idle as surface water delayed backsetting, plowing, and planting. Displeased with the land's wetness, Harbaugh and Lockhart allowed Power to construct a fourteen-mile-long main ditch across the Keystone farm connected to dozens of lateral feeder ditches. Like Power, the *Fisher Bulletin* anticipated that the project, once successful, would shake skeptical farmers from their lethargy and goad them into demanding systematic drainage from the state.[40]

In the spring of 1882, Power's ditches failed abysmally, flooding neighboring farms, the SPM&M's land grant, and federal land. Unwilling to admit defeat, he oversaw the creation of a new twenty-two-mile-long "canal" intended to drain thirty square miles of land into a tributary of the Snake River. Filled with brush, decaying logs, and thick vegetation, the Snake's tributary was unable to carry away the water influx. Clearing watercourses of obstructions had been the second phase of Power's proposed drainage plan, but he apparently ignored it in this instance. Despite these failures, Power pressed ahead. In October, the SPM&M signed a $25,000 contract with the New Era Grading Company to dig more ditches south of Moorhead, Minnesota, and in the vicinity of Grand Forks in Dakota Territory.[41]

Settlers versus the Railroad

Back in Moland, Lommeland, Hogenson, and others sought relief in the courts. On November 14, 1882, sixteen Moland and Morken farmers filed lawsuits against the SPM&M for "wrongfully, unlawfully, and maliciously" diverting water onto their property and ruining two seasons of crops. Their attorney, O. Mosness, worried that his clients' naturalization status imperiled their civil cases. Within weeks of suing, several of the farmers traveled to the county seat of Moorhead to take the oath of citizenship. On November 27, 1882, Arne Ostrem, Peder O. Boen, Ole Syverson, and Jens Pedersen collectively swore their fidelity to the Constitution and pledged that they had lived in the United States for at least five years. Over the next few months, several other settlers followed their lead.[42]

Mosness's prediction proved prophetic, as railroad attorneys used every available legal tactic to delay and dismiss the lawsuits. SPM&M executives alleged that the Moland and Morken farmers were engaged in a nefarious conspiracy to swindle the corporation out of large sums. Convinced that the railroad could not receive a fair trial in Clay County, SPM&M attorneys in late 1882 asked a judge to transfer the "flowage" cases to the US Circuit Court for the District of Minnesota in St. Paul. They also argued that most of the plaintiffs were "aliens" and lacked standing to sue. As the cases awaited hearings in St. Paul, lawyers for both parties agreed to adjudicate the lawsuit of Hogenson, an American citizen, in county court. On April 3, 1883, a county judge ruled against Hogenson and ordered him to pay the SPM&M's court costs. The verdict forced Mosness to obtain a continuance of the other cases while he considered appeal options. In the meantime, a different county judge ruled against the SPM&M in a similar case brought by Andrew N. Forsyth, ordering it to pay him $308 plus court costs for flooding his crops in 1881 and 1882.[43]

The split verdicts ultimately led the Minnesota Supreme Court to agree to hear Hogenson's case. The basic legal issue involved interpretations of diffused surface water law. *Diffused surface waters* were defined as "waters from rain, springs or melting snow which lie or flow on the surface of the earth but which do not form part of a watercourse or lake." Surface water law regulated the use, diversion, and disposal of waters unconnected with a watercourse until they entered and became part of a river, stream, creek, lake, or pond. In an 1881 decision, the Kansas Supreme Court described this transition:

> When surface waters reach and become part of a natural water course, they lose their character as surface waters, and come under the rules governing water courses. . . . And such waters, when they have ceased to spread and diffuse over the surface or percolate through the soil, when they have lost their casual and vagrant character and have reached and come to rest in a permanent mass or body, in a natural receptacle or reservoir, not spreading over or soaking into the soil, forming mere bog or marsh, cannot be regarded as surface waters any more than they can be after they have reached a stream.[44]

In general, states adopted one of two surface water rules: civil law or common enemy. The civil law rule followed the natural law maxim of *aqua currit et debet currere ut currere solebat*, that is, "Water runs and ought to run, as it used to run." Tracing its origins to Roman times and later incorporated into France's Napoleonic Code, the civil law rule imposed liability on landowners when they damaged an adjacent proprietor's property during the process of surface water disposal. The idea of servitude underpinned the rule: An upper landowner could not injure a lower landowner while removing water from his or her property. Conversely, a lower landowner had no right to block the natural flow of water on his land to the detriment of an upper landowner. By constraining landowners' ability to dispose of surface water in any manner they desired, the eighteen states that adopted the civil law rule before 1940 imposed a formidable obstacle on individuals seeking to drain their property, especially before the widespread adoption of drainage districts.[45]

At the other end of the legal spectrum, the common enemy doctrine applied an instrumental approach. The maxim *cujus est solum, ejus est usque ad coelum et ad inferos*, that is, "Whose is the soil, his is even to the skies and to the depths below," underpinned the rule. Often mistakenly identified as the common law, the common enemy rule, as its name implied, treated water lying or percolating across the ground as an adversary that landowners could remove as they desired, regardless of the consequences to others. In the United States, the rule traced its genesis to a series of Massachusetts court cases in 1851, 1859, and 1865. According to the legal scholar Jill Fraley, those cases witnessed the Massachusetts Supreme Court adopt the "long-standing British tradition of treating standing water as an enemy—as something that should be drained." In the District of Columbia and twenty-one states that eventually adopted the rule, Fraley continues, the courts "incorporated the British philosophy of land management by choosing liability rules that prioritized development." As such, the common enemy doctrine imposed little to no liability on proprietors if they damaged a neighbor's property during the process of drainage. Though nineteenth-century courts imposed a host of qualifications and limitations on the common enemy rule, it conformed nicely with prevailing cultural attitudes that prized the conversion of wetlands into healthy and thriving settlements. It also reinforced the myth that wetlands were unnatural, anomalous, and disconnected from hydrological systems.[46]

In the early 1880s, Minnesota adhered to the common enemy doctrine. In his written argument to the court, however, Mosness argued that neither the civil law rule nor the common enemy rule applied to Hogenson's circumstances. "It is a maxim that everyone must so use his own property as not to injure his neighbor's," he argued, "and if the law of surface water is an exception, it violates a maxim as just, as broad and as universal as reason itself." In directly transferring the burden of surface water from its property to nearby farms, the SPM&M's negligence far exceeded any other comparable case. "This case differs from most of those found in the books in that the magnitude of the offense is greater, greater damages are suffered," he continued. "In this case a large stream of water is wantonly thrown upon a farm verdant with growing grain, done at a season when the same are most susceptible of injury."[47]

In their arguments, railroad attorneys massaged existing statutes. First, they contended that surface water law generally applied to disputes between "neighbors," that is, adjacent proprietors. In the case at bar, Hogenson's farm was six miles north of the ditch in question. As a result, he lacked standing to sue. Second, the railroad attorneys emphasized that company engineers had no choice but to convey the surface water onto the prairie instead of the Buffalo River. In the spring, tributaries of the Red River—including the Buffalo—flowed at their maximum capacity. Emptying the ditch into the Buffalo would have caused it to overflow, something expressly forbidden by case law. Nevertheless, SPM&M attorneys conceded that the case essentially boiled down to identifying the most efficient and beneficial use of property: the operation of railroads or wheat cultivation on a few isolated farms.[48]

In November 1883, the Minnesota Supreme Court ruled for Hogenson. Absent an explicit agreement between Hogenson and the railroad, the SPM&M had no right to commit the act in question. Writing for the majority in Hogen M. Hogenson vs. St. Paul, Minneapolis & Manitoba, Justice C. J. Gilfillan concluded that "the right of an owner to improve his land ... does not include the right to gather the surface waters on one's land and turn them upon the land of another, to its damage, even though the former land may as a consequence thereof be improved. In other words, [a landowner] may not in this way improve his own land, by merely transferring to the land of another a burden which nature has imposed on his own land."[49]

The decision signified a dramatic victory for the farmers and emboldened dozens more to file lawsuits. SPM&M executive Allen Manvel complained that the Supreme Court's decision placed the company in legal limbo. As dozens of new lawsuits flooded the courts, the company had to decide whether to contest each lawsuit, and thereby alienate even more settlers, or negotiate out-of-court settlements for potentially dubious claims. Although Manvel "deprecate[d] and dislike[d] any litigation with the farmers along our line," he initially favored fighting the lawsuits "because if we settle with one [farmer] the other [litigants] will think they are as much entitled." In 1885, however, Manvel and attorney Solomon G. Comstock recognized the futility and exorbitant costs of contesting every lawsuit on its own merit, especially given *Hogenson*, and they initiated the long process of reaching out-of-court settlements.[50]

By the beginning of 1887, a county judge had proposed submitting the remaining cases to arbitration. Attorneys for both the plaintiffs and the SPM&M agreed, and over the next several months an arbitration board settled the remaining lawsuits. In addition to providing financial compensation, the SPM&M agreed to fill up and dismantle other problem ditches, emplace culverts underneath its track at various points to alleviate flooding, and ensure that other ditches conveyed water the entire distance to the Buffalo River. In 1893, Hill estimated that the SPM&M paid financial settlements to no less than four dozen Clay County farmers at a cost of $100,000.[51]

Andrew Lommeland also got his day in court. Early in July 1884, his lawsuit went to trial in neighboring Becker County. During the proceedings, corporate attorneys never denied that the Moland ditch flooded Lommeland's farm. After hearing testimony from neighboring farmers, railroad employees, and civil engineers, the jury returned a judgment of $681 for Lommeland. The SPM&M appealed the outcome to the Minnesota Supreme Court based on technicalities related to the jury's calculation of damages and the alleged prejudicial testimony of certain witnesses. On July 12, 1886, the Supreme Court upheld the jury's verdict and drew Lommeland's long ordeal to a close.[52]

Conclusion

In the early 1880s, the Red River Boom transformed the valley into the ground zero of Western settlement. The stunning success of the bonanza farms, rapid population growth, and the emergence of a single-crop economy masked the fact that a growing number of farms, both large and small, suffered from poor drainage. In response, James J. Hill's SPM&M sought to fill the power vacuum created by the failure of federal drainage policy. Seeking to maximize his line's carrying capacity and dispose of its land grant, Hill authorized the construction of dozens of miles of ditches. Power became filtered through intermediary organizations unaffiliated with the state.

After Hill's empire of ditches spectacularly failed, touching off a decade of litigation, he terminated the railroad's direct participation in drainage.

Primarily, the company had attacked a *symptom* of the valley's medley of water resources challenges—water standing on the prairie—rather than the *root cause*: the ill-defined tributaries that spilled out of their banks and did not offer a suitable outlet for surface water disposal. "Under these circumstances," Hill fumed to one valley farmer, the SPM&M "cannot undertake to do any more ditching or improve any ditches we have already made." To avoid new litigation, company engineers filled in most of its forty-five miles of ditches and allowed the remainder to fill up with sediment.[53]

With the *Hogenson* and *Lommeland* decisions, the initial phase of institutional drainage experimentation ended. By 1886, the extent of drainage barely exceeded what it was at the beginning of the Red River Boom, and Hill had proven far more successful at reacting to events than shaping them. Moreover, as long as the SPM&M took the lead in water planning, large and small farmers had little incentive to cooperate. The first phase of drainage also exposed the deficiency of knowledge about wet ecosystems and drainage engineering. In addition to assigning drainage responsibilities to the states with the Swamp Land Acts, Congress completely shunned any responsibility for investigating swamp resources or systematizing drainage knowledge, methods, experimentation, and techniques. National science had not yet developed to a point where it could provide specialized knowledge and administrative solutions for states and local communities, despite policymakers' insistence that the country's wetlands be drained to facilitate the onward march of settler agriculture.

Unwilling to admit defeat, valley settlers redoubled their efforts. Lambasting Minnesota's flagrant disregard of the Swamp Land Acts, they soon rejoined forces with the SPM&M to demand that the state legislature live up to its responsibilities by creating institutions and providing resources to drain the valley's wet prairies and marshes for agriculture.

CHAPTER FOUR

Settler Activism and State Intervention
Draining the Red River Valley, 1886–1900

Elias Steenerson found himself literally mired in the mud. While hauling a load of wheat from his Red River Valley farm to Fisher, Minnesota, in the winter of 1886, some fifteen miles away, his wagon became stuck on the wet prairie. Unable to dislodge it, Steenerson unloaded all eighteen sacks of wheat, disassembled the entire structure, put it back together on dry land, and reloaded it. He arrived in Fisher at dusk and delivered the grain.

The difficult trip reinforced Steenerson's conviction that the Red River Valley would never reach its full potential without a general drainage program that sacrificed individualism to collective action. But what form should a public policy take? Which level of government—township, county, state, or federal—should coordinate and oversee the many interrelated, complex tasks involved in surface water removal? As the owner of 1 million acres of valley land, did James J. Hill's SPM&M have an obligation to participate?[1]

These questions bedeviled the mud-caked Steenerson as he journeyed home. Born on November 4, 1856, in southeastern Minnesota, he was one of nine children raised by Norwegian immigrants Steener Knutson and Birgit Liefson Roholt. In September 1876, the twenty-year-old Steenerson followed his parents to Polk County just prior to the Red River Boom. Soon thereafter, he purchased a quarter section of 160 acres of railroad land and filed for a homestead on another quarter section, establishing a farm near the Sand Hill River he christened "Walhalla." After a brief stint as a public school teacher, he sold farm machinery in nearby Grand Forks. In 1881, he went into business for himself, selling the famed McCormick twine binder and other agricultural implements.[2]

Despite his entrepreneurial instincts, Steenerson's passion remained farming. After arriving home from Fisher that fateful night, he concluded that farmers near the Sand Hill River were at a "disadvantage." The Beltrami Swamp, a marsh that originated where the Sand Hill River lost its course and dispersed during the flattest portion of its descent into the Red River, forced farmers to take a circuitous thirty-mile journey to Crookston, the county seat, which was fifteen miles distant in a straight line. In early 1886, Elias convened a "mass meeting" with his neighbors at the home of his brother Christopher to discuss drainage. "All agreed that something ought to be done," Elias recalled, "but what to do was the question."[3]

After much discussion, the farmers nominated Christopher Steenerson to write James J. Hill in the hopes that "he could help us out." On February 22, Christopher put pen to paper. "The Sand Hill River from Beltrami eastward is quite a large and rapid stream capable of running mills," he explained, until "the chanel disappears and all its waters spreads over the prairies. About 4 miles from the Red River the chanel forms again and is quite deep and wide." In all, the Beltrami Swamp covered "at least 4 or 5 townships of what otherwise would be the most fertile and well settled part of this country."[4]

Christopher's letter reached Hill at the moment he was embarking on his life's "great adventure." Just shy of fifty, he had built the SPM&M into one of the nation's most efficient and best-capitalized lines, capturing the Red River Valley's lucrative wheat trade. By 1886, however, he had reached the conclusion that transcontinental lines with low rates were destined to supplant independent regional railroads. As a result, he persuaded the SPM&M's directors to authorize the extension of the railway all the way to the Pacific. Taking advantage of the era's monetary deflation and low interest rates, Hill oversaw the SPM&M's westward extension from Devil's Lake in northwestern Dakota Territory to Havre, Montana, later in the year. Over the next five years, he orchestrated the completion of the final 800-mile Pacific Extension from Helena to Seattle. When the last spike was hammered into the ground on January 6, 1893, Hill's new Great Northern Railway (GN) became the last transcontinental railroad, catapulting him into the national spotlight.[5]

Absorbed by the dauting task of extending the SPM&M to the Pacific, Hill disavowed any responsibility for drainage. Responding to Christopher's letter, he lashed out, declaring that his company had invested significant sums channelizing the Sand Hill River and building forty-five total miles of open ditches. "After spending several thousand dollars and getting the work well under way," he snapped, "we found ourselves the object of several suits for damage at the hands of parties whose lands were benefitted by the better drainage facilities." Farmers' litigiousness convinced him that "we are not the proper party to move in any enterprise of that kind." Rather, he recommended that Steenerson seek assistance from Polk County.[6]

Elias Steenerson refused to throw in the towel. In May 1886, his perseverance finally paid off when the Polk County Board of Commissioners bowed to public pressure and appropriated $500 for a drainage survey of the townships between the Sand Hill and Red Lake Rivers. The commissioners, however, made the appropriation conditional on the SPM&M kicking in a matching amount. Despite his earlier protests, Hill acquiesced and contributed $1,500. Financial necessity and opportunism explained his renewed interest. Once the SPM&M began its thrust to the Pacific later in the year, he would need a steady flow of cash to pay the interest on the company's bonds. Drainage promised to keep the company's coffers flush with cash as it boosted farmers' annual yields, expanded the company's carrying trade, and enabled the sale of inundated portions of the company's 1 million acres of valley land. Nevertheless, Hill warned that his generosity had limits: "This company does not desire to shirk its share of any improvement, or any common burden that is to be borne on the frontier. At the same time, while we feel that in these matters we have been very liberal in the past, our efforts in that direction have been met by a disposition on the part of some localities to take advantage of any and everything that came their way for their personal benefit, and if this is to continue, the Railway company will have absolutely nothing whatever to do with any of these local enterprises."[7]

The fusion of local and corporate activism energized valley settlers to demand government intervention. Their grassroots mobilization represented just one of many dozens of local outbursts in favor of drainage occurring across the Midwest that collectively comprised a key, yet overlooked, pattern of American state formation. Indeed, government expanded at all levels in response to heightened popular interest in drainage before 1900. As the groundswell of support in the Red River Valley suggests, the establishment of state drainage institutions, county drainage programs, and settler organizations dedicated to land reclamation embedded local communities in environmental policymaking.

Land reclamation also required scientific knowledge and expertise. Across the Midwest, communities sought out expertise and specialized knowledge to pave the way for farm drainage. In this way, the pursuit of water resources development through the application of technical expertise, scientific knowledge, and coordinated planning at the local level preceded the maturation of the science- and efficiency-based federal conservation initiatives during the Roosevelt administration by at least a decade and half. It also led settlers to clamor for more assertive federal involvement and funding.

The Crookston Drainage Convention

Following the meeting at Christopher Steenerson's home, enthusiasm for drainage exploded. On June 5, 1886, a committee of five leading residents, including state senator/future congressman Halvor Steenerson (Elias's and

Christopher's brother) and grain elevator proprietor E. D. Childs, published an invitation in the *Crookston Times* for municipalities and counties to send delegates to a forthcoming drainage convention in Crookston. The meeting's objectives included "the subject of drainage in said section generally, and [devising] a means for the accomplishment of a thoroughly effectual and general system of drainage for said section of country." The organizers also invited Hill, and the SPM&M offered reduced fares for traveling delegates.[8]

On July 1, 250 settlers from Clay, Norman, Polk, Marshall, and Kittson Counties assembled in the Crookston opera house. The two-day convention opened with a lengthy address by C. G. Elliott, an Illinois-based drainage engineer. Born in 1850 in Lowell, Illinois, Elliott received his B.S. in 1877 from the University of Illinois and spent the rest of the century establishing his reputation as the country's foremost agricultural drainage engineer. In 1884, he took a hiatus from agricultural drainage and briefly served as Indianapolis's sanitary engineer. But it was in the inchoate field of agricultural drainage where Elliott made his professional mark. In 1882, he published *Practical Farm Drainage; Why, When and How to Tile Drain*, one of the earliest drainage books to appear in the United States. It described the characteristics of farms requiring drainage, soil properties, drain types and spacing techniques, leveling methods, the determination of grades, tile and ditching technologies, costs, and other related subjects. He did not intend for the book to provide a comprehensive analysis. Rather, his purpose was to "say enough to give the farmer an elementary knowledge of why, when, where and how to drain his farm." Just after the turn of the century, Elliott would also take over the ownership and editorship of *The Drainage Journal*, the country's only monthly periodical dedicated to the subject, which was first published in 1879 in Indianapolis as the clay tile industry's organ. He quickly sold the journal, entered the civil service, and, in the early 1900s, became the USDA's chief drainage engineer, where he garnered wide acclaim as the hemisphere's most talented and sought-after agricultural drainage expert (see chapter 7).[9]

In enlisting Elliott's services, the organizers of the Crookston convention had chosen well. The valley's dormant agricultural resources captivated him. After touring the valley, Elliott announced that he had "never before seen so vast an area, needing as this does, organized systematic drainage." If the Red River's tributaries were not deepened, straightened, and freed from natural obstacles, he feared that they could not carry away an influx of water from drainage districts.[10]

The Illinois engineer recommended digging a series of artificial "main" drains that followed the prairie's coulees and hollows. It was better to "make numerous drains," Elliott explained, than force all of the surface water from bonanzas and smaller farms into the winding, undefined channels of valley watercourses. After the initial step of channelizing the Red River's tributaries and digging artificial mains, groups of farmers organized under state drainage laws would dig ditches that conveyed water from their farms into the outlets. In order to empower farmers to plan, finance, and maintain the construction of ditches and underground tiles, he implored Minnesota to emulate the examples of Indiana, Illinois, and other Midwestern states by passing legislation authorizing the creation of drainage districts or other public entities/corporations under the supervision of county governments.[11]

Elliott thus diagnosed the valley's dilatory pace of drainage as a purely legal, not environmental or technological, matter. The passage of appropriate legislation would unleash settlers, the state, and technology against soggy prairies. With an engineer's instrumental perspective, he consigned all wetlands to a common fate of destruction. He fretted, however, that the large percentage of valley lands owned by absentee capitalists and railroads made it difficult to equitably apportion survey, construction, and maintenance costs. Any successful policy required the cooperation of the valley's largest landowner: the SPM&M.[12]

James J. Hill knew this as well when he took the podium on the convention's second day. Due to ongoing litigation with Clay County farmers, he reiterated that the railroad had abandoned drainage and that he viewed land reclamation as the state's responsibility. Although Hill did not champion a particular policy, he agreed that any proposal to plan and administer a drainage system should "provide for the greatest possible amount of benefit to the [g]reatest number." He then tackled the albatross of cost apportionment. Disclosing that the SPM&M owned 1,013,000 acres in Kittson, Marshall, Polk, Norman, and Clay Counties, he pledged to pay an equitable proportion of any drainage project.[13]

Hill had good reason to cooperate. Half of his company's unsold lands—502,000 acres—were located in the northernmost valley counties of Marshall and Kittson. In 1885, those two counties tallied a tiny settler population of 9,022, which accounted for less than 18 percent of the combined population of the counties represented at the convention. Marshall and Kittson Counties also lagged behind in terms of the land's agricultural utilization. In 1890, Kittson County had only 13 percent of the county's area in farms while Marshall County had 29 percent. For Clay, Norman, and Polk Counties, the percentages ranged from 35 percent to 46 percent. Seeking to dispose of the company's glut of unsold lands in Kittson and Marshall Counties, Hill promised to cover half of the costs of a topographical survey of the Minnesota side of the valley.[14]

The merging of grassroots and corporate advocacy did not overcome the opposition of Kittson County's delegation. Kittson's delegates worried that, as the valley's northernmost (and downriver) county, it was vulnerable to increased flooding (and higher costs) if the winding, flood-prone Red River struggled to carry away the inflow of water from the lower counties. They preferred that the federal government erect a series of reservoirs to help store surplus waters, safeguarding downstream communities against enhanced flooding risks. Kittson County's opposition sparked sharp rebuttals. Elias Steenerson, for instance, shot back that anyone threatening the movement's cohesion and momentum should depart and organize an "anti-drainage convention." Across the United States, as at Crookston, opponents of drainage projects increasingly found themselves vilified, in the geographer Hugh Prince's words, as "reactionaries and obstructionists." The cultural imperative to separate soil and water tolerated no delays.[15]

The convention concluded with the passage of resolutions. The second resolution condemned surface water as a menace to settler progress and prosperity: "A large percentage of land of this valley is practically valueless by reason of obstructed streams ... while an equally large percentage is liable to suffer serious damage by overflow in the spring of the year, the lay of the land being such that the water cannot at present flow back into the Red River." The convention selected one delegate from each county, along with three from the SPM&M, to form a standing executive committee. The committee was responsible for arranging a topographical survey, led by Elliott, and drafting a legislative strategy no later than December 5.

In organizing an expert-led survey, the delegates elevated professional judgment and experience above parochial knowledge, marking the first serious attempt to quantify, classify, and rationalize the valley's drainage problem. In support of the effort, the counties agreed to contribute half of the survey's cost with the SPM&M footing the remainder. Once Elliott wrapped up his survey, Springer Harbaugh, the convention president and co-owner of the Keystone bonanza, would schedule a follow-up meeting in December.[16]

Elliott's Survey, Settler Activism, and Federalism

On July 22, 1886, Elliott returned to Crookston with five assistants and established his base of operations. According to his planning, the topographical survey would begin at the Sand Hill River and cost between $1,000 and $1,500 a month. Over the next few months, Elliott's team made torturously slow progress throughout the valley, progressing only four miles per day. In total, the crew eventually covered five counties, ranged across the equivalent of eighty-two townships, traveled 175 miles north to south and eighteen miles east to west, and in total surveyed 1.7 million acres. Kittson County refused to pony up its portion of the costs, so it was excluded.[17]

In his final report, Elliott attributed the valley's poor drainage to two hydrological factors. First, the Red River's tributaries originated on the higher slopes of the valley's eastern edge (Lake Agassiz's former beaches). As the tributaries snaked their way across the most level portion of the valley, "they take but little of the drainage from either side except that which enters them through small creeks or runs. The banks are higher than the immediate land on each side." This created the phenomenon of the Sand Hill, Snake, Tamarac, and Middle Rivers losing themselves on the prairie. Second, an assortment of "small ravines" and "coulees" unconnected to creeks or rivers punctuated the valley. Similar to the Red River's tributaries, the coulees possessed higher banks than the surrounding land and regularly discharged water onto the adjoining prairie. "Much of the valley has no natural drainage," he concluded. "Farms situated on the immediate banks of streams and coulees are the only ones not subject to injury from Spring and Summer flooding."[18]

Elliott laid out a three-phased approach. First, engineers needed to channelize and open up the Middle, Sand Hill, Snake, and Tamarac Rivers where they dispersed across the prairie. Second, the assortment of ravines and coulees needed to be connected with outlets. Third, additional large ditches needed to be constructed to compensate for the absence of natural outlets. Until those three initial actions were accomplished, Elliott contended, it was "of no use" to consider any other plans, including new legislation. In this manner, drainage in the Red River Valley required a far more comprehensive and layered approach than in other Midwest communities. Before drainage districts or county drainage projects could succeed, settlers needed somewhere to dispose of the water—the key lesson of the SPM&M's abortive drainage program.[19]

Following his receipt of Elliott's report, Harbaugh scheduled a second convention for December 8, 1886. Convening again in Crookston, the meeting opened with Harbaugh's presidential address, where he reminded settlers that they shared a "duty to perform in bringing under the control this vast domain to the uses of man . . . that will pour open its products to the starving millions of Europe and the old world." He then turned the floor over to

Elliott. After summarizing his survey's findings, he estimated that of the half million acres already under cultivation in the five valley counties, only 125,000 of those acres generated an annual profit. Owners of the remaining 375,000 cultivated acres lost money or barely broke even each year. Striking an optimistic tenor, he crowed that proper drainage would, at a bare minimum, double every farm's productiveness.[20]

Following Elliott's briefing, the executive committee assembled a legislative committee of five to prepare and present a general drainage bill to the state legislature. Delegates agreed that the bill's provisions must include the authorization of drainage districts as municipal corporations. Furthermore, the corporations must be empowered to equitably assess damages in proportion to the estimated future benefits associated with drainage, including rising property values and crop yields. The legislation also needed to enable districts to incur bonded indebtedness that could be repaid over a series of annual installments. The likely inclusion of SPM&M lands in drainage enterprises led to a specific provision that corporate property should be assessed in the same manner as private lands. Finally, the legislature needed to create a new "state drainage commissioner" to synchronize the activities of the autonomous districts.[21]

Delegates also appointed a committee of fifteen (three from each county) to lobby for national and state aid. The committee set its sights on petitioning Congress to finance reservoirs at the headwaters of the upstream Red Lake and Otter Trail Rivers to reduce seasonal flooding on the winding Red River. In order to channelize the Red River's tributaries, the committee of fifteen intended to request a state appropriation since intrastate nonnavigable rivers fell under Minnesota's jurisdiction.[22]

The convention thus envisioned a mosaic of complementary institutions at all levels of government working toward a common purpose, exemplifying the manner in which federalism fragmented and broadened authority for water resources planning and management. In the Red River Valley, drainage districts and counties, organized under state law and administered locally, would plan, finance, and supervise drainage works that conveyed surface water into watercourses channelized by the state government and artificial main ditches built by a state agency. Finally, the construction of upstream federal reservoirs would ensure that the influx of surface water did not overwhelm the Red River watershed.[23]

The complexity of the plan, which involved a kaleidoscope of overlapping and competing levels of government, sharply contrasted with the SPM&M's initial ad hoc approach and seized national attention. The editors of *The*

Drainage and Farm Journal described the multicounty undertaking as the largest land drainage project in the nation's history. "The drainage of the Red River Valley, looked at from a purely professional standpoint as an operation in drainage engineering," they boasted, "[is] greater than any covered by one general system, unless it be that which makes of Holland a habitable country." The plan's ambitiousness partly owed itself to Elliott's influence and expertise, but a broader political transformation was also at work. The traditional, parochial environmental politics of the valley was giving way to a broader vision of governance in which settlers sacrificed autonomy, embraced bureaucratization, and welcomed higher taxes to drain wetlands. Extolling drainage as an appropriate government function, the delegates agreed to ask the legislature for "a very liberal appropriation from the state to open up the obstructed river channels in this section."[24]

The delegates justified the request on the grounds that Minnesota had failed to comply with federal drainage policy. J. T. Fanning, a Minneapolis hydraulic engineer who served as Elliott's assistant during the survey, pointed out that under the 1860 Swamp Land Act Congress ceded 249,588 acres of "swamp and overflowed" lands in the six valley counties on the condition that Minnesota sell the lands and invest the revenue in drainage. Instead of following the law, Minnesota sold those lands to pay for the construction of public works and internal improvements in other sections of the state. Justice and equity, Fanning admonished, required Minnesota to return those lands to the counties to facilitate "the object for which the State received them in trust." He also urged the passage of a constitutional amendment authorizing the retrocession of an equivalent amount of state land in those counties that could then be sold by local authorities to defray the expenses of drainage. In his conclusion, Fanning justified the appropriation by pointing to the unique challenges of removing surface water from the floor of a former glacial lake. "A prominent financial difficulty," the final report declared, "is the fact that the heaviest work must be done first, such as the opening of the natural streams where the channels are not well defined."[25]

A Public Policy Revolution:
The Red River Valley Board of Audit

In January 1887, the committee on national and state aid went to work. Cooperating with state representative A. H. Baker, they put a bill before the legislature that appropriated $125,000, divided into five annual installments of $25,000, to improve the Red River's tributaries. The bill would activate a state agency (a "board of audit")—composed of the governor, state auditor, and secretary of state—to plan and coordinate the improvement of the Red River's tributaries in five Minnesota counties. Baker also introduced legislation empowering counties to organize as drainage districts. The proposal authorized the commissioners of every county to act as drainage commissioners, with the county auditor and treasurer respectively serving as clerk and treasurer. Under the bill, the county board of drainage commissioners possessed the authority to judge the merits of petitions requesting the formation of subdrainage districts inside of their county. The minimum threshold for the submission of a petition was the signatures of landowners owning at least one-third of the land within any given tract proposed to be drained. Once the commissioners validated the signatures, they would appoint viewers to make surveys, draw up drainage plans, and estimate costs. Upon approving the plan, the commissioners could then order the work to commence under the direction of a professional engineer. As construction proceeded, the treasurer and auditor would negotiate for bonds and impose assessments, which constituted a perpetual lien on the benefited lands. When the establishment of drainage districts was paired with the state board of audit, Baker bragged that the entire Minnesota portion of the valley could be drained for wheat production.[26]

The 1887 committee's lobbying efforts got off to a rocky start. Constitutional scruples made outgoing governor Lucius F. Hubbard skeptical. Just before leaving office early in 1887, Hubbard conceded that Minnesota had abused the terms of the 1860 Swamp Land Act by "grant[ing] [ceded federal swamplands] as a gratuity to corporations to encourage the building of railroads in sections of the State remote from where [the swamplands] are located." Fidelity to Congress's original purposes would have averted the current imbroglio. "If these lands were now available for the purpose for which they were granted by Congress," he reasoned, "the means would be at hand for the prosecution of the contemplated work." Despite these admissions, Hubbard identified the state constitution as an insurmountable barrier. Section 5 of article 9 prohibited the state from contracting "any debts for works of internal improvement" or acting as "a party in carrying on such works except in cases when grants of land or other property shall have been made to the state specially dedicated by the grant to specific purposes." Section 10 of article 9 imposed an even greater hurdle: "The credit of the State shall never be given or loaned in aid of any individual, association or corporation." Article 9's stipulations against extending state resources to corporations or individuals delivered a staggering blow against the board of audit. As the

St. Paul Daily Globe editorialized, the "restrictive prohibitions of section 5 and 10 of article 9 of the constitution" rendered it a nonstarter.[27]

Over the next few months, Baker met with limited success. On March 8, the bill authorizing counties to organize as drainage districts sailed through the legislature. Modeled after similar Illinois laws and likely written by C. G. Elliott, the legislation empowered county commissioners, upon a simple majority vote, to organize their county into a drainage district with the commissioners doubling as a permanent drainage board. It instructed the commissioners to act as a corporate body that could sue and be sued and to control all drainage matters. Significantly, the legislature also stipulated that the board of drainage commissioners could organize subdrainage districts when a majority of landowners who owned at least one-third of a specified tract petitioned for the formation of a subdistrict. The commissioners would then issue bonds for the subdistricts.

Counties wasted no time organizing themselves into drainage districts. On June 7, 1887, the Clay County board of county commissioners voted to form itself as a drainage district. Norman, Polk, Marshall, and Kittson Counties followed suit in short order. In doing so, however, they departed from Elliott's recommended sequencing. He always maintained that the channelization of the Red River's tributaries must precede drainage district creation, but the legislature took the opposite path.[28]

Although counties could now organize as drainage districts, lawmakers rejected the board of audit. Constitutional qualms, sectionalism, and a belief that the SPM&M should help defray the costs conspired to defeat it. On February 25, the House voted down the board by a tally of 19 to 53. A few days later, Baker floated an emasculated proposal that reduced the appropriation to three annual installments of $25,000, with the SPM&M "contribut[ing] each year an amount equal to one-third of the amount appropriated for that year by the state." Despite the inclusion of a railroad drainage tax, the state senate wavered and the valley's drainage boosters were forced back to the drawing board.[29]

Following the defeat, the valley's committee to secure national and state aid, chaired by lawyer Ezra G. Valentine, continued to meet in St. Paul, organized valley-wide rallies in 1888, and sent a delegation to Washington, DC. In 1889, the committee's perseverance paid off. In his biennial message, Governor A. R. McGill observed, "Unless the valley is properly drained the wheat crop of that country must depend upon the most favorable seasons. An early frost is pretty sure to destroy it on wet lands, while on dry lands it escapes uninjured. If the country were properly drained, the danger from this cause would be remote." The governor's direct intervention did not surmount sectionalism. In one such instance, state senator Frank Arah Day of Martin County objected to an 1891 bill allocating $12,000 annually for the Red River Valley because his southern Minnesota constituents were "as much entitled to drainage as northern counties."[30]

Valentine's committee fared no better with the federal government. A provision in the 1890 River and Harbors Act directed the US Army Corps of Engineers to evaluate the feasibility of constructing two reservoirs at the headwaters of the Red Lake River and at Lake Traverse "for the purpose of diminishing the effects of floods and of storing water for use at low stages in the Red River system." The strategic placement of reservoirs on the Red River watershed, in part intended to diminish seasonal floods associated with an uptick in farm drainage, remained the final element of Elliott's proposed framework. Ultimately, the federal government demurred. In 1892, Brig. Gen. Thomas Lincoln Casey, the chief of engineers, rejected the elaborate scheme. The declining volume of freight on the Red River, as well as the reservoirs' exorbitant $860,000 price tag, rendered the project impractical. Moreover, due to the SPM&M's ongoing expansion, the combined barge and steamboat traffic had plunged from 31,500 tons in 1882 to 3,866 tons seven years later. In Casey's estimation, "The amount of commerce on the Red River of the North does not justify undertaking a project of this magnitude." The future of transportation lay with the iron horse rather than steamboats.[31]

Setbacks at both the state and federal levels dispirited Valentine. In early 1893, he convened the committee in St. Paul to mount one final legislative push. At his own expense, he printed, and circulated among lawmakers, a pamphlet explaining the topographical and hydrological conditions that thwarted agricultural drainage. Dismissing concerns about Article 9, the pamphlet reiterated that valley farmers had asked the state not "to drain the lands of anyone, but simply to put the natural channels and streams in condition to receive and carry off the water." It further highlighted that the disposition of ceded "swamp and overflowed lands" had been inimical to northwestern Minnesota. Policymakers had granted 261,163 acres of the 270,740 acres of Red River Valley federal wetlands ceded by Congress to railroad corporations in other parts of the state. "The unjust course of Minnesota in this matter," the pamphlet asserted, "diverted these lands and the proceeds thereof from the purpose to which they were dedicated by Congress. We now ask that the State, in a small measure, make good to our people the loss sustained by depriving them of these lands."[32]

In 1893, Valentine authored a new bill at the legislature's request that deviated little from previous proposals. It appropriated $100,000 over four years for a board of audit to open up the "closed water-courses leading into the Red River . . . and for opening existing streams in the Red River Valley . . . for drainage purposes" in Clay, Kittson, Marshall, Norman, Polk, and Wilkin Counties. A later amendment added Grant and Traverse Counties. Endowed with the authority to select and design projects, the board of audit consisted of the governor, the secretary of state, one person appointed by the eight counties, and a representative of the GN (the successor of the SPM&M). The bill required the GN to contribute an amount equal to one-quarter of the legislative appropriation as a precondition for the $100,000 appropriation.[33]

On April 17, 1893, the proposal breezed through the legislature. The state senate passed the bill by a vote of 41 to 3, the House by a tally of 85 to 5. What explained the abrupt change in fortune? Valentine's pamphlet apparently convinced skeptical lawmakers that Minnesota's negligent administration of the 1860 Swamp Land Act undermined the valley's agricultural development. As one Duluth newspaper observed, "The general government had made grants of swamp lands to the state, to be used in draining and reclaiming these lands, but the proceeds arising from the sale had been diverted to the southern and eastern part of the state or the lands had been granted to railroads, while the counties in which the lands are located had received no benefit." Equity and prudence demanded that the legislature correct these injustices.[34]

Despite the turn of events, Valentine had committed a colossal blunder. Since 1887, similar bills had imposed a drainage tax on the SPM&M since it stood to benefit handsomely from a state program. In 1889, as he continued extending his line to the Pacific, Hill and the company's board of directors organized a new company called the GN, which in the following year finalized a 999-year lease of the SPM&M. The lease allowed the companies to operate as a single entity although they remained separate corporations governed by two sets of stockholders. Before submitting his bill to the legislature in 1893, Valentine bungled by inserting "GN" in place of "SPM&M."[35]

Upon learning of the miscue, the temperamental Hill flew into a rage. Since he had been away from St. Paul when Valentine drafted the legislation, he was not consulted as in previous years. "The public should understand," Hill scolded Valentine and Governor Knute Nelson, "that it has no more right to appropriate the Company's property or money than it has of any other citizen of the State." Nursing old grudges, he defended his opposition on the grounds that Clay County farmers had responded to the SPM&M's drainage assistance with avarice and hostility: "We have made several efforts to open up these water courses on our own lands, and these efforts have resulted in an attempt through law-suits, to collect heavy damages." Hill's primary source of contention, however, stemmed from the fact that the law taxed the GN to drain the SPM&M's unsold land grant. The GN owned no valley lands. The SPM&M, which had sold almost 650,000 acres of valley land since 1886 but still owned 363,450 acres, represented the appropriate entity to tax.[36]

In the worst traditions of Gilded Age politics, shady backroom deals and cryptic promises sealed the deal. In June 1893, H. M. Donaldson, one of Hill's corporate representatives, met with Governor Nelson. Donaldson extracted promises from Nelson that if Hill agreed to the tax he would receive complete autonomy over the board. As Donaldson crowed to Hill, the governor "gave me to understand in plain language that the board of Audit would be composed of men of your choice and that he would allow their wishes to control the action of the board and in that way YOU would—through your representatives on the board—have the direction of the expenditure of the whole appropriation." Thrilled about the prospects of controlling the board, Hill withdrew his opposition to the $25,000 tax.[37]

A Tool of Corporate Greed: The Red River Board of Audit

The board of audit consisted of Governor Nelson, Minnesota secretary of state Frederick P. Brown, GN chief engineer N. D. Miller, and Valentine (the counties' representative). During its brief four-year existence, the board of audit doubled as Hill's personal drainage company. In 1893 and 1894, the board expended almost $70,000 to build thirty-seven miles of large ditches in five counties. The Sand Hill River Ditch, excavated through the middle plane of the valley where the river spilled onto the prairie and created the Beltrami Swamp, was the initial project. The thirteen-mile ditch drained a substantial block of SPM&M land and accounted for 34 percent of the board's expenditures prior to 1895. Completed in October 1895, the Otter Tail River ditch was the board's fourth ditch, improving another significant chunk of railroad land. Toward the end of the first year, a satisfied Hill rated the board's work as "exceptionally good."[38]

In 1895, the legislature allocated an additional $50,000. The appropriation passed with the caveat that the board spend the money to "reclaim swamp lands granted to the state of Minnesota by [the 1860] act of congress ... and lands owned by the state of Minnesota." After addressing the SPM&M's drainage needs, the board of audit turned its attention to "bring[ing] relief as speedily as possible to the greatest number and at minimum final cost." In all, the board of audit constructed 117 miles of ditches at a cost of just over $162,000 across eight counties. In its final report, it estimated that it had improved more than 1 million acres, drained 130,000 state-owned acres, and increased the median value of drained agricultural lands from $5 to $20 per acre. Provided with suitable outlets, several valley counties appropriated money to build supplementary lateral ditches. After organizing itself as a drainage district under the 1887 law, Polk County spent $100,000 between 1893 and 1901 constructing 350 miles of ditches. Other counties followed suit: Red Lake County (carved out of Polk County in 1896), 110 miles; Marshall County, 60 miles; Kittson County, 50 miles; and Norman County, 30 miles.[39]

In 1897, the board of audit expired. That same year, however, the legislature inaugurated a Board of State Drainage Commissioners responsible for maintaining the ditches built between 1893 and 1897. Appointed by the governor, the board's three commissioners used an annual $500 appropriation to employ a professional engineer to recommend routine maintenance to the ditches, including the removal of debris and obstructions, the prevention of caving banks, and the eradication of flags, reeds, rushes, and other weeds. Once the Board of State Drainage Commissioners notified county governments about the engineer's recommended repairs, they had sixty days to comply. In hoisting maintenance responsibilities upon the counties, the legislature again eschewed drainage responsibilities. In 1901 and 1902, the legislature did not approve further appropriations for the inspection of state ditches. Fed up with the state's creeping austerity, Valentine spent $1,000 of his own money to pay an engineer to inspect the maze of infrastructure and publish a report on its condition.[40]

Meanwhile, in 1901, the legislature created a new drainage commission whose jurisdiction covered the entire state. Consisting of the governor, state auditor, and secretary of state, the commission's governing board oversaw the spending of $25,000 annually from 1901 to 1904 to build ditches predominantly benefiting state-owned lands. By 1915, the state had appropriated over $1 million to excavate seventy-six large ditches across Minnesota to drain additional state lands. The thoroughness and tempo with which the state

Tiling machine on a Minnesota farm, early 1900s. The dramatic explosion of agricultural drainage across the Red River valley and other Midwestern communities required innovative new approaches to laying subterranean clay drainage tile. This photograph depicts a Minnesota farmer using a motorized tiling machine. Reprinted with the permission of the Minnesota Historical Society.

completely reengineered the valley's wetlands proved staggering and represented a harbinger of future events across the nation.[41]

Conclusion

With the Swamp Land Acts of 1849, 1850, and 1860, the federal government delegated drainage to the states. In those laws, Congress provided public land subsidies to fund state institutions that would shoulder drainage responsibilities and expand settler agriculture. The federal government quickly lost interest and exercised no oversight; the states were left on their own to plan and oversee land reclamation, even as millions of acres of arable lands fraudulently passed into the hands of government officials and speculators.

In Minnesota, the ineffectiveness of federal drainage policy, as well as naked self-interest, propelled the SPM&M into local water politics. In his history of the drainage of Midwestern wetlands, geographer Hugh Prince argued that the SPM&M embraced a more interventionist and active role in drainage than any other railroad of its era:

> The lead taken by railroads in initiating ditching and mobilizing public interest in land reclamation was a distinctive feature of relations between corporate interests and states in the northern lakes region. In effect, the states of Michigan, Wisconsin, and Minnesota relied on railroads to promote settlement and economic development and thus expand the tax base. Railroads responded in the conviction that what was good for the [SPM&M] was good for the state of Minnesota. It is remarkable that an ad hoc convention held in Crookston, on the western boundary of the state, should take upon itself responsibility for drafting a general drainage law for the whole state, and that the state legislature should adopt the measure with minor revisions and provide financial backing.[42]

While this was indeed true, the SPM&M's actions were actually reactive and unremarkable. When Congress bailed on drainage, the states were unprepared and unequipped to plan, build, and administer water development projects. With their financial resources, engineering departments, and large workforces, railroads like the SPM&M stepped in, bridging the gap between the federal and local authorities until the states themselves were prepared to lead. Moreover, the role of the SPM&M and other private entities in organizing drainage ventures conformed to the pattern of American associational governance where the federal government relied on intermediary organizations, indirect subsidies, and private entites to achieve environmental policy objectives.

Even more decisively, the SPM&M's involvement in drainage shifted James J. Hill's perspective on the state's role in water management. Once his GN became operational in September 1889, he hustled to finish the country's fifth and most northern transcontinental line by 1893. The onset of the Depression of 1893, the sparse population located adjacent to the GN, and the disintegration of the silver mining industry in Idaho and Montana initially deprived the company of valuable freight revenue. In response, Hill set his sights on promoting dryland farming, agricultural diversification, and especially a federal irrigation program. Owing in large degree to his drainage experiences, Hill in the 1890s rejected any responsibility for irrigation. Rather, he sought to muster public support for federal water initiatives that forged a community of interests between local communities and corporations. Beginning in 1899, he joined the NP, Santa Fe, Southern Pacific, and Union Pacific in subsidizing California attorney George H. Maxwell's crusade for a federal irrigation program for the arid West. Hill's support of Maxwell's National Irrigation Association ultimately succeeded when Congress passed the 1902 Reclamation Act (see chapter 6).[43]

Aimed at securing the stability and expansion of settler agriculture into unsettled areas, drainage revolutionized the relationship between settlers and the state. In the Red River Valley, settlers demanded a state agency to channelize the Red River's tributaries and eagerly welcomed robust government involvement in their lives. Elias Steenerson, whose daunting 1886 trip to Fisher opened this chapter, applauded state intervention and bureaucratization. "I am no socialist," he reflected in his memoirs, "but I have learned from my observation of the development of these times that there are certain enterprises which the State should take hold of, and among them are Drainage." The reclamation of wetlands primed settlers to accept an expansive vision of environmental governance and social organization that elevated expertise, comprehensive planning, technological solutions, and bureaucratization above parochial knowledge and laissez-faire planning. As Steenerson concluded, "State drainage has been and is recognized as one of the proper functions of the State, and has worked untold benefit in developing [Minnesota], especially in the northern part. Large tracts of swamp lands have been reclaimed, and the Sand Hill ditch, and its tributaries have transformed a dismal swamp into cultivated fields."[44]

The political mobilization of valley settlers and corporations exemplified how late nineteenth-century settlers perceived water resource management—especially drainage—as a legitimate state function. Significantly, national science had not yet developed to a point where it could provide knowledge and administrative solutions for states and local communities, despite Congress's insistence that all of the country's wetlands be drained to encourage the dense settlement of dispossessed lands.

Ultimately, this case study of the Red River Valley—one of many dozens of Midwestern communities involved at the time with drainage—demonstrates that draining the nation's wetlands required much more involvement from Washington. Recognizing that the federal government had never surveyed or inventoried the nation's wetlands, the USGS in 1884 hired Harvard geologist Nathaniel Southgate Shaler to spearhead a scientific investigation. As soon as he joined the USGS, Shaler began studying national wetlands resources and responding to settlers' concerns that the federal government had been derelict in assisting drainage-minded farmers. In doing so, he almost singlehandedly unveiled the country's submerged lands to his fellow citizens, but his brand of national science did far more to prop up the social fantasies of eugenicists than to provide practical and technical assistance to ordinary settlers like Elias and Christopher Steenerson.

CHAPTER FIVE

Wasted Lands, Wasted People
*Eugenics, Drainage, and the Legacy of
Nathaniel Southgate Shaler*

America's future looked grim. So warned Nathaniel Southgate Shaler, the head of the United States Geological Survey's (USGS) Atlantic Coast Division. As Red River Valley settlers gathered in the Crookston opera house, Shaler published an ominous article about the looming exhaustion of arable land in the United States. Writing in 1886 for *Science*, he pronounced that the era when settlers could find and cultivate a fertile tract was drawing to a close. He mourned that the country's rising population and dwindling supply of cultivable lands imperiled its future. "The conditions which have determined the occupation of land in the United States differ widely from those which have controlled the settlement of most other countries," he wrote. "In other states there have been political or geographical limits which have greatly restrained the movements of population. In this country there has been from the beginning to the present day, an abundance of good, readily subjugable land awaiting the settler." But that era was passing.[1]

Shaler's bleak prognosis blotted out the recent Indigenous occupation of the continent. The conquest and expropriation of Indigenous homelands had enabled the long-standing availability of "good, readily subjugable land." As a federal government scientist who also held a geology professorship at Harvard, Shaler was complicit in the erasure of Indigenous history and knowledge even as he alleged that the onward march of colonialism had reached a tipping point.

Yet Shaler marshaled statistics to corroborate his gravest fears. By the end of the nineteenth century, settlers were filing claims on 25 million acres of American land every year. From the American Revolution to 1883, they settled 620 million acres of public lands; the next thirteen years witnessed the disposal of a staggering 325 million acres, an area twice as large as Texas. Even more troublesome for government elites like Shaler, with the opening of the Great Plains to dense agricultural settlement in the 1880s, the frontier had finally breached an arid region where agriculture proved a riskier proposition than in the country's eastern humid half, where precipitation remained abundant and more consistent throughout the growing season. Indeed, from 1870 to 1890 the combined population of Kansas, Nebraska, and the Dakotas swelled from less than 500,000 to 3 million.[2]

Joining a growing chorus of intellectuals anxious about the closing of the frontier and the cratering prospects of white land-seekers, Shaler agonized about the social ills that likely accompanied the depletion of cultivable land. In a country bereft of tillable land and increasingly characterized by industrial wage labor, what would galvanize the individualism and self-reliance that differentiated Americans from Europeans? Without a glut of free or cheap land, how could the country reverse its growing farm tenancy rate (the 1880 census recorded that the United States had more tenant farmers than any European country)? In addition, how could it absorb and assimilate immigrants, curb agrarian and industrial unrest, or provide a rural safety valve for city dwellers? Four years before the 1890 census officially declared the frontier closed, these questions bedeviled Shaler. "It is evident . . . that within this decade," he declared in *Science*, "we [will] pass from this old condition where excellent land was to be had for the asking. Before 1890 all such fields will have been occupied."[3]

Land scarcity was only one problem. According to Shaler, the overwhelming tide of immigration in the 1880s threatened to metastasize into a social tumor and destroy the United States from within. From 1871 to 1901, the scale and nature of immigration underwent a tectonic shift. During those three decades, 11.7 million people immigrated to the United States, more than the combined number who had arrived in the 1600s, 1700s, and the first seventy years of the 1800s. Compared with previous immigration waves, the ethnic and religious composition of the "new" immigrants differed markedly. Whereas earlier white newcomers had consisted chiefly of Protestants from northern Europe, the new immigrants hailed from southern and eastern Europe and included a much higher proportion of Catholics and Jews. The 1890s became the first decade in which these new arrivals outnumbered immigrants from northern Europe, persuading many elites, like Shaler, that the *quality* of immigrants mattered just as much as their quantity. His adherence to a neo-Lamarckian evolutionary framework, with its emphasis on the inheritance of acquired characteristics, deepened his pessimism about interbreeding between native-born Anglo-Saxons and the new arrivals. Indeed, he accused them of possessing inferior hereditary characteristics. For Shaler, racial anxieties joined the haunting crisis of landlessness as the country's most implacable foes.[4]

Despite his gnawing pessimism, Shaler—one of the Gilded Age's most popular scientists—rejected the idea that the twin crises of the 1880s spelled the death knell of the United States. While he did not oppose other intellectuals who favored the annexation of Canada, Cuba, or Hawaii to open up new farmland for settlers, Shaler urged Congress to look inward. From his position inside the USGS, he argued that "unoccupied districts," namely Western deserts or Eastern swamps, held the key to escaping the ongoing upheavals. Whitewashing the fact that swamps and desserts were only "unoccupied" due to ongoing dispossession, Shaler observed in *Science* that previous generations of settlers had bypassed swamps as they trekked west. But as the frontier ideal receded into the dustbin of history, Americans would finally be forced to drain Eastern and Midwestern swamplands, such as those in the Red River Valley, or irrigate Western deserts.[5]

Shaler preferred that swamps be "won" to agriculture. But therein lay a dilemma. In the Swamp Land Acts, Congress had not made its wetland resources *legible* to its citizens by inventorying and mapping them, studying their human and nonhuman inhabitants, investigating agricultural opportunities, or probing their geological or hydrological characteristics. As a result, by the mid-1880s the federal government's knowledge about swamps remained limited and imprecise, hampering its ability to assist drainage-minded settlers.[6]

In the last two decades of the century, Shaler embarked on a campaign to fill the knowledge gap. In 1884, he took over the USGS's brand new Atlantic Coast Division, which was charged with conducting a "careful inquiry into the geological history and physical conditions of the swamps and other inundated lands." Shaler spent the 1880s carrying out investigations that culminated with his authorship of the federal government's first two scientific publications on wetlands: "Preliminary Report on Sea-Coast Swamps of the Eastern United States" (1885) and "General Account of the Fresh-Water Morasses of the United States" (1890). Both reports introduced Americans to the diversity, complexity, and varied functions of wetlands. They represented the federal government's first attempt at comparing and quantifying the nation's "areas of approximate inundated lands." Shaler exhorted Congress to drain the fertile wetlands east of the 100th meridian—the longitudinal marker dividing the continent's humid and arid halves—which were situated near urban markets, transportation infrastructure, and social institutions like schools and churches. He boasted that 100,000 square miles of swamps and riparian overflow lands awaited cultivation east of the Mississippi River, enough land to furnish millions of settlers with rural homes.[7]

Shaler's sixteen-year tenure at the helm of the Atlantic Coast Division (1884–1900) elevated him into one of the most significant contributors to the budding American conservation movement. As the protégé of Louis Agassiz, the former head of the Kentucky Geological Survey, and a prolific, nationally acclaimed authority on geological and geographic topics, Shaler deserves a place alongside George Perkins Marsh and John Wesley Powell as the "Big Three" of Gilded Age conservation science. While Marsh's international best seller *Man and Nature* (1864) established *forests* as the foundation of civilization and Powell became the country's most ballyhooed expert on Western *irrigation*, Shaler added a third element: *wetlands*. He repudiated the long-standing myth that watery tracts were wastelands disconnected from broader hydrological webs. And in his USGS reports, he argued that coastal and interior swamps possessed vast biological, botanical, geological, hydrological, and pedogenic diversity and influenced stream flow by storing and then slowly releasing precipitation back into watercourses. No less than forests, rivers, or aquifers, wetlands constituted an integrated part of *watersheds*, a novel concept emerging at the time in Powell's and Marsh's writings. Yet while he saw connections between different environments, he never incorporated Indigenous or Black historical perspectives and knowledge. Nor did he acknowledge limitations to nature's exploitation or question the wisdom of unrestrained economic growth as Marsh and Powell did. As one scholar puts it, Shaler consistently embraced the "maximum utilization of natural resources for the benefit of society."[8]

During the second half of his USGS tenure, Shaler entangled his swamp investigations with concerns about sexual reproduction. In addition to solving the land scarcity crisis, Shaler promised that drained swamps could promote eugenic control, helping native-born Euro-Americans of the highest hereditary quality (e.g., descendants of Anglo-Saxons, Scandinavians, and Germans) avoid interbreeding with the new immigrants. According to his logic, the draining and plowing under of swamps would enable native-born whites to establish homogenous rural communities, live in prosperity, and remain buffered from the temptation of sexual intercourse with the new immigrants—who allegedly preferred living together in urban ethnic enclaves.

Shaler's enlistment of swamps to resolve the country's perceived landlessness and immigration crises signaled how conservation science buttressed the settler colonial agenda. In doing so, he conjoined the goals of the upstart conservation and eugenics movements. Indeed, his categorization and ranking of natural resources based on their relative value to society spilled over into his assessments of different ethnicities and races. But this was not all. Like many Americans who became supportive of eugenic control, he touted the preservation of exemplary species or races, the survival of the fit versus the unfit, the favorability of biological systems in their most pristine forms, and so forth. As exemplified in his approach to drainage, conservation and eugenic goals more times than not comingled, drew upon a similar logic and rhetoric, and were interwoven with broader cultural and racial anxieties.[9]

From Kentucky to Harvard: The Making of a Conservation Giant

Born on February 20, 1841, in Newport, Kentucky, Nathaniel Southgate Shaler grew up alongside the Ohio River's southern banks, a stone's throw away from the bustling metropolis of Cincinnati. As the first surviving child of Nathaniel Burger Shaler and Ann Hinde Southgate, he received no formal education, but childhood hikes alongside the tempestuous Ohio, in tandem with his father's amateur mineralogy hobby, stimulated his love of geology. From local libraries he borrowed and immersed himself in geological classics, including Roderick Murchison's multivolume *Silurian System*, and a lifetime dedication to geology soon ensued.[10]

At age fifteen, Shaler's father, a Harvard Medical School graduate, arranged for a tutor to prep him for college. Johannes Escher, a German-Swiss clergyman and philosophical theologian, instructed Shaler in German, Greek, and Latin literature. After flirting with a West Point appointment in 1859, Shaler followed his father's footsteps and settled on Harvard. From the day he set foot on campus, he enjoyed a front row seat to the buzzing, and often bitter, scientific and theological debates about human origins and the provocative concepts of evolution and natural selection.[11]

Shaler immediately struck up a cordial relationship with Louis Agassiz, a Swiss-born professor. By the 1860s, Agassiz had amassed a reputation as one of the world's most accomplished geologists, naturalists, and paleontologists. An expert on fossil fishes and glaciation, he arrived at Harvard in 1848 on the heels of a hugely popular American lecture tour. His career's latter stages, however, were tinged with controversy. His unflinching opposition to Darwin's theory of evolution, his belief in biblical polygenesis, and his defense of species immutability ostracized him from mainstream science. Moreover, his proclivity for ranking different races and even classifying them as distinct species, particularly Southern Blacks, sullied his legacy. Agassiz's reputation as Darwin's fiercest foe was also well deserved. Unable to reconcile his fervent conviction in divine design with natural selection, he went on to spar with Harvard botanist Asa Gray and other scientists in highly publicized debates up until his 1873 death.[12]

Despite Agassiz's antipathy toward evolution and sponsorship of racial polygenesis, the Swiss-American's inviting personality and intellectual credentials appealed to Shaler, who enrolled in Harvard's Lawrence Scientific School to study under him. Agassiz assigned him to investigate fossil brachiopods and, after Shaler momentarily returned to Kentucky when the Civil War erupted, he went on to receive his B.S. in geology summa cum laude. His initial work on brachiopods resulted in his first publication in 1861, marking his decision to specialize in geology.[13]

Shaler eagerly soaked up the evolutionary origin of species. In this manner, he was hardly unique. Within two decades of Charles Darwin's *On the Origin of Species* (1859), the theory of evolution had won almost unanimous acceptance from the American scientific community even though many scientists, including Shaler, quibbled about the specific mechanism behind species transmutation. In dismissing the Genesis accounts of creation and a global flood based on the fossil record, post–Civil War biologists and geologists elevated natural law above the supernatural. In Shaler's case, his youthful estrangement from orthodox Christianity may have contributed to his zealous embrace of evolution. He later recalled:

> My father never went to any church, and . . . he attached little importance to what was taught there. My mother was in a limited way a church-goer and kept a pew in the Episcopal church, though she often went to the Methodist meetings, taking me to one or the other. Of these churches, both of the orthodox type, I remember only the tedium of the performance and the development of an intense hatred of the being who, with the power to arrest Satan and his works, permitted him to torment men. . . . Against the devil himself I had no such rage, for it was clear that he was only a bigger kind of bad man, such as I saw about me.

In another passage from his *Autobiography*, Shaler mocked fire-and-brimstone sermons, as well as the belief that eternal punishment awaited unrepentant sinners, as a "lie."[14]

Notwithstanding his early disaffection from Christianity, he never shrugged off misgivings about natural selection and clung to a neo-Lamarckian evolutionary framework. First crafted by the French naturalist Jean-Baptiste Lamarck (1744–1829), neo-Lamarckism held that the characteristics an organism acquired from interacting with the physical environment were passed from generation to generation. Until strict hereditarianism supplanted it in the 1910s, neo-Lamarckism's emphasis on the inheritance of acquired characteristics deeply influenced eugenicists and scientists. Later, it prodded Shaler to wade into immigration and racial politics.[15]

The Civil War interrupted Shaler's life after his 1862 graduation. Returning home, he received a captain's commission in the Fifth Kentucky Battery, but persistent ill health forced his departure from the army. Although he had to resign his commission, his second return to Kentucky proved eventful as he married Sophia Penn Page. In the summer of 1864, the newlyweds returned to Harvard, where Agassiz appointed him as a paleontology assistant at the Museum of Comparative Zoology and tasked him with cataloguing the institution's fossil collection. Half a decade later, at the age of twenty-eight, Shaler was promoted to professor of paleontology, a position he held until 1888, when the university changed his title to "professor of geology." By 1872, with Agassiz's health faltering, Shaler took control of both Harvard's geology and zoology programs. Nineteen years later, he became the Lawrence Scientific School's dean, a position he held until his 1906 death.[16]

As Shaler's star ascended at Harvard, he burst onto the scene as one of the Gilded Age's most visible polymaths and a "pioneer of early modern geography." Endlessly curious about the natural world's relationship to social and cultural phenomena, he helped popularize geological and geographic knowledge, making complex scientific theories digestible for ordinary Americans. Moreover, from 1872 onward he published on a dizzying array of topics: earthquakes, hurricanes, volcanoes, tornados, soils, forests, weather forecasting, pest infestations, natural gas, nature contemplation, the oscillations of continents, the history of warfare, the evolution of dogs and rattlesnakes, geological deposits, glaciation, race relations, and, of course, swamps. Trained as a geologist, he was increasingly enticed by historical geography and treated the field of geography as a subordinate branch of geology. In doing so, he cemented himself as a transitional figure as the scientific enterprise professionalized, knowledge balkanized, and the lens for evaluating the place of man in nature shifted from natural theology to natural law.[17]

A Voyage to England and the Kentucky Geographical Survey

In 1872, the year Shaler assumed control of Harvard's geology and zoology programs, he set sail for England, hoping to rub elbows with Britain's greatest scientific minds. During the multiple-month journey, he engaged the likes of Darwin, Thomas Huxley, Charles Lyell, John Tyndall, and others, but he also stumbled upon historical efforts to drain the English fenlands. His *Autobiog-*

raphy, posthumously published in 1909 by his wife, credited the 1872 trip for crystallizing his curiosity about swamps. "This fen region, the reclaimed Marshlands (some two thousand square miles of the best corn land in England)," the *Autobiography* noted, "set Mr. Shaler to thinking what might be done by draining the vast acreage of swamp in the United States. His imagination also played about the political consequences of the appropriation of this land to farming uses." Although the *Autobiography* never clarified the remark or pinpointed the source of his interest in drainage's "political consequences," within a decade he had concluded that the federal government must do more to bring the unexploited resources of wetlands into the market.[18]

Later that year, Shaler returned to Harvard and capitalized on his first big break outside of academia. By the mid-nineteenth century, rapid industrialization and the substitution of coal for wood in iron production were leading many states to search for buried mineral wealth by authorizing geological surveys. In 1860, for instance, the California legislature launched a statewide geological survey. A decade later, Kentucky, Missouri, Ohio, and Wisconsin followed suit. Mindful that Shaler coveted an opportunity to lead his home state's geological survey, Agassiz wrote Kentucky governor Preston H. Leslie on his behalf. Leslie wasted no time extending a job offer; in August, Shaler accepted it while staying on at Harvard. In the first year, his $10,000 appropriation allowed him to conduct a general reconnaissance of every county, study plant and animal life, create maps, and publish a report. In 1875, he received additional funding to conduct a triangulation survey. Later that year, he submitted his first annual report, which offered a glimpse into his empirical approach to nature, as well as the unmistakable influence of George Perkins Marsh.[19]

In the appendix to his 1875 report, Shaler included forty-one pages lifted verbatim from Marsh's 1874 *The Earth as Modified by Human Action* (the second edition of his classic *Man and Nature*, the nineteenth century's most popular, groundbreaking conservation text). In May 1864, Scribner published *Man and Nature*. The book sold over a thousand copies in its first few months, prompting an immediate reprinting. With his admonition that "man is everywhere a disturbing agent," Marsh dissected how prehistoric Mediterranean and Middle Eastern civilizations had vanished after overexploiting the natural environment, particularly forests. The same fate awaited contemporary nations, he prophesied, unless they curbed forest destruction and restored clear-cut landscapes. In addition to selling forests as the foundations of civilization, *Man and Nature* delved into the interconnectivity of different ecosystems,

Nathaniel Southgate Shaler. Reproduced from *The Autobiography of Nathaniel Southgate Shaler with a Supplementary Memoir by His Wife*, 1909.

including the relationship between forests and stream flow. By doing so, Marsh theorized the concept of a "watershed," a specific, hemmed-in area where the functions and activities of living things influenced one another. When the book's second edition appeared a decade later in 1874, it was already an international best seller, and it kickstarted the conservation movement, shattering the cultural ethos that uninterrupted economic progress, technology, and expansionism yielded no environmental repercussions.[20]

Marsh's ideas captivated Shaler. In his report's appendix, he hailed Marsh's book as a "great masterpiece [that] should be read by all who desire to understand the effects of man's action on our earth's history." The appendix featured extracts from the second edition of *Man and Nature* that juxtaposed the forest resources of Great Britain, continental Europe, Russia, and the United States; examined the destructiveness of floods; and heaped praise on woodland and forest preservation efforts. Although Shaler conceded that cherry-picking selective portions of Marsh's book presented "a very inadequate idea of the author's treatment of his subject," he judged that the subject was so fresh and compelling as to warrant a partial reprinting.[21]

Although Shaler borrowed liberally from Marsh (especially the linkage between forest cover and stream flow), he crafted his own ideas about the relationship between man and nature, namely the dangers of unchecked population growth and the geological and hydrological functions of wetlands. In his 1875 report, he highlighted the "terrible inundations" of France's Garonne River as a reminder "of the dangers which [will] menace this country if it is ever stripped of its timber." Shaler continued that "with each succeeding year of increased forest clearing ... we may reckon on far more sweeping floods, acting upon a more and more densely peopled region." On this latter point, he proclaimed that the scale of humanity's environmental destructiveness was proportional to its population density. As he sounded the alarm about rampant population growth, he despaired that "we must watch all the dangers which will come from a rapid increase of population, and the consequent disturbance of the old relations of the forces of nature."[22]

Since Shaler assumed the inevitability of population growth, Kentucky's future would include deforestation and flooding on the Ohio River and its tributaries. Consequently, he pleaded for more assertive federal intervention in flood control. In this regard, *Man and Nature* again influenced him. As historian Donald Worster has pointed out, by "introducing a new sense of limits, Marsh's book also pushed many to begin rethinking the role of government in the United States." Shaler identified the federal government as the only entity capable of coordinating the efforts of multiple states in curbing flooding in the face of population pressures and forest clearing. Specifically, he endorsed the strategy trumpeted in civilian engineer Charles Ellet Jr.'s 1852 report to Congress on reducing flooding on the lower Mississippi River. Ellet proposed constructing a series of dams, reservoirs, and artificial lakes upstream to retain water during the wet season, while selectively releasing it during the dry season. In doing so, the hydrological cycle's seasonal vicissitudes could be regulated, achieving both flood control and navigation improvements. In foreshadowing the concept of multiple-use planning that so enthralled Progressive Era conservationists, he boasted that the storage reservoirs could supply irrigation water: "Within fifty years this method of fostering crops will be made an important element in the agriculture of this country."[23]

Shaler was just warming up. In addition to dams, reservoirs, and artificial lakes, he argued that certain classes of swamps aided in diminishing downstream torrents. Pointing to the geology of New England's "high-lying swamps" in his 1875 report, he observed, "In New England a soil resting on sand and gravel, a great system of ponds and lakes, and great areas of high-lying swamps, all serve to retain the greater part of the flood-water and to discharge it slowly." In Shaler's estimation, these "high-lying swamps" were just as indispensable in promoting navigation and flood control as storage reservoirs and forest cover. Although he did not identify the specific hydrological instrument at work, the assertion that high-lying swamps sequestered excess moisture before methodically releasing it back into watercourses anticipated future ecological understandings of wetlands. Whereas Marsh had hammered out the relationship between *forests* and watersheds, Shaler grafted *swamps* onto the concept, suggesting that the constituent parts of each water basin were complex, integrated, and interdependent.[24]

Settlers since the onset of colonization had demonized swamps as anomalous phenomena unconnected to forests, rivers, or other hydrological systems. Now, Shaler teetered on the brink of a breakthrough in his culture's understanding of the interdependencies inside of watersheds. He rejected the dominant view that watery tracts might be the vestiges of Noah's flood or some other mythical event, fixing swamps' origins in ancient geological events. Geology, not supernatural forces, offered the clue to unraveling swamps' hydrological functions.

Outside of high-lying swamps, Shaler asserted that other watery tracts, such as bottomland and riparian swamps, must be drained for agriculture and to protect settler bodies. In far southwest Kentucky, he pointed to the Reelfoot Lake region as being "too productive to be left to waste and fever-breeding." With the adoption of levees, dikes, or the "Holland system" of windmill-power pumps, he predicted that the "waste"inundated tracts adjacent to the Mississippi or Ohio Rivers would dry out and become prime farmland for settlers. Moreover, the confinement of floodwaters through the combination of dams, levees, or windmill pumps would eradicate malaria. "The experience of Holland shows conclusively that the institution of a permanent water level is the one effective bar to the production of malaria," he contended.[25]

In appraising Kentucky's swamps, as well as the state's timber resources, Shaler displayed his faith in meticulous empirical observation. Specifically, during his leadership of the Kentucky Geological Survey, which expired in 1880, he always sought to categorize, judge, and then rank natural resources based on their usefulness to society and industry. With watery tracts, he ranked "high-lying swamps" in their natural condition as superior to swampy bottomlands and riparian swamps because they normalized streamflow, mitigated downstream flooding, and did not poison adjacent communities with malaria. Merchantability and market considerations likewise informed his assessment of timber resources. In his annual reports, he rated black walnut, hickory, tulip, and white oak as "more valuable," "noble," "precious," and "by far the most valuable." Over the 1870s, the steady replacement of white oaks with "comparatively worthless" black and red oaks alarmed Shaler because "it may cease to furnish the basis for some very important industries which now quite depend upon it." Fearful that the white oak faced "extinct[ion]" within a century because of a paucity of saplings across the state, he recommended more investigations to forestall the species' decline.[26]

In time, Shaler's penchant for grouping and judging species crept into his assessments of humanity. When in the late 1880s he assailed immigration from southern and eastern Europe, he grouped and rated ethnic and racial groups as he did black, red, and white oaks. Significantly, his proclivity for categorizing and ranking tree species and swamps, as well as his concerns about preferred species' reproductive capacity and potential for extinction, mirrored the future agenda of eugenicists. In making policy recommendations, conservationists and eugenicists drew on the same rhetoric and obsession about reproduction, extinction, and the protection of "noble" species.

At the Helm: Shaler, the USGS's Atlantic Coast Division, and New England's Coastal Wetlands

As Shaler's tenure in Kentucky wound down, events at the national level were reshaping the face of federal science. After the Civil War, Congress looked to facilitate the economic growth and integration of the American West with the rest of the country by reviving Western scientific exploration, which had stalled since the 1863 abolishment of the US Army Corps of Topographical Engineers. In a four-year period after the war, Congress authorized four major Western surveys: Clarence King's Geological Exploration of the Fortieth Parallel (1867); Ferdinand V. Hayden's survey of Nebraska, which later expanded into Utah, Wyoming, Montana, Colorado, and Idaho (1867); John Wesley Powell's great Western survey for the Smithsonian Institution (1869); and Lt. George Wheeler's Geographical Surveys West of the One Hundredth Meridian (1871).[27]

By the late 1870s, the four simultaneous Western surveys were troubling policymakers. In 1878, Congress formally asked the National Academy of Sciences to investigate. On November 6, 1878, the academy recommended that Congress discontinue the Hayden, Powell, and Wheeler surveys and activate a new "Geological Survey" to investigate "geological structure, natural resources, and products." On March 3, 1879, President Rutherford B. Hayes signed legislation that created a new Department of the Interior entity—the USGS—responsible for the "classification of the public lands, and examination of the geological structure, mineral resources, and products of the national domain."[28]

After languishing as a scientific backwater during its first two years, the USGS entered a period of rapid expansion under its second director, John Wesley Powell. As the director of the Smithsonian Institution's Bureau of Ethnology and a national celebrity, Powell had garnered acclaim for his daring 1869 passage down the Colorado River and through the Grand Canyon, as well as his famous *Report on the Lands of the Arid Region of the United States* (1878). In that publication, he famously urged Congress to abandon the Jeffersonian rectilinear method of land subdivision in the West.[29]

From his first day on the job, Powell harbored grand ambitions for both himself and the USGS, eventually engineering a stunning feat of bureaucratic aggrandizement. By 1885, the survey's annual budget had ballooned from $106,000 to more than $500,000, the number of its employees and field staff had jumped from a few dozen to 283, its initial decentralized structure had been replaced with a streamlined Washington-based organization, and "general geology" and less utilitarian forms of science had become institutional prerogatives. One of Powell's greatest achievements, however, lay in steering the USGS's growth into a transcontinental institution. As part of his initial 1882 budget request, Powell sought an increase of $100,000 for work in the West and an extra $100,000 to expand into the Appalachians and the Mississippi Valley. Even more notably, he worked behind the scenes to attach an amendment to the 1882 General Sundry Bill instructing the USGS "to continue the preparation of a geological map of the United States." When that twelve-word amendment became law, Powell had liberated the survey from its Western orbit. As one writer observed, "To prepare a geological map of the United States he had first to prepare a topographical map.... And to make a topographical map of the United States he had to go outside the [Western] public lands." The USGS was poised for a meteoric phase of growth.[30]

Despite his long association with the West and irrigation, Powell never overlooked the Eastern United States. He organized new USGS divisions in the upper Mississippi Valley, Appalachia, and New England and signed cooperative agreements with several Eastern states. Starting with Massachusetts in 1884, he committed the USGS to produce 1:125,000 scale maps of every New England state. And also in 1884, he tapped Shaler to lead the USGS's new Atlantic Coast Division. It was a superb selection because Shaler brought a wealth of practical and academic experience to the task.[31]

In his first year, Shaler hurried to complete his coastal swamp report. Published a year later in the USGS's *Sixth Annual Report*, the forty-five-page "Preliminary Report on Sea-Coast Swamps of the Eastern United States" classified and inventoried wetlands along the New England seaboard. In it, Shaler demolished long-standing myths in two ways. First, he laid bare seacoast swamps' variety and complexity. Second, he challenged the long-prevailing view that landscapes characterized by an abundance of surface water were anomalous. Rather, they possessed unfathomable diversity and were connected to other environments through dynamic hydrological and botanical webs. Moreover, he located their origins in ancient geological processes: "The development of the shore swamps of New England is intimately connected with the glacial history of this district during the last ice period. The forces at work in that period determined the shape of the surface and thus fixed the outline of the shore. The oscillations of level attending on the coming and going of the ice sheet did much to determine the history of the coast line. It will therefore be necessary to preface the account of the New England coast swamps with a brief statement concerning the effects of glacial action on its territory."[32]

"Preliminary Report on Sea-Coast Swamps" opened with a rundown of coastal New England's glacial and geological history. After the wide-ranging overview, Shaler separated coastal swamps into three groups: saltwater, freshwater, and estuarine. Salt water swamps owed themselves to the interaction of salt water plants, various invertebrates, and detrital matter washed in from tides. The second category, freshwater swamps, formed where watercourses crisscrossed and meandered through level coastal lowlands, which hindered natural drainage and led to surface water deposits. Finally, estuarine swamps "formed where essentially fresh water is lifted and lowered by the tide, composed in the main of grasses and the alluvium of rivers." Ultimately, Shaler's three-tier scheme represented the government's first classification system for the country's wetlands. And while settlers continued to interchangeably use "swamp," "marsh," "bog," "inundated land," and other related terms, Shaler contended that coming to grips with inundated landscapes would demand more scientific and semantic precision.[33]

Shaler never intended for his report to catalyze an appreciation of seacoast swamps' intrinsic worth. Indeed, he appreciated the sublimity of many natural landscapes, but swamps were another matter. He sought to persuade Americans, and perhaps even Congress, to emulate European countries by "winning these marshes to agriculture." In the past, the biggest obstacle to drainage, he reasoned, had been a glut of arable, available land. Yet the imminent closing of the frontier, which he publicized in his *Science* article, mandated the reclamation of "unoccupied" swamps. After consulting the USGS's map collections, he put the amount of reclaimable wetlands "east of the one hundredth meridian" at between 20,000 and 40,000 square miles.[34]

Shaler hyped drainage as a superior land reclamation option to irrigation. "The great advantage of the more northern marsh areas," he continued, "is found in the fact that they are generally near the large centers of population of the country." Given this proximity, farmers would enjoy easy access to roads, railroads, urban markets, schools, and houses of worship. The conversion of coastal swamps into agricultural fields, he crowed, would lift land prices to $200 per acre. Even better, swamp soils' "practically inexhaustible" character promised heftier yields and slower rates of exhaustion. In the Northeast, he put the amount of reclaimable coastal swamps between Maine and New York at more than 200,000 acres. Desirous of promoting a white settler agenda, Shaler finally rounded out his report with a detailed catalogue of "the larger salt marshes of New England and Long Island"—unsettled soggy areas in need of government intervention to create farms.[35]

Significantly, Shaler excluded centuries of accumulated knowledge about swamps from Indigenous and Black perspectives. The omission signified his belief that environmental knowledge was only useful to the extent that it advanced the government's preferred methods of production: settler agriculture, timber harvesting, mineral development, and so forth. The exclusion served as a form of epistemological policing that solidified whose knowledge mattered (Shaler's and the federal government's) and whose was expendable and dismissible (Indigenous communities and Blacks). It also erased centuries of Indigenous and Black history, reinforcing the myth that swamps had always been unoccupied or undeveloped and thus exclusively reserved for settler occupation and exploitation.

The Nationalization of Shaler's Investigations

Shaler's "Preliminary Report on Sea-Coast Swamps," 1886 *Science* article, and subsequent publication about New England's fluviatile swamps elevated him into a preeminent authority on wetland resources. As his association with wetlands deepened, the media sought out his opinions on controversies involving swamps. Queried by one journalist about the folk tradition of "mysterious lights seen hovering over swamps at night," he balked. Having spent many years investigating swamps, he had never witnessed a "will-o'-the-wisp" or experienced other "luminous experiences." "The reports about moving lights visible above swamps," he responded, "may be due to subjective impressions induced by gazing into darkness." Empirical science—not folklore or superstition—explained swamps' inner workings.[36]

In June 1885, Shaler approached Powell about broadening his investigations. "I propose that you put the task of studying all of the marshes of this country in my hands," he appealed. "The matter is one of much scientific interest and great commerce importance." Powell immediately approved the request. Given his unbending conviction that very little of the West could be irrigated, Powell had once promoted the creation of a centralized federal agency to plan, administer, and oversee drainage and flood control projects. Curious about the untapped wealth of swamps, he eagerly awaited the results of Shaler's investigation.[37]

Shaler hustled to recruit a team of assistants. Alfred C. Lane, a Harvard mathematics instructor, was one of the first to come on board. In early 1886, Lane moved to Europe, enrolled at the University of Heidelberg, and spent the next two years traveling across Great Britain and continental Europe, collecting data on European drainage methods and procedures. Two years later, he returned to Harvard, completed his PhD, and served with the Atlantic Coast Division before becoming the Michigan Geological Survey's petrographer. Shaler next brought on Spencer Penrose, one of his geology students. Early in 1886, Shaler boasted to Powell that "as both [Lane and Penrose] have private means, I shall be able to secure their services for their bare expenses in the field." Another of Shaler's protégés, Collier Cobb, also joined the effort. Within a decade, Cobb won acclaim as the country's top shoreline geologist as the chair of the University of North Carolina's geology department.[38]

Initially, Shaler operated with a $5,000 annual budget, but the USGS bumped it up to $7,500 in 1888. In addition to conducting field studies and examining maps, Cobb mailed questionnaires to professors, state chemists, newspaper editors, prominent landowners, and attorneys up and down the Atlantic Seaboard. The questionnaire centered on four topics: the quantity of undrained swamps near their residences, the public health impact of drainage, available swampland timber resources, and the value of drained plots.[39]

As Shaler's team engaged settlers up and down the East Coast, visited the Great Dismal Swamp, and carried out fieldwork, Congress in 1888 authorized a simultaneous irrigation survey in the West. The much-studied Powell Irrigation Survey, also under the USGS's direction, gobbled up far more resources and publicity. In fiscal year 1889 alone, it received $100,000; a year later, it worked with $250,000. Politically, it became a lightning rod of controversy as Powell verbally jousted with Western policymakers in congressional hearings. While Shaler worked in relative anonymity and freedom, Westerners dissected every detail of the irrigation survey, always fearful that sinister motivations explained the USGS's dilatory pace. Nevertheless, the existence of concurrent drainage and irrigation surveys signified that public policy options for federal water development were never hitched solely to the West—even after Congress abruptly dismantled the irrigation survey in 1890.[40]

That same year, Shaler published his final study in the USGS's *Tenth Annual Report*. Entitled "General Account of the Fresh-Water Morasses of the United States, with a Description of the Dismal Swamp District of Virginia and North Carolina," the lavishly illustrated eighty-four-page report consisted of two parts. The first part provided an overview of wetland hydrology, paleontology, plants, and soils across the country's humid half; detailed swamps' geological origins; and unveiled a new classification system for "inundated lands." It also showcased a state-by-state list of "areas of approximate inundated lands," assessing that upwards of 105,000 square miles were "winnable" to settler agriculture. In the second part, Shaler turned his attention to "the Dismal Swamp District of Virginia and North Carolina," describing the area's fossil record, geology, hydrology, vegetation, healthfulness, and drainage prospects.[41]

Shaler opened "General Account" by defining *swamps* and probing their origins. He explained that swamps "developed wherever there is a condition of embarrassed drainage which serves to retain a sufficient amount of the water to prevent the complete decay of the vegetable waste which accumulates on the soil as a result of the entire or partial death of the plants which occupied the district." Three specific conditions contributed to swamps' "embarrassed drainage": the shape and contours of the land, specifically "the gradients by which that rain-fall descends to the sea"; the quantity and seasonal timing of rainfall; and annual temperatures. The physical attributes of any given swamp, Shaler generalized, hinged on the interaction of all three conditions in a specific locality.[42]

The interplay of land gradients, rainfall, and temperatures yielded an assortment of variations. Reflecting on his six years of investigations, Shaler deemed it "impossible to introduce any single classification of swamps which will adequately designate their several characters in a way to make it of any value." His investigations also uncovered dramatic disparities in botanical, climatic, geological, hydrological, and pedogenic characteristics. These variations led him to announce a new "tripartite system" in 1890.[43]

The tripartite system was Shaler's "most satisfactory" attempt at a classification system. At the top, he divided inundated lands into two groups: first, "those which are visited by salt-water or are beneath the surface of the sea," which he deemed *marshes*; second, "those which are formed above the ocean level," which he named *swamps*. Next, he split "marine marshes" and "fresh-water swamps" into three subcategories based on surface characteristics: (1) marine, (2) "those which are formed in fresh-water basins," and (3) "those which are formed above the level where there is a permanent standing water." Finally, he used those subcategories to generate even more subclassifications for marine marshes and fresh-water swamps. This was a departure from earlier publications. In "Preliminary Report on Sea-Coast Swamps" (1885), he had designated three kinds of swamps: salt water, freshwater, and estuarine marshes. In *Science* a year later, he split "American inundated lands" into four broad geographic groupings: tide-water marshes, lacustrine swamps of the glaciated district, the Mississippi's Delta swamps, and "the class of wet lands or upland swamps where the marshy condition is due to the action of plants in retaining water under the surfaces of considerable districts." The inconsistencies no doubt reflected the fact that he was the first American scientist to attempt a classification system for a still misunderstood and vilified landscape.[44]

As with all of Shaler's USGS publications, his "General Account" underwrote the settler agenda. In one of its most suggestive sections, "The Economic Uses of Morasses," he isolated two broad economic categories: "first, the resources afforded by the swamps in their natural state, and second, the advantages which may be won through the drainage of these areas."[45]

Economically, Shaler trotted out six possibilities for swamps: silviculture, peat production, cranberry farming, iron ore harvesting, diatomaceous earths, and flood production/streamflow stabilization. Leveraging the data assembled by Collier Cobb, he first boasted about the quantity of timbered swamps in New England, Virginia, and the Mississippi River floodplain. While he singled out New England swamps for their prodigious white cedar yields, the Great Dismal Swamp, south Atlantic Seaboard, and Mississippi River Delta offered even bigger prospects. In those areas, companies could reap windfalls from cypress and juniper trees. He was particularly bullish about the Great Dismal Swamp. He projected that at least one-third of it could support the intensive cultivation of juniper and perhaps other tree species. In closing, the overall silviculture possibilities mesmerized Shaler: "A considerable part of our fresh-water morasses may advantageously be used for forest culture, because they generally lie near to deep water and can be made accessible to transportation by very cheap canals." Ultimately, the enormous amount of unharvested timber in American swamps belied alarmist rumors that the nation faced an imminent timber famine, one of the catalysts for the upstart conservation movement. Speaking at an 1894 meeting of Boston's Beacon Society, for instance, Shaler hammered home the point: "We are in no danger of suffering from treelessness.... The swamps, those nurseries of the forest, amount to 115,000 square miles in the United States."[46]

Peat provided another untapped prize. Although the first commercial attempt to manufacture peat in the United States did not occur until 1902, Shaler briefly touched on the subject, although he expressed pessimism about its prospects. As long as coal deposits remained abundant and experiments with packaging peat into "a more compact and serviceable source of heat" faltered, Shaler held out little hope. Moreover, he detected a "popular prejudice" against peat as a heating element. When set ablaze, it possessed an ashy character, emitted a strange odor, and failed to generate intense heat. He inferred that economic and social conditions did not merit a domestic peat industry. Agricultural fertilizer offered another prospective utilization of peat, but he judged that the chemical variations of peat soils rendered such usage inconsistent and difficult to predict. He proved prescient in this regard; until 1908, peat was not used in the United States commercially as a fertilizer.[47]

The promise of cranberry farming remained brighter. According to Shaler, 100,000 acres of Northern swamps could support cranberry growth. Among that acreage, bogs offered the most optimal conditions. He went on to explain that preparing a bog to support cranberry cultivation involved stripping away the upper vegetation, dissecting and removing the top layer of peat, placing a layer of sand three inches to a foot in depth, and finally erecting dams to manage water levels. All of this could be accomplished for an initial outlay of $300 per acre and, on average, sixty days of labor per year. Bogs also had the built-in advantage of enabling their nutrient-dense waters to fertilize the cranberry roots. His analysis centered on ushering cranberry cultivation into unexploited areas. "The above-described form of tillage," he insisted, "has an especial importance for the reason that it makes avail of areas which have hitherto contributed nothing to the economic interests of the country and gives employment to a large amount of labor." Specifically, he singled out cranberry farming around Cape Cod for improving rural conditions, maximizing agricultural output, and opening up economic opportunities for new settlers.[48]

The abundance of iron ore in New England's small marshes likewise proved alluring. Commonly known as "bog ore," the residues originated due to glaciation and other geological processes. During the eighteenth and early nineteenth centuries, companies heavily exploited bog iron, but the discovery of new, cheaper sources near commercial centers resulted in industries abandoning rural areas. Nevertheless, Shaler praised the potential aggregate quantity of bog ore in New England and other Northeastern states, which he tabulated at sixty square miles (at a depth of at least one foot). In total, he placed the amount of bog ore deposits in New England at 100 million tons; the entire country probably contained 500 million tons.[49]

Diatomaceous earths were the fifth economic possibility Shaler explored. These earths' commercial value stemmed from their employment as a "protecting envelope" for heavy industry's boilers and steam pipes. Fortunately for manufacturers, this grayish, compact material was widely distributed across the waters of American swamps. Although Shaler did not speculate about the quantity or distribution of recoverable diatomaceous earths—or the means of recovering them—he recognized that their usage as a protective coating in manufactured heavy items made them a critical resource for industry.[50]

One of the most notable portions of "The Economic Uses of Morasses" section analyzed the "economic value [of riparian wetlands] from their effect on the flow of the steams into which they discharge." In it, Shaler built on his previous publications by reaffirming the role of swamps in watersheds. Since the European colonization of North America, settlers had castigated watery landscapes as wastelands whose only redeeming virtue was their availability for settler agriculture. Now, Shaler countered that wet tracts that stabilized stream flow and reduced flooding might deserve to be left in their natural condition. "Our swamps have an important economic value from their effect on the flow of the streams into which they discharge," he explained. "They act as natural reservoirs of the rain-water, storing it in times of much rain and pouring it out slowly into the streams. They thus bring about the relative steadfastness of the flow which is characteristic of all rivers which rise in swamp districts." Many white Americans, including George Perkins Marsh, had mostly portrayed wetlands as unnatural landforms that resulted when soil from denuded forests washed downhill and accumulated in malarious, submerged tracts. Chafing at the notion that riparian wetlands or "upland swamps" owed themselves to anything other than geological forces—and were thus divorced from the hydrological cycle—Shaler argued that swamps merited inclusion in watersheds alongside forests, rivers, soils, and aquifers.[51]

Shaler never fashioned an ecological agenda for shielding swamps from agricultural exploitation. Indeed, his approach to water resources administration was utilitarian, instrumental, and dedicated to the settler agenda. Unlike Marsh and Powell, he seldom acknowledged limits to land exploitation and downplayed the consequences of unbridled economic growth. Despite identifying some of the *ecological* attributes of wetlands, he did not represent a protoecologist, and he touted settler agriculture as the land's highest purpose. And as in his 1885 report, Shaler's 1890 "General Account" made no effort to integrate Indigenous and Black knowledge, which had accrued from centuries of corporeal and experiential immersion as well as their efforts at modifying nature. These exclusions worked to patrol the boundaries of knowledge creation. Shaler's silencing of nonsettler voices denied them a stake in the future management of national swamp resources while he elevated himself into the subject's premier expert, purveyor of knowledge, and scientific gatekeeper.

Under Shaler's guidance, the USGS commissioned conservation science to satisfy the bottomless appetite of settlers for farmland. The ultimate aim of Shaler's 1885 and 1890 reports was to streamline and systematize knowledge about swamps and inundated lands to facilitate their conversion into farms. Parroting antebellum farm reformers, he emphasized that swamp soils, enriched by the accumulation of centuries' worth of animal and vegetable matter, could withstand the "tax which husbandry puts upon [them]." Even though drainage seldom stirred the public's imagination with the same intensity as irrigation, Shaler's work with the USGS, which ended in 1900, elevated him into the country's intellectual patron of drainage.[52]

Both of Shaler's USGS reports dissolved all doubt that wetlands were just as indispensable for economic growth as forests and irrigation. Yet, which settlers stood to benefit from the reclamation of America's inundated lands? Large investors, the middle class, or the dispossessed? New immigrants or native-born Americans? Moreover, what constituted the ultimate social and cultural aims of drainage? Capitalism and profits? Preserving democracy, building communities, or perpetuating the family farm?

Shaler provided few answers to these questions in his USGS reports. In other publications, however, his scientific racism, anxiety about immigration from southern and eastern Europe, and captivation with eugenic control spurred on his fascination with swamps. With Shaler more than perhaps any of the government's other conservation scientists, the subject of drainage could not be divorced from Anglo-Saxon fears of race mixing and the coming eugenics revolution.

Fusing Conservation and Eugenic Aims:
The New Immigration and the Promise of Swamps

In 1891, at fifty years of age, Nathaniel Southgate Shaler ascended to the deanship of Harvard's Lawrence Scientific School and embarked on his life's final stage. He led the Atlantic Coast Division for another decade, publishing on a range of topics in USGS annual reports: the nature and origins of soil (1892); the geological history of harbors (1893); the geology of common roads (1895); the origin, geographic distribution, and commercial value of peat (1895); the geology of Cape Cod (1898); and Virginia's Richmond basin (1899). Outside of the survey, his publications sought to popularize topics related to geology and historical geography in widely read periodicals and books. He also pivoted to cultural and social subjects. Toward the end of the 1880s, his endless slate of publications in *The Atlantic Monthly, North American Review, Scribner's Magazine, Science,* and other liberal venues increasingly veered into immigration, race, and imperialism.[53]

Shaler's personal brand of Anglo-Saxon supremacy fueled his eagerness to weigh in on immigration. As a transnational ideology of whiteness and race-making, "Anglo-Saxonism" exploded in popularity in the United States and Great Britain at the end of the nineteenth century. Particularly in New England, Anglo-Saxonism assumed a regional flavor, evoking contradictory responses from patrician intellectuals. On the one hand, Brahmin elites like Shaler swaggered that the settlement of New England had improved and refined their ancestors' racial pedigrees, elevating them into the most innovative and enterprising Anglo-Saxons on earth. Their boasts about New England serving as the epicenter of Anglo-Saxon greatness, however, were tempered by a growing disillusionment about racial decline. Brahmins were aghast about New England's plunging birthrates, outmigration, poor diets, and overcivilization. Such factors portended the race's inevitable decline, at least in New England. As the scholar Bluford Adams argues, the region's patrician intellectuals shared an "enormous pride ... in their dual identities as Anglo-Saxons and New Englanders and their belief that those identities were mutually defining."[54]

By the 1890s, Shaler joined other Brahmins to cast the new immigration as the gravest threat to both Anglo-Saxonism and New England. Dismissing immigrants from Hungary, Poland, Russia, and southern Italy as hereditarily inferior to native-born Americans of northern European ancestry, he asserted that the deluge of new immigrants beginning in the 1880s would erode democratic governance since they were "as permanently foreign to our institutions as are the Africans." The rising number of immigrants arriving on US soil from outside of northern Europe posed an existential threat. "The course which we are to take with reference to the incursion of these peoples, who are so radically foreign that we cannot hope to Americanize them," he forecasted, "is perhaps the gravest problem which our nation has now to face."[55]

In 1893, Shaler grappled with the consequences of the new immigration in an article that has been dubbed "the single most important anti-immigration screed of the period." Appearing in *The Atlantic Monthly*, "European Peasants as Immigrants" drew on mainstream evolutionary biology, physical anthropology, environmentalist human geography, and the now-discredited Teutonic school of history. In identifying ethnic and racial homogeneity (e.g., Anglo-Saxonism) as the bedrock of cultural and national identity, Shaler despaired that the country teetered on devolving into a "race oligarchy." As "the most vicious and persistent form of despotism," race oligarchies emerged anywhere superior and "inferior races" lived together. The despotic characteristics of race oligarchies sprung from the fact that inferior races could *live* in a civil society but not become *part* of it. In Shaler's rendering, a division between "masters and servants" characterized such oligarchies, and they were the antitheses of equality, freedom, civil society, and other democratic ideals.[56]

In the Western Hemisphere, Shaler cited Latin America and the American South as examples of race oligarchies. In the latter, the coexistence of unequal races propped up a two-tiered society at odds with democratic citizenship:

> There are many things which go to show that the oligarchic form of society in our Southern States, brought about by the essential diversities of the white and black races, is already affecting the system of their government. The negro has little or no more place in the body politic than he has in the social system. One third of the population in that part of the country is excluded from the most educative duties of the citizen, those which should come to him through the trusts which his neighbors confide to his care.... My reason for noting the facts above mentioned is that ... the presence of any considerable mass of alien people (alien, though they may have been born upon the soil) is, to a democratic state, a danger of the most serious sort.[57]

After laying out his race oligarchy concept, Shaler rebuked lax immigration enforcement. "We have suffered grievously from the folly of our predecessors in recklessly admitting an essentially alien folk into the land," he lectured readers. "Alien folk" encompassed two groups: Africans and European peasants. Shaler lamented that the majority of Africans trafficked into the United States originated in the Guinea Coast, whose inhabitants were allegedly far less hereditarily robust than other Africans. As a consequence, American Blacks were intellectually deficient, bereft of "all the instincts of a freeman," and unprepared for democracy—as demonstrated by the failure of Southern Reconstruction. "It was a more desperate and immediate evil to have the Southern commonwealths converted into mere engines of plunder," he sermonized, "as was the case during the so-called period of reconstruction, when the blacks controlled these States." With no short-term prospects for integrating Blacks into civil society, Southern communities where Blacks outnumbered whites had already collapsed into an oligarchic condition: "Where the black population becomes dominant, only the semblance of a democracy can survive; the body of the people will, as in Hayti, shape their society and their government to fit their inherited qualities." In such a circumstance, the only alternative would be for whites, "by their intellectual superiority and their cooperation with the abler negroes," to impose order in "a forcible way." Race oligarchies birthed violence and authoritarianism.[58]

Shaler next turned his attention to European peasants. Drawing on his neo-Lamarckism, he denigrated southern and eastern European immigrants as "a separate lower estate" that were "in essentially the same state as the Southern negro." Just as he had categorized and ranked the relative value of red, black, and white oak species for the Kentucky Geological Survey, he now juxtaposed and compared racial groups. Since these European peasants displayed no "upward striving," they lacked a "large sense of citizenly motives" as well as "social or political longings."[59]

Shaler attributed the peasants' hereditary shortcomings to nurture as well as nature. Over the centuries, the peasant underclass's "abler youths" had been siphoned off and prevented from reproducing by two entities: the army and the Catholic Church. After leaving his family to pursue a career as a mercenary, a peasant often died in battle or returned to his village too old to procreate. Since only a bare minimum of peasants possessed the intellectual and physical abilities to improve their social standing, their exclusion from the collective gene pool depleted their villages' hereditary makeup. In addition to military service, the Catholic Church had unleashed "a far more efficient means of impoverishing the peasant blood." Celibacy sterilized tens of thousands of peasants who enlisted in papal service. During Catholicism's 1,200-year existence, Shaler accused it of robbing no fewer than 100,000 of the laboring class's best talents. "The result has been," he expounded, "that while the priesthood and monastic orders have systemically debilitated all the populations of Catholic Europe, their influence has been most efficient in destroying the original talent in the peasant class." Put simply, peasants were hereditarily deficient and irredeemable.[60]

The scourge of celibacy had not wrecked northern Europe. "So far as my experience goes," he continued, "the peasantry of Germany and the Scandinavian countries is in a much higher state than that of southern Europe; there, is, indeed, a distinct improvement visible as we go northward." Even more so than the German and Scandinavian nations, England possessed only a remnant of a peasant underclass due to the limited influence of Catholicism and monasteries: "In the northern part of Europe, owing to the development of those forms of Christianity in which the clergy is not celibate, and in which the monastic order finds no important place, the greater part of the population has been, for many generations, exempt from this destructive influence."[61]

Drawn from the southern and eastern European underclass, the new immigrants arrived in America hereditarily incapable of participating in civil society. Even worse, the new immigrants threatened to enfeeble the high eugenic quality of native-born Anglo-Saxons through interbreeding. Since scientists at the time assumed that the mixing of races resulted in the worst hereditary characteristics being passed to the offspring, the interbreeding of native-born Americans and new immigrants could hasten the debilitation of the only race capable of sustaining democratic citizenship, triggering the onset of a race oligarchy in New England.

Shaler's scientific racism and Anglo-Saxon chauvinism prompted him to join other Boston Brahmins in spearheading efforts to curb immigration. In 1894, he became one of the founding vice presidents of Boston's new Immigration Restriction League. Led by John Fiske, the league supported a literacy test for immigrants, emphasized assimilability differences between *old* and *new* immigrants, linked racial concepts to an overarching view of national destiny, and became nativists' leading voice for immigration restriction. On a personal level, Shaler endorsed educational and health standards to bar undesirable peasants. If he had been a "reluctant and unlikely nativist" in the 1880s, as one scholar has claimed, he became a full-throated supporter of strict restrictionism by the century's end.[62]

What would happen if the new immigration continued unfettered? Gravely frightened about its scale and tempo, Shaler advertised swamps as eugenic sanctuaries. He obliquely advertised land reclamation—particularly drainage—as one of the only remaining strategies to ensure that Anglo-Saxons did not engage in sexual intercourse with peasant immigrants. He floated the idea in his enormously popular *Nature and Man in America*. In that 1891 book, he took solace that "the immigrants from continental Europe have in the main betaken themselves to the cities of New England, and have shown little disposition to obtain control of the soil." Recruited by urban industrialists, the new immigrants flocked to cities for jobs, built insular urban neighborhoods, and avoided rural environs.[63]

The settlement patterns availed native-born Anglo-Saxons of an opportunity to fashion homogenous rural enclaves. "Although much of the strength of New England has gone West to found new States," he surmised, "enough remains to ensure the perpetuation of the original stock, so that ... the towns will be largely composed of descendants of foreigners of alien race, and the country districts of folk of English blood." Converting swamps into sanctuaries for Anglo-Saxons theoretically provided the means of putting this vision into effect. His USGS investigations had identified 10,000 square miles of swamps in the New England and Laurentian districts available to be "won to the plough." Composed of the "richest soils," the swamps were destined to provide rural homes for future generations of Anglo-Saxon settlers. "With the inevitable crowding of our American population which the next century is to bring about," he concluded, "these swamps will be drained, and by their drainage a vast area of excellent land will be won to tillage." Drainage could foster eugenic control by geographically separating superior and inferior races, curbing opportunities for interbreeding.[64]

But that was only half of Shaler's racialized vision. In 1900, as the war in the Philippines dragged on, he implored the government to recruit "abler negroes" to serve as infantrymen. His reasoning was simple: "They would endure tropical climates better than whites." Once the war concluded, the government could then encourage the Black soldiers and their families to remain "permanently and contentedly established in Luzon." Drainage, imperialism, and settler myths about Black immunological superiority were thoroughly entangled in Shaler's vision, which fused the fates of foreign policy and domestic conservation. While the conquering of hostile *foreign* lands required the deployment of Black soldiers as surrogates for immunologically vulnerable whites, it also ensured that more Anglo-Saxons could remain stateside and raise their families on *internal* lands made healthy and productive via government-assisted drainage programs.[65]

Conclusion

By injecting swamps into the day's most pressing social and racial issues, Shaler mobilized the state's resources to offer environmental solutions. On one day, drainage could postpone a future of land scarcity for white settlers; the next day it could create an exclusionary space for Anglo-Saxons, allowing them to procreate and raise families together. In this way, the social basis of land reclamation could never be divorced from patrician scientists' obsessions with race, eugenic control, and immigration.

Ultimately, Shaler's appropriation of conservation science for racial purposes foreshadowed how the emerging prophets of irrigation, including George H. Maxwell, Frederick H. Newell, Elwood Mead, and William Ellsworth Smythe, drummed up support for a federalized irrigation program in the West whose primary purpose was to perpetuate the white family farm and densely pack dispossessed tracts with settlers. By the time his tenure with the USGS ended, Shaler never wavered from his belief that the rewards and riches of reclaimed swamp soils should exclusively benefit a chosen racialized few: native-born whites.

Shaler's conservation legacy thus remains just as complex as the wetlands he investigated. His "Preliminary Report on Sea-Coast Swamps" and "General Account of the Fresh-Water Morasses of the United States" symbolized achievements for government-sponsored conservation science, despite their glaring omissions of Indigenous and Black knowledge and experiences. In less than six years, Shaler pointed to the diversity and varied hydrological functions of wetlands; established their interconnectedness with forests, watercourses, and aquifers; studied their soils and vegetation; inaugurated the nation's first wetland classification system; and liberated Americans from supernatural explanations for the origins of swamps. Nevertheless, he departed from the contemporary examples of George Perkins Marsh and John Wesley Powell in never fully appreciating the limits involved in the exploitation of nature or articulating the environmental consequences of uncontrolled and unregulated capitalism and economic growth.

Shaler never doubted that swamps existed to enrich settlers. He marshaled conservation science to generate options to avert the exhaustion of arable land and the flood of new immigrants. The nation's "land-reserve" of swamps, his USGS investigations suggested, offered an escape valve for both crises. The drainage and occupation of millions of acres of swamps would stymie the nation's descent into landlessness while also offering a method of eugenic control, isolating native-born whites in homogenous rural communities. In this manner, the promise of swamps bonded the aims of conservationists and budding eugenicists. Small wonder that the conservation and eugenics movements, particularly after 1900, enjoyed such broad crossover appeal as proponents of both movements sought to avert the decline of the most valuable species while inventorying, classifying, and ranking the relative value of others.

During Shaler's USGS tenure, he accommodated settlers' demands that the federal government do more to provide drainage support. However, he approached conservation science from racist and moralistic perspectives while displaying little appetite for addressing the practical matters of land drainage: governance, technology, law, clay tiles, the securing of outlets, and the rivalries between large and small landowners over tax assessments. As a result, the federal government would forcefully intervene in drainage during the Progressive Era to dispense specialized knowledge, educate farmers, and provide other drainage-related resources to communities.

CHAPTER SIX

A Construction Agency
The Reclamation Service and Settler Home Creation, 1902–1910

The embattled settlers of northern Minnesota faced financial ruin. So warned Republican Rep. Halvor Steenerson. Early in 1906, the Minnesota congressman blamed the spiraling crisis on swamps and federal Indian policy. Back in 1889, Congress had imposed allotment on Minnesota's twelve Anishinaabe reservations, forcing the transfer of 3 million acres of Indigenous land to the federal government and coercing all members of the tribe to relocate onto either the White Earth or Red Lake reservations. The "ceded" Anishinaabe lands then became available for purchase to white settlers under the terms of the Homestead Act. Revenue generated by the land sales was deposited into a "Chippewa in Minnesota Fund," which would draw interest and subsidize annual payments to tribal members. In 1904, the Red Lake Anishinaabe band ceded another quarter million acres to the federal government in order to avoid another round of allotment and maintain communal control over a diminished land base. With the Anishinaabe herded onto two reservations and millions of acres of their homelands opened to prospective settlers, Steenerson anticipated the swift occupation of the expropriated lands.[1]

To Steenerson's disappointment, the Anishinaabe cessions never sparked a land run. In fact, by 1906 at least half of the over 3 million acres of "ceded" land remained unsold and a third of the new settlers had fallen into default on their payments. According to the congressman, the settlers' hardships stemmed from northern Minnesota's swampiness and its patchwork of private, state, federal, and Anishinaabe lands. Boundaries were once again scuttling drainage attempts. Since unsold "ceded" lands remained under federal control, they could not be incorporated into drainage districts or taxed and the state drainage agency lacked the authority to build ditches across them. As a consequence, the crop fields of many recent settlers remained swampy and waterlogged. Thousands fell behind on their payments. "Some of these people who have gone into that region and agreed to pay to the Indians are in default," Steenerson lamented, "because . . . by reason of the country there being wet they cannot get their crops in, and they have a failure of crops, and it retards the development of that whole region." Small wonder that he portrayed settlers as far bigger victims of federal Indian policy than the dispossessed Anishinaabeg.[2]

As an architect of the 1886 Crookston drainage convention and former attorney in drainage cases before the Minnesota Supreme Court, Steenerson understood that when mobile nature straddled or moved across political boundaries, it obliged government intervention. The disorderliness of surface water—its defiance of property and political boundaries—made it difficult to guarantee the permanent presence of settlers on Anishinaabe homelands without federal assistance. In response, the congressman drew up legislation that put the federal government in charge of draining northern Minnesota.

On January 2, 1906, Steenerson unveiled his plan. During a meeting with USGS director Charles D. Walcott, he reminded Walcott that in 1902 Congress had created the Reclamation Service in the USGS to build irrigation projects in the sixteen Western states and territories (initially excluding Texas) using revenue generated from public land sales. He argued that his district's swamps were no less deserving of federal aid than Western deserts. "There are millions of acres in Northern Minnesota ... [which] will remain worthless until the state and federal authorities take hold of the matter and furnish drainage outlets," he declared. "It is just as meritorious to reclaim swamp and overflowed lands and make them productive [with] drainage as it is to reclaim arid lands and make them productive by irrigation." Moreover, the Reclamation Service was the only entity capable of synchronizing the drainage of private, state, federal, and Anishinaabe land. Without federal leadership, northern Minnesota's "waste and useless" swamps and peatlands could not be made "productive and suitable for homes of the home seeker." Days later, Steenerson introduced legislation empowering the Reclamation Service to drain swamps across *all* of Minnesota.[3]

Steenerson's proposal coincided with the high tide of the well-known Progressive Era conservation movement. Responding to cultural fears about the closing of the frontier, unprecedented immigration, a potential timber famine, the chokehold of industrial monopolies, urban growth, and imperial expansion, a small coterie of Washington scientists and engineers used the embryonic administrative and regulatory state to implement federal natural resources initiatives. Promoting centralization above localism, science above politics, and rational planning above laissez-faire economics, these conservationists, aided by Republican President Theodore Roosevelt's enthusiastic support, sought to transform Americans' relationship with nature. They nationalized Western irrigation, dramatically expanded the system of forest

Congressman Halvor Steenerson. Courtesy of the Library of Congress.

reserves and national parks, transferred control of forest reserves to the USDA, imposed grazing fees on public lands, authorized the leasing of coal and waterpower sites, and passed the 1906 Antiquities Act. Disdainful of limited government and political compromise, conservationists championed bureaucratization as the antidote for inefficient and wasteful local governance.[4]

Drainage appealed to federal conservationists no less than irrigation, forest protection, or national park creation. Leaders of the Reclamation Service agreed with Steenerson that the country's approximately 100 million acres of undrained wetlands testified to the inability of settlers, drainage districts, and states to convert swamps into rural homes in a timely and efficient manner. Moreover, the existence of inundated tracts spanning state lines, including the Great Dismal Swamp (Virginia and North Carolina), Kankakee Marsh (Indiana and Illinois), and the Mississippi River bottomlands, also demanded federal attention. Without hesitation or delay, Walcott and Reclamation Service chief engineer Frederick H. Newell threw their weight behind Steenerson's bid to drain Anishinaabe homelands, but they quickly commandeered it for their own purposes.

Newell supported draining Minnesota's "ceded" Indigenous lands, but he encouraged Steenerson to expand his legislation so that it covered wetlands in *any* state or territory. In doing so, he pursued two statist aims. First, he sought to undo the tradition of local control over drainage that had prevailed since the Swamp Land Acts. Second, he conceptualized drainage as the linchpin in his vision of bureaucratic aggrandizement, that is, a deliberate effort to seize water management responsibilities, including drainage, from rivals such as the USDA and the US Army Corps of Engineers. If the effort succeeded, it would solidify a nationwide constituency for the Reclamation Service, curry favor with Southern Democrats, and institutionalize drainage as a legitimate federal activity.[5]

The Reclamation Service's pursuit of drainage led to the bitter contestation and politicization of wetlands. During Theodore Roosevelt's administration, federal conservationists and technocrats adopted two approaches to drainage. As this chapter shows, the Reclamation Service promoted *constructing* large federal swamp drainage projects that would be settled and farmed by a new generation of settlers. As the service rallied congressmen and conservation organizations to throw their combined weight behind a federal drainage takeover, it eyed launching new federal reclamation projects on public and "ceded" swamplands outside of the West. The service's construction model, however, clashed with a *consulting* approach pioneered by the USDA's Office of Drainage Investigations (ODI). Unlike the Reclamation Service, the ODI never received congressional authorization to construct drainage projects; rather, its engineers disseminated specialized knowledge across the rural periphery and orchestrated experimental drainage work in conjunction with states and drainage districts (see chapter 7).

The Reclamation Service and the ODI pursued the same end (a drained and densely settled rural landscape) using competing models (construction versus consulting). In the end, the service's ambitions fell victim to a series of contingent events that Newell neither foresaw nor could control. Federalism, sectionalism, personal rivalries, and a growing rift between Republicans and Democrats over the degree to which the federal state should intervene in water planning and management obstructed his nationalizing reform. As the failure suggests, settlers ultimately preferred that the federal government consult, coordinate, and guide—rather than co-opt and control—drainage. Once the Reclamation Service jettisoned drainage, Congress steadily expanded the ODI's personnel and budget during Roosevelt's second term (1905–09), preserving the tradition of local sovereignty over drainage that continues to the present day. It also extended this tradition to "ceded" Anishinaabe lands. Once the federal drainage crusade had stalled by 1908, Steenerson and fellow Republican Rep. Andrew Volstead convinced policymakers to pass legislation empowering state agencies and drainage districts to incorporate federal and "ceded" Indigenous lands in Minnesota into their projects and impose taxes on them. One way or another, conservationists intended to drain, populate, and domesticate the nation's wetlands to expand the reach of settler agriculture.[6]

The Origins of Federal Reclamation and Progressive Era Water Politics

The Reclamation Service traced its origins to the previous decade's economic and political upheavals. The crippling 1893 depression, skepticism about the assimilation of southern and eastern European immigrants, population growth, violent labor unrest, nostalgia for the frontier (which Frederick Jackson Turner declared closed in 1893), and nagging doubts about rural America's vitality whipped up Western enthusiasm for a federal irrigation program. Moreover, the admissions of Idaho, Montana, North Dakota, South Dakota, Washington, Utah, and Wyoming as states between 1889 and 1896 magnified the West's political clout, giving the region 30 percent of Senate seats by 1900. Since the Civil War, Westerners had excoriated the inequitable distribution of river and harbor appropriations, which disproportionately benefitted Eastern communities. In 1895, for instance, Nebraska Governor Silas A. Holcomb complained to his legislature that since Congress had invested millions of dollars on levees it should also "direct its efforts toward turning the waters of the western tributaries of the Mississippi River into great reservoirs, and thence into irrigation ditches for the development of sections of the country which produce very little." The majority of Westerners agreed. In 1901, their chorus of discontent reached a crescendo when Montana Republican Senator Thomas H. Carter filibustered to death the annual river and harbor bill after a conference committee stripped provisions favorable to the West. Carter's filibuster portended the region's dawning political prowess and eagerness to share a greater portion of federal water development money.[7]

Western transcontinental railroad corporations, which stood to benefit handsomely from a federal irrigation program, also prodded Congress to act. The transcontinentals saw irrigation as an instrument to develop and sell their large unsold federal land grants, promote dense rural settlement, increase their freight, and reverse the migration of Westerners to the East and to Canada (which from 1896 to 1914 exceeded 600,000 people). In 1899, James J. Hill's GN, the Northern Pacific, the Santa Fe, and the Southern Pacific each pledged $500 per month to California lawyer George H. Maxwell's campaign on behalf of a federal irrigation program and his new National Irrigation Association (NIA). In 1902, Maxwell's feverish lobbying and slick publicity campaign paid dividends when Congress passed the Reclamation Act, which created the Reclamation Service.[8]

The terms of the federal reclamation program were generous and straightforward. Settlers on government projects only paid for their portion of the costs of constructing irrigation canals, reservoirs, or other structures in ten annual installments. Those payments, as well as the proceeds of public land sales across the entire West, were deposited into a revolving "reclamation fund," which supporters anticipated would ensure a perpetual stream of revenue. Uncle Sam had entered the irrigation business.[9]

Frederick H. Newell, the Reclamation Service's first leader, had assembled an impressive resume by the time the Reclamation Act became law. Born on March 5, 1862, in Bradford, Pennsylvania, he lost his mother to an untimely death when he was an infant, leaving him to live with different relatives throughout childhood. A diligent pupil, he excelled in school and, in 1880, enrolled at the Massachusetts Institute of Technology. He graduated five years later with a degree in mining engineering, gained employment at one of his father's businesses, and then joined the Ohio Geological Survey. In 1888, his career's trajectory tilted upward when he met USGS director John Wesley Powell. Impressed with Newell's credentials and talents, Powell hired him to lead a group of engineering graduates to study the volume, velocity, and features of Western rivers. Newell quickly rose to become the USGS's chief hydrographer, and he ingratiated himself with Washington scientists. In the 1890s, he delivered lectures at the Cosmos Club, the National Geographic Society, the American Geographical Society, and the American Forestry Association. He also befriended the forester Gifford Pinchot, who introduced him to New York's rising governor and future vice president, Theodore Roosevelt. By the time an assassin's bullet killed President William McKinley and Roosevelt entered the Oval Office in September of 1901, Pinchot and Newell were the new president's closest natural resource advisors.[10]

Newell favored a deliberate and cautious approach to building federal irrigation projects. Although he had predicted wildly that his Reclamation Service would eventually irrigate 60–100 million acres (an area the size of California), in his initial slow approach he sought to manage public expectations, to learn from mistakes, and to accumulate experience. This strategy proved stillborn. Barely a month after the ink dried on the Reclamation Act, Roosevelt strong-armed the Reclamation Service into bolstering the Republican electoral agenda. Concerned that Nevada Rep. Francis G. Newlands (federal reclamation's congressional sponsor) and other leading Democrats would claim credit for the program, he hyped irrigation as a Republican public works program. In July 1902, with midterm elections looming in the fall, Roosevelt directed Secretary of the Interior Ethan A. Hitchcock to launch as many reclamation (e.g., irrigation) projects as possible. The president expected that irrigation development would drum up much fanfare and cement large Republican majorities. In March 1903, Hitchcock complied. He authorized the first five federal reclamation projects: Gunnison (Colorado), Milk River (Montana), Salt River (Arizona), Sweetwater (Wyoming), and Truckee-Carson (Nevada). Over the next four years, the Reclamation Service kept up the torrid pace, launching twenty-three projects across sixteen Western states and territories, elevating irrigation into a cornerstone of Roosevelt's conservation agenda.[11]

From the beginning, Newell put a premium on constructing large dams and irrigation structures. As the historian Donald J. Pisani observes, the Reclamation Service billed itself as a *"construction agency,* and it did nothing to provide western farmers with the capital needed to clear land, grade it, fence it, purchase livestock, and build houses and barns." Nor did the service design model towns, attempt to organize settlement, or dispense specialized knowledge about desert agriculture to farmers. This approach owed itself to Newell. As Pisani points out, "Frederick Newell was not a hydraulic engineer, and he never designed an irrigation project—either before or after 1902." Predictably, Newell remained focused on constructing large reclamation projects and never prioritized arming settlers with the knowledge, resources, or capital to succeed.[12]

As the Reclamation Service raced ahead, several developments hinted that the political landscape was becoming ripe for Congress to expand federal reclamation beyond irrigation. Early in 1905, South Carolina Democratic congressmen approached Rep. Charles Q. Tirrell of Massachusetts, the Republican chairman of the House Committee on Irrigation of Arid Lands, about how they could borrow money from the reclamation fund to drain Southern swamps. Although Tirrell took no action and nothing came of the request, it suggested that Congress might have an appetite for tackling drainage.[13]

Swamp drainage greatly appealed to Newell. On Western lands owned by the federal government, the Reclamation Service already claimed that it could drain swamps in tandem with irrigation. As Newell explained to a Sacramento audience in 1906, "The National Reclamation Act covers the reclamation of swamp land when taken in connection with an irrigation project."

As a result, he envisioned transforming the Reclamation Service into a truly national entity that planned and coordinated a range of water projects—irrigation, drainage, navigation, and perhaps even flood control—on public lands and waters in every state in the country.[14]

Halvor Steenerson's proposal to drain Anishinaabe territory afforded Newell a fortuitous opportunity. On January 4, 1906, the congressman introduced legislation that would deposit the revenue generated by the sale of Minnesota's few remaining public lands into a "Drainage Reclamation Fund," which the Reclamation Service would use to drain the state's public and "ceded" Anishinaabe lands. In the prior year, Minnesota's public land sales had topped $340,000, enough money to initiate surveys for limited areas. The proposal authorized the secretary of the interior to organize surveys and examinations, withdraw public and "ceded" Indigenous lands from entry, designate project locations, and condemn lands. To curb land speculation, the measure directed settlers to pay one-fifth of their share of construction costs at the time of entry and their remaining expenses in four annual installments.[15]

The proposal thrilled Newell and his boss, USGS director Charles D. Walcott. If passed, it would unlock access to communities outside of the arid West and make the reclamation of federal deserts and wetlands coequal priorities. With an unflinching faith in federal experts' capacity for planning and administering complex projects, Walcott in February 1906 implored Steenerson to broaden his proposal. He instructed the congressman that it was "inadvisable to enact legislation of this kind which would be applicable to but one State when the provisions of the bill might be applied to a number of other States." Creating a second federal reclamation program for drainage would galvanize a national constituency, engender support from Southern Democrats, and pave the way for the drainage of "ceded" lands.[16]

Steenerson's actions touched off a deluge of competing proposals. Drainage especially intrigued North Dakotans. Although North Dakota was one of the states covered by the Reclamation Act, its political leaders lambasted the leisurely pace of federal irrigation development in their state. Lacking deep canyons and gorges, the state offered few suitable big dam locations. The state's congressional delegation, however, nursed darker conspiracy theories. Between 1902 and early 1906, public land sales in North Dakota had generated $5 million, or about one-sixth of the reclamation fund, yet the Reclamation Service had spent less than $500,000 inside of its boundaries (far below the 51 percent mandated by law). North Dakota Republican Senator Henry C. Hansbrough, a sponsor of the Reclamation Act, speculated that Newell harbored a bias against the Northern plains.[17]

Hansbrough's fearmongering resonated with Lewis W. Hill, the vice president of the GN and James J. Hill's son. Having bankrolled George Maxwell's publicity crusade on behalf of federal irrigation, the Hills expected the Reclamation Service to erect multiple irrigation projects near the GN corridor. "I think [Newell's] interests are all in the southwest," Louis grumbled in 1904. "Either this is the case, or he knows nothing about North Dakota when he [claims] ... that there is no considerable irrigation project practicable in North Dakota, except possibly in connection with the Yellowstone River." In late January 1906, Hansbrough and fellow Republican Rep. Asle Gronna, following the recommendation of a statewide drainage convention in Grand Forks, introduced twin bills to transfer the next $1 million generated by North Dakota's public land sales to draining wet prairies in the Red River Valley. Since the USDA had spearheaded a drainage survey of the valley's Cass, Traill, Grand Forks, and Walsh Counties on October 13, 1905, both bills entrusted the USDA with leading the new federal drainage program.[18]

The events in North Dakota and Minnesota also excited Southerners. In 1902, Southern Democrats had supported the Reclamation Act by a wide margin of four to one; three years later, congressmen from the region demanded that Rep. Charles Q. Tirrell grant their communities access to the reclamation fund. They interpreted the flood of proposals as the precursor of an enlarged federal reclamation program. The moment for expansion had arrived. In March 1906, North Carolina Democratic Rep. John Humphrey Small introduced a bill to divert $3 million from the reclamation fund for the USDA to drain the Dismal Swamp. As he told *The Washington Post*, "I was not aware that the reclamation fund could be turned to such a practical use." A short time later, South Carolina Democratic Senator Asbury Latimer demanded a similar diversion to drain his state's abandoned rice fields.[19]

Schemes to tap the reclamation fund or assign drainage responsibilities to the USDA thrust Newell into turbulent political waters. Although he sought to unify the country under a single reclamation policy, the Reclamation Service could ill afford to alienate its primary constituents: Western farmers. Indeed, the period of 1906–7 witnessed a surge of Western antipathy toward the Roosevelt administration's conservation agenda due to the leasing of coal and waterpower sites and the imposition of grazing fees in national forests. Proposed reclamation fund diversions to Eastern or Midwestern states infuriated Westerners. The Walla Walla, Washington, *Evening Statesman*, for instance, skewered Hansbrough as an "enemy of [the] reclamation service" who had unleashed a "death blow" on federal irrigation. The *Idaho Statesman* howled in protest that the legislation "open[ed] the way for a general onslaught on the reclamation fund, and will lead to the demolition of the reclamation service." *Forestry and Irrigation*, the nation's leading conservation periodical, bristled that Hansbrough had handed the West a "mighty serious setback" because reclamation fund diversions threatened to delay the completion of unfinished reclamation projects or the commencement of new ones.[20]

Gross mismanagement had already made the financial situation dire. Less than four years after its creation, the Reclamation Service spiraled into a fiscal crisis. In April 1906, the influential conservation organ *Irrigation Age* published an ominous article entitled "Reclamation Fund Exhausted." In it, Reclamation Service consulting engineer C. E. Grunsky admitted that the hasty launching of twenty-three reclamation projects in less than four years to satisfy Roosevelt's partisan agenda had pushed the reclamation fund, with its $28 million coffers, to the brink of insolvency. Despite his eager support of expanding federal reclamation into wetlands, USGS director Walcott could not shake off doubts about the prudence of subsidizing drainage with irrigation dollars. "The condition of the reclamation fund," he conceded, "would not permit the setting aside of any such amount either from the funds now on hand or from the proceeds to be expected for a considerable number of years." Simple arithmetic rendered reclamation fund diversions for drainage a nonstarter.[21]

Undeterred, Steenerson rolled out a new bill modeled after the Reclamation Act. Introduced in March 1906, the legislation would create a revolving "drainage fund" from public land sales in non-Western states for the USGS director to reclaim wetlands in *any* state or territory and on "ceded" Indigenous tracts. The proposal prioritized "projects in the States in which the proceeds of sales and disposal of public lands have been greatest." Finally, it empowered the government to accept land cessions from states and to incorporate drainage districts, country drainage boards, and other private lands into projects—if the plots did not exceed 160 acres.[22]

Newell and his allies applauded Steenerson. First, his bill generated a new funding stream for drainage that left irrigation dollars untouched. Second, it kept drainage responsibilities in-house. Indeed, the rivalry between the USDA and the Reclamation Service over water planning and administration had grown more contentious since 1900. Newell and his supporters in the influential NIA, specifically George H. Maxwell and Guy Elliott Mitchell, in private correspondence to Walcott chided the North Dakota and Dismal Swamp measures as "evil schemes" that unwisely "split the Reclamation work between the Department of the Interior and the Department of Agriculture."

Efficiency—the ubiquitous buzzword of the federal conservation movement—supposedly demanded that all reclamation work (e.g., irrigation and drainage) remain under one roof. In March 1906, Mitchell published a *Forestry and Irrigation* article advertising the Reclamation Service as the appropriate agency to drain the nation's swamps. The service, he argued, already designed drainage systems on federal irrigation projects and collaborated with local farmers in organizing water user associations and assessing costs. Moreover, the USGS's Topographic Branch, to which the Reclamation Service belonged, had "already run its lines over many of the great swamp areas of the eastern states ... [and] will be ready to launch out into immediate activity in drainage." The Reclamation Service provided the best home for drainage.[23]

Crucially, Steenerson's drainage crusade summoned the Progressive Era's romanticism and nostalgic pining for a timeless rural past. Between 1870 and 1900, the population of the United States, swollen by immigration and natural increase, doubled while the industrial workforce expanded to encompass a third of the population. Foreign-born Americans accounted for a third of the population growth from 1860 to 1900; in the period between 1870 and 1920, immigrants represented one out of every three industrial laborers. By 1920, the population had jumped to 105 million and the census reported that, for the first time in US history, more people lived in urban areas than on farms.[24]

The ongoing demographic upheaval haunted Newell. He bemoaned that the concentration of Americans in cities, as well as their wage dependence on venal industrial bosses, eroded independence, individualism, and self-reliance. Nothing less than the wholesale relocation of millions of urban families to new rural homes on former swamps could revitalize the family farm and inculcate into Americans the hardiness and personal responsibility that he fetishized. As such, the government shouldered a sacred obligation to "promote land-owning citizenship."[25]

The symbol of the settler home permeated the rhetoric and logic of drainage boosters. In an editorial published in dozens of newspapers across the country, NIA secretary Guy Elliott Mitchell hailed Steenerson as "the man who can provide homes for the industrious and strong-armed citizens." The Minnesotan's efforts to convert public and "ceded" Anishinaabe swamps into homes, Mitchell announced, crowned him as "a benefactor to the race."[26]

Newell was no less committed to expanding rural white home ownership. "In any scheme the part played by the United States and its employees is that of initiating, guiding, and securing ... the best use of the [drainage] fund in securing what is vital to the life of the commonwealth, namely, the extension and perpetuation of a class of small landowners," he later observed. "The success of democratic institutions is tied up with this class of citizens and the largest menace to the maintenance of the institutions of a free country," he evangelized, "come[s] from the concentration of citizens in a few large industries, crowded in big cities, and under the control of a few capitalists." Drainage was the panacea for escaping a future of urban violence and mobocracy.[27]

Newell's smug moralism exposed one of the biggest contradictions at the heart of federal conservation. Newell, Steenerson, and other conservationists were essentially elite romantics who worshipped a preindustrial rural America. Ironically, recreating an ageless rural America by populating the nation's swamps first required assembling a modern bureaucratic superstructure in Washington, a fact that eluded all of the latter-day Jeffersonians.

Partisan Divisions and the Road to Oklahoma City

In a span of five months, Halvor Steenerson managed to put swamps onto the center stage of national environmental politics. Publicity and support for his drainage campaign came from a diverse cast: Maxwell's NIA, conservation periodicals, the Reclamation Service, romantic Jeffersonians, and a group of policymakers and bureaucrats concerned about the plight of settlers and the fate of "ceded" lands. As the summer of 1906 dawned, however, the momentum behind federalized drainage slowed due to the combined forces of federalism, partisanship, and warnings about the unchecked growth of federal environmental powers.

In May 1906, Western Republicans killed Steenerson's bill in committee, in large measure because the available revenue from Midwestern public land sales amounted to a mere $5 million. Wyoming Republican Rep. Frank W. Mondell, the influential chairman of the Committee on Irrigation of Arid Lands, echoed the prevailing sentiment when he declared, "It is idle to talk of starting this great work with this small amount of money. It would be all frittered away in making surveys and nothing would be left for actual work." Single-state drainage proposals fared little better. Robust dissent from Nebraskans and Idahoans in the House Committee on Public Lands cast a dark pall over any attempt to divert irrigation dollars to drainage. In a blistering minority report, they pilloried drainage bills for potentially delaying the completion of unfinished Western irrigation projects, launching a new and intrusive federal program, championing private over public land development, enriching existing property owners rather than aiding new settlers, and violating federalism by suppressing local control over drainage. The same committee approved Small's Dismal Swamp bill. In doing so, it invited "those States which expect such great benefits from irrigation [to] join with other States and sections in the reclamation of our fertile swamp lands." Attempts to nationalize drainage had balkanized supporters of federal reclamation and jeopardized the Republican Party's distributive program, prompting Speaker of the House Joseph G. Cannon, an Illinois Republican, to prohibit a floor vote on any drainage bill.[28]

The setbacks convinced some policymakers—namely, South Carolina Democratic Rep. Asbury F. Lever—to revisit the government's woefully inadequate knowledge about wetlands. In June 1906, he ridiculed the rank amateurism of drainage proponents. "It took ten years of constant agitation to enact irrigation legislation," Lever told a local newspaper. "To accomplish anything a campaign of education must first be inaugurated." At a bare minimum, Congress needed to build upon Nathaniel Southgate Shaler's 1885 and 1890 USGS swamp investigations and authorize a comprehensive scientific inventory of the nation's wetlands. Lever floated a joint resolution appropriating $20,000 for the USDA to survey wetlands in the Carolinas, Georgia, and Virginia; study the quality and richness of wetland soils; and estimate the per-acre costs of drainage. The resolution also directed the USDA to gather, study, and disseminate a list of European policies related to drainage and quantify the amount of wetlands that had already been reclaimed and cultivated. On June 26, 1906, South Carolina Democratic Senator Asbury Latimer introduced a similar resolution in the upper chamber.[29]

The resolutions widened a partisan rupture. Favoring the Reclamation Service's construction approach, Republicans greeted them with suspicion and incredulity. Apparently oblivious to events during the first half of 1906, Maine Republican Senator Eugene Hale protested that Democrats' endgame was to commit "the Government to an additional scope of duty by undertaking to drain swamps and wet lands throughout the United States." He encouraged senators to refer the proposition to the Committee on Agriculture and Forestry. Oregon Senator Charles William Fulton slapped down the resolution for violating the spirit of federalism. "The swamp lands have been given to the several States by Congress," he asserted. "Surely, it was not intended that Congress should go to work and reclaim the swamp lands."[30]

California Senator Frank P. Flint offered the most searing criticism. Unlike his fellow Republicans, Flint never objected to the interventionist environmental state accumulating additional duties. Rather, his disapproval centered on Democrats' concession to the USDA. Unfamiliar with the magnitude of the USDA's experimental and cooperative drainage work, Flint queried Latimer on the Senate floor if his resolution "[did] not mean the providing of an entirely new engineering service" within the USDA. Latimer retorted that the USDA already employed a corps of agricultural drainage engineers that dispensed expert technical and legal advice to states and drainage districts, to which the Californian quipped, "They ought not to have it in my opinion." Efficiency had taken a backseat to partisanship, with Democrats lauding the USDA's consulting approach and Republicans lining up behind the Reclamation Service's pursuit of construction responsibilities.[31]

With Congress deadlocked, the epicenter of drainage activism quietly shifted to Oklahoma Territory. Like North Dakotans, Oklahomans had grown weary of the plodding pace of federal irrigation development. In 1905, J. B. Thoburn, the secretary of the territorial Board of Agriculture, argued in his biennial report that Oklahomans would be best served if the territory's share of the reclamation fund subsidized drainage rather than irrigation. At the very least, he encouraged farmers living in the low-lying and flood-prone watersheds of the Washita River, the Little River, and the Deep Fork Branch of the North Canadian River to demand that the USGS survey those lands and distribute topographical maps.[32]

In August 1906, the NIA joined the fracas. During a speech to the Deep Fork Drainage Association in Chandler, Oklahoma, Thomas L. Cannon, the head of the NIA's St. Louis branch, lambasted the Reclamation Service for ignoring Oklahoma. Public land sales in Oklahoma had contributed $4 million to the reclamation fund, but not a penny had been spent inside of the territory. Cannon encouraged Oklahomans to clamor for the revision of the 1902 law so that it covered drainage as well as irrigation. "By stretching the text of the act," he argued, "land might also be reclaimed by drainage. If the reclamation of land by irrigation is constitutional and right, then the reclamation of land by drainage is equally so." According to Cannon's estimates, more than 1 million acres in Lincoln, Oklahoma, and Pottawatomie Counties—many of which were likely owned by the railroad corporations Cannon represented—stood ready to benefit from drainage.[33]

By October, the Oklahoma City Chamber of Commerce had caught wind of Cannon's publicity tour. Teaming up with Governor Frank Frantz, the Chamber of Commerce voted to host the country's first national drainage congress. It invited state governors, mayors, agricultural societies, commercial and industrial organizations, county and township representatives, drainage districts, railroad executives, and federal natural resource agencies to a December summit. The chamber's circular promised to initiate "a general movement for the reclamation of lands by [drainage] ... and to start a campaign of education."[34]

On December 5, 1906, the first national drainage congress kicked off. For three days, seventy-five delegates from seventeen states, multiple railroad corporations, and three federal agencies (the USGS, the USDA, and the Forest Service) presented lectures on agricultural drainage. Minnesotans were heavily represented and applauded themselves for shining the national spotlight on drainage. As A. G. Bernard, the president of the Minnesota Drainage League, explained, "Minnesota, I think, has set the example for the drainage movement. Our state was followed by North Dakota, and then the call for the national drainage congress was issued."[35]

Bernard pledged his support for new legislation drafted by California Senator Frank Flint. Born on July 15, 1862, in Middlesex County, Massachusetts, Flint moved to San Francisco as a child and later studied the law. In the 1890s, he served as an assistant US attorney, judge of the Los Angeles County Superior Court, and finally the US district attorney for the Southern District of California. His law career catapulted him into the political spotlight and, in 1905, he won election to the Senate as a Republican. He served a single term and did not seek reelection.[36]

Flint exerted a disproportionate influence on Progressive Era water politics. As a close friend of USGS director Walcott and the chairman of the Committee on Geological Survey, he was positioned to advance the Reclamation Service's agenda since it remained subordinate to the USGS until 1907. As chairman, he first waded into national water politics by maneuvering to put the Reclamation Service in charge of tackling the Sacramento River Valley's ensemble of water-related challenges. Along with the Columbia River, the Sacramento River, rising in the Trinity Mountains, was the Far West's only other navigable intrastate watercourse. The river's broad, fertile, and level valley claimed an average width of forty miles. As a navigable watercourse, the Sacramento River boasted an impressive velocity and volume, ranging from 5,000 to 600,000 cubic feet per second. During these flood stages, the river regularly overflowed its banks and transformed the valley into a vast "inland sea." In the valley's lower portions, the seasonal floodwaters fed sprawling stretches of towering tule (large bulrush) marshes that sometimes exceeded fifteen feet in height. Early in the 1900s, the USGS calculated that the valley contained 600,000 acres of tule swamps. The Sacramento's status as an intrastate navigable river significantly meant that federal engineers would only have to deal with a single state in adjudicating water rights. Thus rather than viewing the Sacramento River Valley's water challenges (irrigation, flood control, and drainage) as an insuperable obstacle, the Reclamation Service hyped the valley as a "staging area" to showcase its managerial capabilities and the potential effectiveness of "multiple-purpose" development.[37]

The multipurpose concept dazzled Progressive Era conservationists. It envisaged the comprehensive, holistic, and simultaneous development of *all* natural resources across *entire* river basins. Forests positioned at the headwaters of streams, the argument went, would counter runoff, decrease soil erosion, moderate stream flow, and reduce sedimentation. When paired with forest reserves, reservoirs would store water for irrigation, reduce the incidence and intensity of floods, and starve downstream riparian wetlands of water, allowing them to be drained and cultivated. Flint heralded the Sacramento River Valley, with its interrelated water challenges, as the perfect testing area for the concept.[38]

In March 1906, Flint had introduced a measure that appropriated $5 million for the USGS to initiate surveys and build a range of reclamation projects across the Sacramento and San Joaquin Valleys. The legislation further empowered the USGS to build storage reservoirs with the aim of capturing floodwaters to prevent downstream floods, irrigate adjacent lands, and dry out tule lands "for agricultural purposes." It required that at least three-fourths of landowners agree to repay their share of surveying, construction, and administrative costs over two decades. Furthermore, the law directed the USGS to ensure that the landowners organized themselves into drainage or irrigation districts under California law.[39]

The proposal overjoyed the Reclamation Service. Consulting engineer C. E. Grunsky, a towering figure of the engineering profession who had just wrapped up a stint on the Isthmian Canal Commission, effusively praised Flint. An unyielding proponent of centralized planning and administration, Grunsky disdained the Swamp Land Acts. "The United States did not pursue a wise policy . . . when it gave the swamp and over-flowed lands in Arkansas and other states to these states to be reclaimed," he later wrote. Furthermore, a "diversity of local interests, [and an] unwillingness to substitute organized effort for individual effort," he explained in his assessment of Flint's bill, "makes it appear hopeless to attempt to secure the adoption and execution of general plans . . . by cooperative action unless the same can be had under provisions of the Reclamation Act. . . . It is a proper and wise policy for the Federal Government to aid in . . . drainage . . . along the lines laid down in the Reclamation Act."[40]

Despite Grunsky's denigration of local sovereignty, Secretary of the Interior Ethan A. Hitchcock in April 1906 killed the Reclamation Service's flirtation with the multipurpose concept. He fretted that the service's support of Flint's proposal, when juxtaposed with its opposition to single-state drainage bills in Minnesota and North Dakota, evinced "inconsistency" and "discrimination." In response, Flint scrapped his multipurpose strategy.[41]

In late November 1906, Flint reemerged from the shadows and drafted a new drainage bill almost identical to Steenerson's updated legislation. It dedicated, retroactive to July 1, 1905, the revenue generated by public land sales in states not associated with the 1902 Reclamation Act to a "drainage fund" to be used by the Secretary of the Interior to drain "swamp and overflowed lands" in any state, territory, and "ceded" Indian tracts. On December 5, the first day of the Oklahoma City national drainage congress, he introduced it with much fanfare in the Senate.[42]

Back at Oklahoma City, the Reclamation Service's chief statistician and assistant chief engineer lined up delegates behind Flint. Although it is unclear if there was prior coordination, on the last day of the meeting delegates "emphatically endorse[d]" Flint's proposal. In addition, they passed a series of resolutions praising Steenerson for bringing drainage to the forefront of national conservation politics, encouraged states to pass effective drainage district statutes, and congratulated the USDA's "surveys and investigation of special drainage problems." In endorsing a second federal reclamation program, the delegates downplayed the promise of multipurpose planning and showed their appetite for enshrining drainage a as legitimate conservation activity, one that maximized nature's output and inaugurated millions of new settler homes.[43]

Kansas v. Colorado, Federalism, and the Albatross of Privatization

Frederick Newell's pursuit of federalized drainage entered its second and decisive phase after Oklahoma City. During this period, his ambitions to convert the Reclamation Service into a national water resource agency were dashed, never to be revived again. Just as he had miscalculated the popularity of a top-down managerial program for drainage, he proved tone deaf at recognizing the multitude of legal and political factors that made federal reclamation expansion a nonstarter. In addition, a landmark 1907 Supreme Court decision slammed the door shut on his aims.

The fateful 1907 case of Kansas v. Colorado stemmed from a prolonged clash between Kansas and Colorado over interstate water rights on the Arkansas River. Originating in Colorado, the Arkansas flowed through Kansas,

Oklahoma, and Arkansas before it joined with the Mississippi River. Kansas argued that irrigation and manufacturing diversions in Colorado diminished its flow to the detriment of western Kansans, and it asked the Court for relief. Kansas based its suit on riparian rights, the legal doctrine reserving an undiminished flow of water to riparian landowners. Claiming the right to appropriate water within its own borders, Colorado invoked state sovereignty.[44]

Critical of both states' reasoning, the federal government intervened on behalf of the Reclamation Service, which attacked Kansas's position since it imperiled the government's authority, under the property clause, to improve and dispose of its arid public lands by threatening the legality of upstream dams. Riparian rights, the Reclamation Service contended, conflicted with the agricultural development of arid regions that required the storage and conveyance of water over long distances. They also made it difficult to evaluate the quantity of water available for irrigation and undercut the federal government's monopolization of surplus water in interstate streams. US attorneys balked at Colorado's claim of sovereignty, arguing that it violated the theory reserving control over unappropriated waters to Washington. One of the Reclamation Service's broader goals in the case, however, centered on persuading the court to annul riparian rights, the doctrine predominating in the Eastern United States. If this occurred, the road would be open for building drainage projects outside of the West.[45]

In May 1907, Associate Justice David J. Brewer delivered the court's unanimous decision. From the perspective of drainage, Brewer's decision to uphold the Colorado diversions proved far less important than his dismissal of the federal government's position and admonishments against expanding federal reclamation. Espousing a narrow view of federal power, Brewer revisited the court's "constant declaration ... from the beginning that this Government is one of enumerated powers." He conceded that the commerce clause conferred broad authority on Congress to regulate navigable interstate streams, as he himself had recognized in United States v. Rio Grande Irrigation Company (1899), but the case at bar differed because the entirety of the Arkansas River was not navigable. Brewer next denied that the property clause lent Congress a blank check to override or invalidate state laws in pursuit of public land reclamation. Though the court had never fully ruled on the property clause's reach, Brewer emphasized that "it does not grant to Congress any legislative control over the States, and must, so far as they are concerned, be limited to authority over the property belonging to the United States." Brewer hinted at, but did not explicitly declare, the unconstitutionality of the 1902 Reclamation Act. But he left no room for confusion when he signaled that the court would frown upon the extension of federal reclamation outside of the West: "While arid lands are to be found, mainly if not only in the Western and newer States, yet the powers of the National Government within the limits of those States are the same (no greater no less) than those within the limits of the original thirteen, and ... in the absence of a definite grant of power, the National Government could enter the territory of the States along the Atlantic and legislate in respect to improving by irrigation or otherwise the lands within their borders. Nor do we understand that hitherto Congress has acted in disregard to this limitation."[46]

Brewer's narrow reading of federal authority trounced any legal justification for a federal drainage program. Most significantly, the ruling rejected that the national government possessed any right, either real or implied, to improve private land, even if a project also involved the simultaneous drainage of some quantity of public (or "ceded" Indigenous) lands. In conjunction with *Kansas*, the Swamp Land Acts extinguished any prospects for federal drainage because those policies had privatized the nation's "swamp and overflowed" lands. In 1907, the USDA estimated that the federal government held title to only 5 percent of the estimated 77 million acres of wetlands east of the 115th meridian. Since the drainage of the scattered pockets of public and "ceded" wetlands depended upon compelling the participation of adjacent private landowners, the Reclamation Service's chief legal advisor Morris Bien grew pessimistic: "So far as the public lands of the United States are concerned, the question [of drainage] presents but little difficulty from a legal standpoint, but where a large portion of a project is not owned by the United States, the power of the Federal Government to engage in construction must be regarded as doubtful." A federal drainage program was a nonstarter.[47]

National newspapers agreed. Quoting from Brewer's ruling, *The Washington Post* editorialized in September 1907 that "the authority of Congress within the States is limited to control over the property belonging to the United States within their limits. Congress cannot enact laws overriding State laws relating to reclamation and it cannot assume control of lands owned by the States." Similarly, an internal USGS memorandum stipulated that the federal government could not directly construct drainage projects unless the states retroceded their Swamp Land Act donations. This was unlikely because the bulk of their grants had already passed into private ownership. The Swamp Land Acts, the memo declared, fortified local control: "Unfortunately, many years ago the Legislature enacted and gave these [swamp] lands to the states provided the latter drained and reclaimed them. In practically no instance has this been done.... While it is the belief that the Government is in a position to demand the recession of these swamp lands when it may be ready to undertake to drain them[,] it is doubtful if any such course will be desirable or necessary."[48]

The Department of the Interior's General Land Office (GLO) adopted a similar stance. During the previous year, the deluge of drainage bills in Congress had perturbed the GLO. As the Department of the Interior office responsible for federal wetlands cession under the Swamp Land Acts, the GLO bitterly opposed any measure that infringed on the status quo. In March 1906, GLO commissioner W. A. Richards had denounced a second reclamation act for transgressing federalism. As Richards concluded, "Congress has made ample and liberal provision for the reclamation of swamp lands in the various laws heretofore enacted on that subject." In January 1907, he doubled down, recommending that any federal drainage legislation be struck down.[49]

Allies of the Reclamation Service refused to budge. Over the next twelve months, Guy Elliott Mitchell and USGS employees orchestrated a media blitz to drum up support. They flooded *National Geographic*, *Hearst's International*, the *American Review of Reviews*, the *Journal of the American Peat Society*, and other periodicals with articles that showcased drainage as a wiser investment than irrigation and highlighted its significance in building permanent and prosperous settler communities. At an estimated cost of $5–6 per acre, drainage was far cheaper than irrigation's $30 price tag. As always, the articles propagandized drainage as a palliative for modern society's social ills, including urban overcrowding and fears about food shortages, and highlighted its role in keeping settler bodies healthy. Given the richness of swamp soils, drainage gifted new farmers with windfall harvests and soaring land values (estimated at $50–$200 per acre).[50]

The articles dismissed constitutional scruples. Since many of the nation's largest undrained swamps crossed state lines, the onus for their development fell on Congress. The hydrological interconnectedness of swamps and river stream flow likewise justified federal intervention. Hinting at a dawning, if halting, ecological understanding of watershed hydrology, Mitchell emphasized how the drainage of a local swamp or marsh "affect[ed] the river flows or flood problems in another State." Hydrological webs joined the multitude of state boundaries to render drainage districts and states inadequate for the task. Halvor Steenerson had always been correct: Drainage welcomed centralized oversight and control. The Reclamation Service's construction approach was the only viable public policy option for draining federal and "ceded" swamps.[51]

Localism Triumphant: The Demise of the Reclamation Service's Construction Approach

Despite Justice Brewer's ruling and a hardening conviction that federalism reserved drainage as a local prerogative, the Reclamation Service pressed ahead. On November 26–27, 1907, the National Drainage Association (NDA), formed the previous year in Oklahoma City, hosted its second annual meeting on the campus of Johns Hopkins University in Baltimore. New Secretary of the Interior James A. Garfield, congressmen, USDA and Department of the Interior personnel, prominent entomologists, US Army Corps of Engineers officers, and state legislators attended the proceedings. The meeting adjourned after two days without endorsing a specific policy. Unlike at Oklahoma City, the USDA resisted the Reclamation Service's agenda. According to USDA engineer R. P. Teele, "Our friends in this Department succeeded in controlling ... and preventing any endorsement of the work of the Geological Survey [e.g., Reclamation Service] by the Committee on Resolutions, and the Flint bill dropped out of sight entirely."[52]

Napoleon Bonaparte Broward—the Democratic governor of Florida, new NDA president, and zealous proponent of draining the Everglades—had much to do with the NDA's policy of neutrality. Seeking to solidify a diverse coalition, he persuaded delegates to avoid championing any proposal that favored either the Reclamation Service or the USDA. "It would be manifestly unjust and unwise for the National Drainage Association, or any of its members[,] to attempt any influence whatever in the legislative action of Congress," he stated. Broward, too, encouraged the NDA to "confine ourselves to urging national drainage as a general policy and not as any particular project." Vexing constitutional questions, however, dampened delegates' outlook. Endorsing the opinion of the Reclamation Service's legal staff, Democratic South Carolina Senator Asbury Latimer confided to *The Charlotte Observer* that "I fear some trouble on constitutional grounds. All the swamp lands held by the Federal government have been turned over to the States and the States have ... disposed of them to corporations or individuals. ... In the reclamation service[,] some private lands are being reclaimed, but the title to much of it still remains in the Federal government; while the government has title to none of the swamp lands." The Swamp Land Acts remained a legislative albatross.[53]

Despite this pessimism, national policymakers spent the next two months introducing no fewer than nineteen drainage bills. The Senate also took the step of authorizing the inventorying and quantifying the nation's wetlands for the first time since Nathaniel Southgate Shaler's investigations. On December 9, the chamber approved Latimer's resolution directing the USDA to submit a report on the location and quantity of the nation's undrained and drained "swamp and overflowed" lands; the impact of drainage on public health, cultivation, and rural communities; and a summary of state and foreign drainage laws.[54]

The surge of activity cloaked how the Reclamation Service's construction approach had already lost steam. Kansas v. Colorado, GLO opposition, intense partisanship, and the tiny amount of wetlands owned by the federal government hampered its efforts. Yet the allure of bureaucratic aggrandizement persisted. As a consequence, Newell enlisted new Secretary of the Interior James A. Garfield to make one final attempt at forging a bipartisan consensus. On January 25 and February 1, 1908, they hosted two meetings with Democratic and Republican policymakers. The bill that emerged from the negotiations, which Flint introduced in the Senate in early February, retroactively dedicated, starting on June 30, 1901, the proceeds of public land sales outside of the West to a "drainage fund" to be used to reclaim public or "ceded" wetlands in any state. The bill gave 160-acre plots of public land to settlers, requiring them to cultivate half the land and then repay their share of construction costs in ten annual installments. The bill's sections 5 and 6, inserted to mollify Southern Democrats, authorized loaning money to states, municipalities, drainage districts, or corporations engaged in reclaiming private wetlands through bond purchases. A first lien on the land would secure the loan, which had to be repaid within a decade. The conferees drafted these provisions to attract support from North Carolina and Alabama congressmen, whose respective states prohibited the issuance of bonds for drainage or had not passed drainage district legislation.[55]

In subscribing to the classic Progressive Era governance playbook, Garfield and Newell exhibited a skeptical and paternalistic view toward local governance. Local efforts at drainage, they claimed without citing specific evidence, had resulted in "confusion" and "delay." Only federal experts insulated from the corruption and self-interest of local politics possessed the appropriate temperament to "make general surveys and broad plans of development, involving all drainage issues from the headwaters down to the navigable portions of streams." Revealing their elitism, they concluded that the bill in question accomplished these goals by "leav[ing] to qualified men the solution."[56]

On February 24, 1908, the Senate Committee on Public Lands favorably reported Flint's bill with minor amendments. The committee rhapsodized that the proposal promised to drain 80 million acres in thirty-four states and territories, an area the combined size of Ohio, Indiana, and Illinois. If the secretary of the interior subdivided that entire area into forty-acre plots, the committee projected the rapid availability of rural homes for 12 million settlers.[57]

Riding a wave of momentum, Newell and Garfield anticipated that the legislation would sail through Congress. An untimely death soon soured its prospects. On February 20, South Carolina Senator Latimer unexpectedly died. Over the previous years, Latimer had served as the most vocal Democratic supporter of federalized drainage. While participating in the negotiations in Garfield's office, he had initially criticized Flint's new bill as a "Republican measure." By the end of the negotiations, however, he had patched up his relationship with Republicans, thrown his support behind the legislation, and asked Florida Governor Broward to write his gubernatorial counterparts across the country to support the proposal. The moment Latimer died, Newell and the Reclamation Service lost their biggest Democratic ally.[58]

When Flint's bill reached the Senate floor in April, Democrats immediately assailed its constitutionality. Colorado's Henry M. Teller cited Kansas v. Colorado as proof that "the Government could not go into a State and improve lands that do not belong to the General Government." Georgia's Alexander Clay piled on as well, chiding the legislation for going beyond the original Reclamation Act. "The reclamation act," he argued, "did not contemplate that the Government funds should be utilized for the purpose of loaning money to private individuals to develop and reclaim their lands." Georgia's other senator, Augustus O. Bacon, unloaded on section 5 as "one of the most stupendous and unlimited projects for emptying the Treasury that I have ever heard of" and dismissed explanations of why the word "corporation" appeared in the final text. James Clarke of Arkansas feared that the measure heaped too much authority on the secretary of the interior, lacked a provision guaranteeing states contributing to the fund a proportionate return of their contribution (such as the Reclamation Act's 51 percent mandate), and departed from the precedent of "using the public lands in a locality for the improvement of that particular locality." Clarke worried that the proposal erected a liberty-usurping leviathan that substituted the will of one bureaucrat (the secretary of the interior) for the sovereignty of hundreds of drainage districts.[59]

Enraged, Flint accused Southern Democrats of duplicity and of throwing the floor debate into a partisan fracas. Yet he misidentified the source of their alienation. They blocked nationalization less out of a desire to embarrass Republicans or subvert the Reclamation Service's construction-based agenda than out of a conviction that the USDA's approach to drainage, which provided expert advisory *consultation* to settlers rather than *construction* services, offered a more popular and viable alternative (see chapter 7). On Bacon's recommendation, the Senate struck out sections 5 and 6, emasculating the bill so that it applied only to federally owned wetlands. Bacon then offered an amendment to kill the entire bill, but the Senate rejected it by a near party-line vote of 15–37, with nearly half of the body abstaining. The amended bill never came up for another vote, and the Reclamation Service's dream of securing drainage responsibilities died a quiet, inglorious death.

Following the defeat of federalized drainage, a handful of congressmen, most notably Mississippi Democratic Senator John Sharp Williams, periodically revisited the subject. From 1912 until World War I, he introduced legislation in every congressional session to create a federal drainage program, but his intentions were symbolic and partisan. Writing to Secretary of the Interior Walter L. Fisher in 1912, the Democrat admitted that "I have a specific purpose in view...the reclamation of swamp and overflowed lands as a counterpoise in appropriations to the immense sums which have been spent for irrigation." The rural homebuilding idealism that had jump-started the Reclamation Service's pursuit of drainage had vanished, replaced by a bitter struggle over dwindling funds earmarked for land reclamation. The statist attributes long attributed to conservationism—efficiency, the elevation of science above localism, favoring of executive action, and decision making by dispassionate experts—by 1912 had run aground on the shoals of federalism, localism, and bureaucratic territorialism.[60]

With the death of the federal drainage crusade, the outstanding question of how to drain expropriated Indigenous territory in Minnesota lingered. On that matter, Rep. Halvor Steenerson soon burst back onto the scene.

State Power, Land Reclamation, and the Fate of the Anishinaabe "Ceded" Lands

From the moment Steenerson in January 1906 set foot in USGS director Charles D. Walcott's office, he believed that only the federal government could coordinate the reclamation of northern Minnesota's private, state, federal, and "ceded" Anishinaabe lands. Although he favored a second federal reclamation program for drainage, he embarked on a parallel campaign to secure legislation applicable only to Minnesota.

His efforts eventually bore fruit. On June 21, 1906, Steenerson and Minnesota's congressional delegation persuaded Congress to appropriate $15,000 for the USGS to conduct a "drainage survey of lands ceded by the [Anishinaabe] Indians in the State of Minnesota which remain unsold and are wet, overflowed, or swampy in character, with a view to determining what portions thereof may be profitably and economically reclaimed by drainage." During the summer and fall, USGS personnel carried out the survey, and they issued a preliminary report on November 7. Since the initial appropriation proved far too small to survey the entire Anishinaabe cession of over 3 million acres, the USGS limited its investigations to a 630-square-mile swath adjacent to the Mud River, located northeast of Thief River Falls, Minnesota. The preliminary report, published in early 1907, projected that the so-called Mud River region's 266,750 acres could be drained at a per-acre cost of $3.23. Overall, the report tabulated the total cost of drainage improvements at $1.1 million.[61]

The report's optimistic tenor armed Steenerson with enough ammunition to keep the Anishinaabe drainage question before Congress. Over the next two years, Congress appropriated another $20,000 to survey the remainder of the "ceded" tract. In an interview in late 1907, Steenerson bragged that an overwhelming majority of settlers wanted Native homelands quickly drained and converted into homes. "I have never found but one man who opposed the drainage of these lands," he maintained, "and he was a game sport who said that if the Red Lake lands were drained it would prevent duck and wild goose hunting."[62]

For Steenerson and fellow conservationists, settler agriculture served as the most enlightened and noble form of land usage, and he denied that wetlands possessed intrinsic value or should remain in their natural state to provide habitat for waterfowl and wildlife. Nor did he show any concern for how agricultural drainage obliterated long-standing Indigenous connections with the land and water, including harvesting wild rice or gathering other medicinal or edible aquatic plants. In an era of rampant urbanization and a declining farm population, it behooved Congress to transform every wet tract into rural homes. In fact, when Steenerson broke with the national conservation coalition in 1913 by opposing the damming of Hetch Hetchy Valley in Yosemite National Park, he did so to protest "the eternal drawing upon the Federal Government resources . . . to make cities more attractive at the expense of the country." Jefferson's romantic vision of an agrarian republic was still alive and well in the early twentieth century.[63]

As the push for a second reclamation program was losing steam in early 1908, Steenerson redoubled his efforts. On January 9, he sent a letter to Recla-

Minnesota farmer laying drainage tile at the base of a trench, 1910. Reprinted with the permission of the Minnesota Historical Society.

Hovland tile ditcher crew at work in Minnesota, 1910. The explosion of drainage across the United States during the Progressive Era witnessed corresponding bursts of technological innovation and development with dredgers, ditchers, and tile laying machines. Reprinted with the permission of the Minnesota Historical Society.

mation Service chief legal officer Morris Bien about how the federal government could facilitate the drainage of Anishinaabe homelands. Specifically, he floated rewriting an earlier bill so that it would "harmonise with the state law, in case we should want the ditches laid out under that law." Bien, who remained steadfast in the belief that federalism buffered drainage from a federal takeover, worked behind the scene with Steenerson and his fellow Minnesota Republican Rep. Andrew Volstead to craft a solution. Their new proposal authorized Minnesota to treat federal and "ceded" lands in the same manner as state or private lands when it came to drainage; that is, Minnesota's state drainage agency could build outlets across lands owned or held in trust by the federal government and assess the costs of doing so against future settlers. A decade before Volstead dried out the nation's taverns and saloons by shepherding national prohibition through Congress, he set his sights on drying up Anishinaabe homelands to solidify the indefinite presence of settlers in north-central Minnesota.[64]

When the so-called Volstead Act reached the House floor on April 20, 1908, it sailed through Congress. During the debates, Volstead and Steenerson unspooled a story of white settler victimization in explaining how half of the over 3 million acres of "ceded" Anishinaabe lands still remained unsettled. "There are about a million and a half acres that have not been entered at all, because nobody could make use of them for homes," Steenerson groused. The whites who had settled the land, he continued, were "poor" and "unable to make any payments." A harsh, unforgiving environment had conspired with patchwork landownership patterns to undermine settlers' attempts to cultivate their fields in a profitable and timely manner. Queried by a colleague about an ownership breakdown of the "ceded" lands, Steenerson responded by emphasizing the legal and environmental obstacles plotting against settlers: "The situation confronting them was that a majority of these lands could not be assessed [drainage taxes] because they are Government lands. You could not extend the drainage improvements against them, and consequently all drainage improvements failed." With a flair for sarcasm, Volstead returned to the symbol of the home and mocked opponents: "If you want to preserve [the swamps] for all time for the frogs, good and well; but it seems to me that here is an opportunity to furnish homes for 100,000 people. We are not asking near as much as the people that go upon the arid lands of the West. The National Government gives the work and advances the money to reclaim those lands. We do not ask that. We expect to go down into our own pockets to pay for it, and we will do it cheerfully if you will only give us the right." On the following day, the House passed the bill by a margin of 241–20, with 109 members not voting. The Senate quickly approved an amended version with a voice vote, and the full House gave its final nod of approval in mid-May. On May 20, 1908, President Roosevelt signed the Volstead Act into law.[65]

Conclusion

With the Volstead Act's passage, the brief national drainage movement came full circle. When the federal government first considered taking up drainage, it did so as an errand in settler colonialism, seeking to secure the lasting presence of new settlers on dispossessed Anishinaabe homelands in Steenerson's congressional district. Conservationists in the Reclamation Service, however, had different ideas. They signed on to Steenerson's legislative agenda by making it a pretext for institutionalizing drainage as a federal activity. When federalism, localism, and partisanship thwarted the Reclamation Service's statist agenda, the best Congress could do was to empower Minnesota to treat federal and "ceded" lands no differently than state or private property and equip it, and not the Reclamation Service, with construction responsibilities.

In only eight years, the Volstead Act enabled the construction of approximately 2,000 miles of drainage ditches on the "ceded" Anishinaabe tract. The law symbolized how the interventionist environmental state proved far more skilled at dispersing power than wielding it in centralized bureaucracies. In Progressive Era water politics, the national served the local. Legislative attempts to consolidate authority at the center yielded centrifugal results that more often than not preserved local sovereignty by empowering intermediary institutions at the expense of Washington technocrats and bureaucrats.[66]

Environmentally, the result was the same. Whether Congress or local institutions drained and domesticated Indigenous homelands, the act of land reclamation dismantled and erased centuries of Indigenous relationships with the land, soil, and water. In replacing Native flora and fauna with settler agroecologies—dry crop fields, roads, fences, ditches, and so forth—land drainage became one of the most consequential, and decisive, weapons in the settler state's arsenal.

Ultimately, Newell and the Reclamation Service proved unable to persuade Congress to dedicate the same amount of resources to swamps as it did to deserts. Drainage remained a cherished public policy objective, but the failure of nationalization revealed that federal bureaucrats and engineers often reacted to circumstances and events far more than they shaped them. Despite the Reclamation Service's failure, however, the scale, intensity, and pace of drainage exploded from 1905 to 1920, mocking federal conservationists' claim that only the central government could spearhead water and landscape engineering projects on a national scale.

CHAPTER SEVEN

Consulting Advisors
Federal Drainage Engineers and Settlers, 1902–1913

The expert guided and the citizen decided. Such was the advice of C. G. Elliott, the head of the USDA's Office of Drainage Investigations (ODI). In November 1909, A. D. Crooks, the secretary of the Wilson County (Kansas) drainage committee, asked Elliott to provide assistance for his county's beleaguered farmers. Heavy downpours during the previous year had led the meandering Fall and Verdigris Rivers to repeatedly flood, swamping thousands of acres of lowland crop fields. Galvanized by the inundations, Crooks and other landowners launched a movement to establish a drainage district, which under state law required the signatures of two-fifths of the taxpayers in the designated area. Once the signatures were collected, the landowners could petition the county to authorize a district, which then imposed taxes and issued bonds to pay for levees, ditches, and/or river channelization.[1]

From the outset, the proposition struggled to gain traction. Early in January 1909, a group of landowners organized a drainage committee, drew up a tentative reclamation plan, and circulated throughout the valleys rallying farmers to sign the petition. The disposal of surface water, a transient natural resource that defied property boundaries, always required some degree of cooperation, coordination, and individual sacrifice. Yet getting Wilson County settlers to work together proved daunting. By mid-January, the committee's efforts had faltered; too many landowners complained about levees, canals, spillways, and other improvements being built on their properties. Most of the settlers supported a district, but they recoiled at government intrusion onto their private property, regardless of the future benefits. State Senator F. M. Robertson, an advocate for the district, mocked opponents' perspective in a scathing editorial: "Will they put up a levee somewhere to back the water up on my land? Will they dig a spillway across the country and cut my land in two and put a big ditch between my house and the public road? Will they, in straightening the river, destroy a wide streak of my best land and compel me to go to a bridge (which they may not keep up) in order to reach my farm across the ditch?"

Rebuking the naysayers as "stubborn" and "selfish" for wanting benefits without bearing the costs, Robertson pleaded for unity and sacrifice. "The promoter can only answer that such may be the case and that they ought to be willing to sacrifice something to their own and the general good, especially since the land cut off will be made enough more valuable to offset the inconvenience. . . . All admit that something should be done; that an absolute guaranty from overflow would vastly enhance the value of their lands, and all would sign the petition if they could be persuaded that all the incident damages would be inflicted on the other fellow," he concluded.[2]

In spite of the festering opposition, Frank Prunty and S. M. Singleton, on January 15, 1909, submitted a petition to the county signed by dozens of Wilson County farmers, calling themselves the Benedict Drainage District, for the formation of a 6,000-acre district. Two weeks later, the county board of commissioners rejected it, unable to verify that enough signatures had been collected. The setback left the settlers "hopelessly divided," and the drainage movement plunged into acrimony and inaction.[3]

Two months later, the local *Fredonia Daily Herald* weighed in on the gridlock. "This subject needs a more liberal and broad-minded consideration," it editorialized on March 11, 1909. "Our private opinion is that this matter will never be settled until taken out of the hands of the farmers." In particular, the paper inveighed against the absence of coordination and comprehensive planning. Summoning the classic Progressive Era governance formula, the *Herald* recommended that the issue be placed "under the supervision of a competent civil engineer." Professional engineers—not self-interested settlers—should oversee hydrological and topographical surveys and then use those findings to identify the appropriate locations for levees, ditches, spillways, and other structures. In theory, expert-led planning would finally supplant naked self-interest and ad hoc measures.[4]

But therein lay a dilemma. Who bore responsibility for paying for the examinations and surveys? Moreover, by 1909 the agricultural drainage engineering profession remained in its adolescence; few universities included agricultural drainage in their civil engineering curricula and the field itself lay outside of the mainstream of the civil engineering profession. As a result, many district organizers, especially those outside of the Midwest, lacked access to competent engineers and specialized knowledge.

Fortunately, C. G. Elliott's recently created ODI put forth a solution. Since its creation, the ODI had championed an approach to drainage that sought to marry federal expertise with local autonomy. It preached that farmers like those in Wilson County, when guided by the technical wisdom of disinterested federal engineers, should control their own political destinies. The ODI assisted settlers who chose to organize a drainage district but for whatever reason could not seal the deal. After settlers petitioned the office, it judged the proposal's merits, dispatched personnel to the requesting community to conduct topographical and hydrological surveys, and provided legal advice. Unlike the Reclamation Service, it did not construct water management projects; rather, ODI engineers branded themselves "consulting advisors." They interacted with landowners befuddled by intractable challenges or disagreements, carried out surveys, studied soil conditions, estimated costs, offered legal guidance, and proposed the most feasible plan, including the locations of district boundaries and underground tiles and ditches. The ODI then submitted a report to the farmers or local drainage committee, who were then free to decide whether to proceed.[5]

The ODI's activities in Wilson County followed this script. On November 5, 1909, the local drainage committee formally requested ODI assistance. Two months later, federal drainage engineer W. J. McEathron arrived in Fredonia, Kansas, where he engaged the county drainage committee, studied local agricultural conditions, investigated the soil, and spent four days carrying out a topographical survey. Prior to departing, he assured the committee that drainage constituted a "very easy proposition" and reminded them of the ODI's consulting mission: "The government does not do the work of drainage nor require it to be done, that is with the people themselves."[6]

Upon returning to Washington, McEathron drew up a preliminary report, which he forwarded to the drainage committee in February. It recommended a combination of main diversion ditches, smaller ditches positioned at surface depressions, the removal of dead trees and other debris from river channels, and a series of low levees. Emphasizing the report's preliminary nature, McEathron encouraged the committee to arrange for a comprehensive survey at a future date.[7]

The next summer, in July 1911, the ODI sent engineer D. L. Yarnell to Kansas to check in on the proposed Benedict Drainage District. Yarnell quickly discovered that prodrainage sentiment had evaporated. "After two dry seasons without overflowing streams," *The Wilson County Citizen* reported, Yarnell "did not find the interest in the work of the drainage department particularly intense." With his site visit completed, Yarnell packed his bags and departed. So ended the ODI's advisory consulting mission in Wilson County.[8]

Throughout the Progressive Era, the ODI served as the state's clearinghouse for professionalized drainage knowledge. Led by Elliott, the nation's most distinguished and experienced drainage expert, it deployed national science to mediate and balance the interaction between peripheral communities and the federal state. By 1910, the office's tiny staff of thirty engineers and administrative personnel had also published and disseminated instructional bulletins about drainage techniques, procedures, and policies, as well as engaging in a range of experimental work with universities and state governments. More significantly, the ODI's allegiance to a federal-local partnership reconciled the Progressive Era's tension between autonomous individualism and the impulse for modernization and expert-led efficiency.

In promoting collaboration with farmers, the ODI exemplified the USDA's burgeoning role as an "advocate of a 'modernizing associationism.'" As the historian David E. Hamilton has argued, USDA leaders from the 1880s to the New Deal consistently "perceived a need for national systems to unite public and private institutions on behalf of its vision of agricultural progress. By building an associative sector, the Department could cut across partisan, community, and regional boundaries to forge institutions dedicated to its own goals. . . . Its associative machinery, so it insisted, made possible a means of modernizing without centralizing, of rationalizing without coercing."[9]

The ODI's advisory consulting mission embodied the USDA's modernizing associationism, but this mission also had its vices and drawbacks. As federal drainage engineers fanned out across the country and interacted with land companies and speculators, there was ample opportunity for them to self-deal and misappropriate federal resources for personal gain. By the end of the decade, the ODI found itself embroiled in a number of highly publicized scandals that sullied its reputation as a neutral dispenser of professionalized knowledge.

The ODI's consulting approach likewise put the ODI at odds with the Reclamation Service. Although both entities paid homage to traditional notions of Progressive Era governance—the elevation of specialists above generalists and coordination above laissez-faire—they pursued entirely different approaches. The ODI never envisioned itself as a construction agency. "The Government has never assisted in the construction of drains for the benefit of lands in private ownership," Elliott explained in 1908. The office's budget only supported "making investigations of lands requiring drainage, developing plans, and assisting in various way[s] to promote the work, but no part of this appropriation is available for the construction of the work."[10]

As a result, the ODI mostly limited itself to *farm drainage*, that is, dispensing technical and scientific assistance to improve existing farms for material and public health purposes. This clashed with the Reclamation Service's broader emphasis on *swamp drainage*, which emphasized draining large public swamp tracts to create new rural homes. Before World War I, the vast majority of communities seeking drainage-related assistance, including those from Wilson County, solicited the ODI for farm drainage support. In Wilson County in April 1906, the *Fredonia Daily Herald* praised drainage-minded settlers in a neighboring county for "not waiting for the appropriation by congress to do something toward preventing the overflow of the Neosho river." Small wonder that almost no archival evidence suggests that communities solely placed their faith in federalized drainage.[11]

Ultimately, the ODI helped facilitate the largest single outburst of land drainage in American history. Due to soaring farm incomes, elevated precipitation levels, high demand for agricultural commodities, and advancements in steam dredging technology, the Progressive Era witnessed an unprecedented explosion of drainage districts. By 1920, the date of the federal government's first drainage census, 65.5 million acres fell under the jurisdiction of drainage districts and municipal drainage projects, with more than three-fourths of that total (78 percent) organized since 1900. The nationwide proliferation of districts, many of which benefited from ODI assistance, revealed that settlers wanted the government to furnish expert knowledge and assistance rather than build swamp drainage projects. The ODI accomplished this by tethering intermediary organizations—drainage districts and county drainage boards—to federal expertise and knowledge. By the end of Roosevelt's second term, the interventionist environmental state was proving far more skilled at consulting than constructing, at coordinating than controlling, signaling that the nationalization of water resources tasks was neither inevitable nor triumphal—even as federal experts became increasingly integrated into farmers' lives.[12]

The Origins and Growth of the USDA and Federal Water Wars

The emergence of a state drainage apparatus occurred as the USDA entered a period of meteoric expansion. Created in 1862 during the Civil War, the USDA's original purpose was to "acquire and to diffuse among the people of the United States useful information on subjects connected with agriculture" and to disseminate "new and valuable seeds and plants." Initially, many policymakers interpreted the USDA's creation as a simple transfer of the Patent Office's mission of providing seeds to congressmen for distribution to constituents. Within two years, however, the USDA was publishing monthly crop reports and operating small scientific facilities.[13]

During the last quarter of the nineteenth century, the USDA grew at a dizzying pace. In 1877, Congress appropriated $18,000 for hiring entomologists to study boll weevil outbreaks. In 1883, it added a Veterinary Bureau; a year later, it activated a new Bureau of Animal Husbandry to "prevent the exportation of diseased cattle, and to provide means for the suppression and extirpation of pleuropneumonia and other contagious diseases among domestic animals." In 1887, the USDA created quarantine stations to inspect and cull infected livestock. More significantly, the Hatch Act (1887) authorized the department to cooperate with land grant universities in applied agricultural research. Two years later, the department achieved cabinet-level status. In 1890, it added a Weather Bureau. In the century's last decade, its Divisions of Chemistry, Forestry, and Biological Survey led research efforts ranging from food adulteration to soil analysis to forest conservation. Also by 1890, the USDA employed 2,400 Americans, many of them graduates of land grant universities, who within two decades cemented its reputation as "the principal scientific agency of American government" and one of "the world's leading research institutions."[14]

The department's accumulation of responsibilities entangled it in irrigation and drainage politics. Indeed, the ODI traced its lineage to the 1898 creation of the USDA's Division of Irrigation and Reclamation of Arid Lands. In July of that year, Wyoming state engineer Elwood Mead became the chief of the division, which the USDA soon rechristened the Office of Irrigation Investigations (OII). The OII's birth catapulted it onto a collision course with the USGS's Hydrographic Division. Led by Frederick H. Newell, the Hydrographic Division operated over 100 streamflow gauging stations on the Arkansas, Colorado, Columbia, Missouri, Sacramento, and San Joaquin Rivers, as well as a bevy of Eastern rivers. As the campaign for a national irrigation program reached a fever pitch by 1900, Mead and Newell jockeyed for position. At the time, Mead floated his plan for Congress to cede 5 million acres of public grazing lands to each Western state, subsidizing the creation of state-level irrigation departments that would act as laboratories of professional and technical knowledge. Newell, as discussed in chapter 6, harbored different ambitions. Aligning himself with Nevada Democratic Senator Francis G. Newlands, he helped draft legislation putting the USGS in charge of irrigating Western public lands. When the Reclamation Service was created in 1902, it fell under the Department of the Interior's USGS and Newell became its first chief engineer.[15]

Newell and Mead at first shared a cozy rapport. The relationship, however, quickly turned sour. After the OII's activation, Newell vilified Mead not only as a meddler who sought to seize authority over irrigation from the USGS, but as someone who promoted state water rights to the federal government's detriment and duplicated his division's work. Conversely, Mead discounted Newell as a dilettante who recklessly inflated the number of public acres available for reclamation, knew little about water rights, and ignored irrigation's drainage aspects.

California became ground zero of the feud. In November 1899, the California Water and Forest Association pledged $10,000 to defray the costs of a comprehensive study of water rights, irrigation, and policy led by Mead's office. Published in 1902, the OII's *Report of Irrigation Investigations in California* trotted out a list of recommendations for reforming the state's water laws. It also propelled Mead's rise as an irrigation expert, prompting the University of California to hire him to head a new Irrigation Department at Berkeley. Mead, who remained employed with the OII, saw the appointment as a springboard toward building a "Bureau of Irrigation" in the USDA to administer a federal irrigation program. Congress's 1902 placement of the Reclamation Service in the Department of the Interior devastated Mead, but the OII retained a foothold in California. In addition to his academic responsibilities at Berkeley, Mead's alliance with Governor George Pardee ensured that the OII received annual state appropriations of $10,000 to study irrigation methods, pumping systems, and various aspects of seepage, drainage, and the improvement of alkali-choked irrigated tracts. Disdainful of Mead's political connections and his preference for state sovereignty over water, Newell and his close ally George H. Maxwell chafed that the OII was undermining federal reclamation. Fuming in February 1903, Maxwell berated Mead as "a viper in the bosom of the national irrigation movement[,] professing always friendship ... [but] aiding every effort that was being made to knife and destroy it."[16]

Following Mead's *Report*, the California Water and Forest Association organized a commission to draft a new model water law by January 1903. Composed of Mead, Newell, and Judge John D. Works, among others, the committee recommended the passage of a bill mirroring the *Report*'s findings. The commission similarly proposed a four-member board of civil or hydraulic engineers to adjudicate legal disputes among irrigators, bestow new water rights, calculate the quantity of water required to raise various crops, and curb inflated water claims. Maxwell and his allies lashed out at the commission's proposal—the "Works Bill"—as a backdoor attempt to subvert federal reclamation. He worked behind the scenes to counteract Mead's "malicious and Machiavellian methods," ultimately securing the proposal's defeat. In the end, the California imbroglio churned up the lingering bad blood between Mead and Newell/Maxwell and fragmented the interventionist environmental state into warring factions.[17]

The Institutionalization of Agricultural Drainage Engineering and the Rise of C. G. Elliott

In 1903, as Californians battled over the Works Bill, at the national level Congress directed the OII to take up drainage. The abrupt institutionalization of drainage investigations owed itself to two unrelated factors. First, the accelerating scale and tempo of Western irrigation had triggered waterlogging problems on many federal irrigation projects, including soil alkalization, canal seepage, and rising water tables. The assumption that all desert soils were one and the same had led irrigators to overwater broad swaths of the West. Overirrigation turned federal and private irrigated plots into soggy, mosquito-laden fields as water seeped from unlined supply canals and groundwater levels surged upward. In April 1904, Mead estimated that one-third of California's irrigated lands had "already been converted into semi-swamp area and the boundaries of these areas are extending each year." He further claimed that 10 percent of Colorado's older irrigation districts required drainage. Many policymakers linked the future of irrigation to fresh studies of alkalization and seepage, as well as to new experiments focused on relieving irrigated tracts of excess water.[18]

Second, curiosity about the drainage of Midwestern river bottomlands and other wetlands to support the expansion of agriculture spurred on Congress. In 1903, a spate of floods on the lower Illinois River prodded Rep. Henry T. Rainey, an Illinois Democrat, to task the OII with drainage responsibilities. Rainey wanted federal assistance for his constituents, who were in the midst of leveeing the lower Illinois River, so they could expand their corn and wheat production on rich bottomland soils. Like their counterparts in other parts of the country, early twentieth-century Illinois farmers benefited from the era's robust farm economy as commodity prices generally exceeded farm costs, affording them more purchasing power than other workers. When this was combined with technological advancements, farm mechanization, and expanding federal scientific support to agriculture, the Progressive Era (notably 1909–14) was one of the best times to be a farmer in American history. The healthy rural economy buoyed Rainey's efforts. In 1903, Congress instructed the OII to take up drainage; the office immediately renamed itself the Office of Irrigation and Drainage Investigations (OIDI). C. G. Elliott, the country's most experienced agricultural drainage engineer, became the OIDI's engineer in charge of drainage investigations. He remained at the pinnacle of federal drainage for the next decade.[19]

Born on June 8, 1850, in Lowell, Illinois, Elliott attended Illinois Wesleyan College and Oberlin College before enrolling in 1873 at the University of Illinois in Urbana. Deeply involved in campus affairs, he edited *The Daily Illini*, served as the president of the sophomore class, led the Student Governmental Group, and later guided the Philomathean Literary Society. It was Elliott's immersion in the university's civil engineering program, however, that sparked his lifelong passion for drainage. Within a decade of completing the program in 1877, he emerged as a trailblazer in the methods, policies, and technologies associated with the fledgling field of agricultural drainage.[20]

Glimmers of Elliott's sophisticated understanding of drainage first shined through in his junior-year academic journal. In the mid-1870s, the University of Illinois's civil engineering program required students to keep summer journals that detailed work opportunities, personal travel, and observations of engineering structures. On September 17, 1875, Elliott submitted his journal, which included reflections on meteorology, geology, a frustrating fishing trip ("1 Bite per 6 hrs"), a bridge spanning the Vermillion River, farm drainage, and the strenuousness of life on his father's farm. On the last of these subjects, he condemned the unusually stormy and wet harvest season. "For the last few weeks all of my skill and especially my muscle has been exerted in the hay and harvest fields," he complained. "This part of farming has been particularly difficult this year because of the raging weather." As he grew older, the memory of toilsome summers on his father's farm made Elliott sensitive to the daily grind associated with settler agriculture. Small wonder that he dedicated the remainder of his professional life to developing the specialized knowledge, engineering principles, policies, and technologies to reduce the burdens of farming.[21]

Elliott's summer journal showcased his wealth of drainage-related talents. On his father's soggy farm, he conducted land surveys, took levels, and designed a small drainage project. "There are several ponds on my father's farm that are of no value without drainage," he recorded. In June, he spent several days using surveying instruments, including water levels and rods, to calculate surface elevations. Once his measurements were completed, he sketched out a ditch and underdrainage system that would dry out the ponds and other prairie sloughs. Although it is unclear if his father ever carried out the drainage plan, the episode highlights Elliott's keen understanding of determining proper elevations and surface deviations, a prerequisite for successful agricultural drainage engineering. The written description of the project, composed in his impeccable cursive handwriting, was accompanied by clear drawings and sketches. Indeed, Elliott mused in one of the journal's concluding sections that competent engineers must keep every drawing "as presentable as possible," a later prerequisite for engineers he hired while leading the USDA's drainage investigation.[22]

Elliott's name was synonymous with agricultural drainage consulting and engineering by the time the USDA put him on its payroll. After earning his B.S. in 1877 from the University of Illinois, he opened a thriving drainage engineering practice in Tonica, Illinois, in the late 1870s. (He would go on to earn a C.E. from the university in 1895.) In 1882, he published *Practical Farm Drainage; Why, When and How to Tile Drain*, one of the earliest American books about drainage techniques and procedures, and he wrote regularly for the *Drainage and Farm Journal* and other Midwestern agricultural journals. He also became a minor celebrity at the annual conventions of tile manufacturers and civil engineering societies, where he delivered well-received lectures on the benefits, methods, and cost assessments of drainage.[23]

With drainage, all roads lead back to the valley of the Red River of the North. In 1886, delegates at the Crookston drainage convention chose Elliott to lead a topographical survey of the valley's entire Minnesota side, described at the time as the country's largest drainage survey (see chapter 4). A decade later, he prepared the USDA's first drainage bulletin. Three years later, the Division of Soils republished it as *Farm Drainage*; in 1904, the Office of Experiment Stations published an enlarged edition as *Farmers' Bulletin 187*, "Drainage of Farm Lands." In 1901, Elliott bought the *Drainage Journal* (previously the *Drainage and Farm Journal*), which hyped drainage as the essential ingredient of settler land management. The following year the USDA hired him as the federal government's first agricultural drainage engineer.[24]

Elliott entered federal service at a time when the civil engineering profession had not fully integrated agricultural drainage. Despite the budding public interest in drainage, American universities paid it little heed. At the University of Illinois, Elliott's alma mater, only a handful of baccalaureate theses focused on drainage-related civil engineering topics were completed before the mid-1900s. Over the ensuing decade, Elliott moved swiftly to systemize techniques and practices, arrange experimental work with universities and individuals, and mentor the first generation of federal drainage engineers.[25]

In the second edition of Elliott's popular *Engineering for Land Drainage* (1912), he synthesized the drainage engineer's repertoire of skills and knowledge: a working knowledge of soil characteristics, subsoils, and strata; surveying abilities; drawing skills; an appreciation of hydrological and hydraulic principles; mapmaking experience; math skills; and a baseline understanding of the law. In addition, the period's growing technological sophistication required comprehension of how to lay underground tile lines, position levees, channelize rivers, erect pumping stations, and utilize steam-powered dredges.

Moreover, engineers required proficiency in translating technical concepts and complex mathematical formulas into jargon-free reports digestible by settlers. As Elliott put it, "The facts should be presented in a logical order, and expressed in terse and clear language, and should, as far as possible, be free from technicalities." Epistemological specialization magnified the need for rhetorical generalization.[26]

Objectivity also mattered. Elliott prized disinterestedness. When interacting with a local drainage committee or local drainage boosters, the engineer summoned science, education, and dispassionate expertise. Engineers on the federal payroll should remain unmoved when cajoled by local officials to take a particular position. This became the linchpin of the OIDI's advisory consulting: The expert guided, but the citizens decided. As he expounded, "[The engineer] should not be a tool in the hands of the [drainage] board or any interested party, but an honest counselor and director of the undertaking for which he has been employed. His plans should possess such merit that they will appeal to his clients and any differences of opinion or judgment should be courteously discussed. Such a course calls for the exercise of a high order of common-sense, good judgment and integrity in addition to the technical skill[s]."[27]

Elliott positioned drainage at the heart of the federal conservation movement. In a 1910 letter, he ranked drainage as no less foundational to the nation's future as irrigation, forest conservation, and public lands management. "As to our connection with the conservation movement," he elaborated, "the utilization of these waste lands and the betterment of fields which are already in cultivation, come properly within the scope of the work. Not only does drainage increase the revenues of a country and encourage its settlement, but is usually essential in its healthfulness." Concerning the latter, etiological breakthroughs during the period, most notably British physician Ronald Ross's 1897 discovery that the female anopheles mosquito transmitted malaria, prompted many states and municipalities to organize mosquito abatement organizations that promoted draining land and/or dumping kerosene into wetlands to kill mosquito larvae. Ross's discovery also reinforced the cultural misperception that wetlands were sickly and in dire need of a medical diagnosis and cure (e.g., drainage), as reflected in in the Washington press corps' pet name for Elliott: "Swamp Doctor."[28]

Compared with most of the era's federal civil engineers and conservationists, Elliott's career remains cloaked in obscurity. The OIDI chief did not leave behind a personal paper collection, he lacks a biographer, and his prolific writings seldom strayed beyond intricate technical and legal minutiae.

He also lacked wealth and social status. A technocrat, he did not publicly embrace the euphoric nostalgia for a timeless rural past that enthralled so many of his generation's conservationists, including Newell and Maxwell; nor did he fetishize Jeffersonianism, social Darwinism, or eugenics fantasies in his correspondence and publications. Elliott thus favored drainage to improve existing farms rather than celebrating it as an instrument to carve out new homes for urban factory workers or immigrants. Pure and simple, the expansion and betterment of settler farms was his ideology and life's work.

The Convergence of Federal Expertise and Local Governance: The OIDI's Advisory Consulting

Elliott tackled his new responsibilities with alacrity and gusto. In 1903, the inaugural year of federal drainage work, the OIDI headed investigations and plans to reclaim seepage- and alkali-damaged irrigated tracts in Fresno, California; the Ahtanum and Yakima Valleys in Washington; and Wyoming's Greybull Valley. In the country's humid half, federal engineers studied wetlands drainage in Hancock County, Iowa. Finally, the office dipped its toes into the fledgling field of soil erosion prevention by installing a series of experimental underdrains on terraced "hillside land" in northern Georgia. An office press release boasted that the initial round of investigations constituted "the first work of the kind" and that "the wide range, specific differences of treatment, and financial and sanitary importance of drainage in its relation to the improvement of farm lands, make it one of the most unique subjects which engage the attention of agriculturalists in many localities." The only good swamp, the release emphasized, was one that had been drained, plowed over, and occupied by settlers.[29]

In his introduction to the OIDI's first annual report, Elwood Mead (serving both as a Berkeley professor and the office's director) described the many classes of land in need of drainage. Seizing on long-standing cultural anxieties, he brooded that "fertile, dry prairie land which could be brought under cultivation with no expenditure except for breaking the sod" was racing toward exhaustion and that a food crisis was imminent. Only three options existed for feeding the nation's citizens: boosting yields on existing farms, improving "waste lands," and irrigating Western deserts. "Drainage," Mead emphasized, "plays an important part in each of these."[30]

In terms of the first two options, Mead referenced Nathaniel Southgate Shaler's 1885 and 1890 USGS reports. To the east of the 100th meridian, Shaler had placed the area of wetlands available for drainage at between 67 and

84 million acres (roughly equivalent to the amount of Western deserts the Reclamation Service hoped to irrigate). Mead argued that drainage was settlers' best reclamation option because of wetlands' prolific soils and proximity to transportation corridors and cities. In fact, he forecasted that the drainage of the nation's wetlands would boost the collective value of American farms by $1.5 billion. But drainage also aided Western agriculture. The West's reputation as a uniform desert belied its geographic diversity. In California alone, 1.75 million acres of wetlands awaited drainage and cultivation. Outside of the Golden State, pockets of wetlands in North Dakota's prairie pothole section, the valley of the Red River of the North, and southeastern Oregon's Malheur Basin were also prime targets. Moreover, the OIDI homed in on preventing the abandonment of irrigated tracts suffering from canal seepage, rising water tables, and soil alkalization.[31]

During the next two years, the OIDI carved out a role in streamlining drainage techniques, standardizing knowledge, and sharing expert advice. Mead delegated all of the office's drainage work to Elliott, who methodically expanded the OIDI's footprint and clarified its purpose and role. He emphasized that his office confined itself to "all problems relating to the removal of surplus water from lands adapted to agriculture, their protection from overflow, and determining experimentally methods of regulating the quantity of water in soils used for agriculture." His engineers and staff researched, planned, and consulted, but they did not construct. His federal experts, in tandem with local farmers, united to make better policies and broaden participation in land conservation initiatives. In Mead's apt description, government engineers doubled as "consulting advisors," applying science and specialized knowledge to mediate local disputes. The OIDI sought to forge a new relationship between expert and citizen based on cooperative dialogue, engagement, and trust. Federal drainage engineers, temporarily loaned to drainage districts, converted complex technicalities into social and political action at the local level.[32]

The OIDI's dissemination of expert knowledge and shepherding of grassroots political development stretched the environmental state's tentacles into rural areas heretofore untouched by the federal government. Between 1902 and mid-1905, for instance, it consulted with and advised drainage districts encompassing a modest territory of 300,000 acres. In immersing farmers in the process of water planning, the OIDI sought to rise above the cauldron of parochial loyalties by "unit[ing] the people" and deploying national science to enhance "more harmony in their execution." The office appointed itself as a mediating chaperone, guiding communities of farmers through the early process of district formation and then nurturing and monitoring their progress from afar.[33]

The OIDI also did not shy away from policy formation. As a coordinator of social and political action, it endorsed uniform legislation at the state level, which made providing legal assistance easier. In the summer of 1906, for instance, Georgia state representative and future governor Lamartine Griffin Hardman approached the OIDI about drafting a bill authorizing drainage district formation. He further inquired about the flurry of drainage bills in Congress. In response, Elliott remained noncommittal about pending legislation, merely stating that it "brings up a new feature of legislation which Congressmen realize must be very carefully considered." Nevertheless, he implored Hardman not to wait for direct appropriations: "Whatever is done along this line in the future does not obviate the desirability and even necessity for each State to secure a good working law by means of which landowners can cooperate under legal restrictions to carry out the drainage necessary to improve the swamp lands of the State." Elliott enclosed copies of a 1904 Iowa drainage statute, which he had coauthored. Since Iowa's circuit and supreme courts had upheld its constitutionality, he brandished it as a template for future legislation. In Elliott's view, state legislators—like settlers—profited from federal assistance in transforming proposals into meaningful policies.[34]

Beyond studying the drainage of irrigated land, equipping farmers with specialized knowledge, and promoting consistent state laws, the OIDI presided over experimental activities intended to systemize knowledge and validate new technology. One of the office's largest experimental undertakings occurred in tandem with the University of Minnesota's Northwest Experiment Farm. Established in 1895, the farm occupied 475 acres north of Crookston, Minnesota, in the valley of the Red River of the North. Suffering from poor surface drainage, the farm struggled to prosecute its educational and experimental missions. As a result, in 1903 the state legislature appropriated $5,000 for the installation of a drainage system, but it failed to deliver results. Two years later, the legislature appropriated an additional $4,000 to finish the job.[35]

With the new funding in hand, Professor William Robertson, the farm superintendent, requested OIDI assistance. In October 1905, the OIDI dispatched drainage engineer John T. Stewart to conduct preliminary surveys and draw up a general plan. Three months later, the University of Minnesota and the USDA entered into a cooperative agreement "to construct the drainage system in accordance with the plans under the supervision of an [OIDI] engineer." Subsequent agreements obligated Robertson to oversee observation wells composed of four-inch drain tiles and to report regularly on "the effect of the underdrains upon the soil and upon the growth of crops compared with adjoining land which is not drained." The OIDI also tasked him with sending soil samples to the Washington office, monitoring precipitation levels, and measuring flowage rates in main outlet ditches. By June 1908, the new underdrainage system had proven successful, enabling the superintendent to cultivate 20 percent of the farm with cereal grains.[36]

In addition to its experimental endeavors, the OIDI engaged in pioneering research on soil erosion. Although the USDA's Bureau of Soils and state agencies had launched the National Cooperative Soil Survey in 1899 to map national soils and classify land usage, soils languished as an ancillary field of inquiry during the Progressive Era. In large measure, soil erosion's marginalization owed itself to the towering presence of Milton Whitney, the chief of the Bureau of Soils until 1927. Reductively fixated on the concept of soil texture, Whitney denied the severity of soil erosion despite troves of contrary evidence, including the prevalence of eroded and gullied agroecosystems in the South. His dominance over federal soil science ensured that the state's soil-related activities remained limited to soil texture until the New Deal.[37]

Despite the prevailing soil texture obsession, Elliott unveiled a novel soil erosion prevention and mitigation trial. The OIDI's first annual drainage report described a cooperative experiment undertaken with Lamartine Griffin Hardman, the same Georgia state legislator who corresponded with the office in 1906 about drainage legislation. On Hardman's farm near Commerce, Georgia, one of his fields had become "deeply gullied and given over to briers and wild grass." According to Elliott, the severe erosion stemmed from natural climatic processes rather than anthropogenic factors: "The steeper slopes, when cultivated, are easily eroded by the heavy rainfall which occurs in the South during the winter. The surface freezes slightly in the winter, which renders the soil all the more easily moved by subsequent rains. Many hillsides have been abandoned to gullies, broom grass, and briers on account of surface washing." Throughout the nineteenth century, many northern Georgia farmers resorted to terracing, and a vigorous debate ensued about whether level or graded terraces constituted the proper type. Elliott worried that level terraces, due to gravity, pulled water down from the upper slopes and into lower natural depressions, which were susceptible to seepage. "Several such seeped spots give rise to corresponding small ditches," he reasoned, "which soon become deep gullies and eventually necessitate the abandonment of the field."[38]

To thwart additional gullying and hillside erosion, the OIDI installed an experimental underdrainage system on a gullied field. Government engineers buried four-inch tiles at a depth of two and one-half feet underneath the slopes of eroded ditches. They then connected the subterranean tile lines to an outlet, filled in the gullies, graded the surface, and seeded the land with wheat. Elliott anticipated that underdrainage may slow or halt further hillside erosion. "The drains were laid in this way for the purpose of intercepting the water which oozed down the slope underneath the surface and cropped out at some point," he observed in his first annual report. "It is expected that the carrying away of the seepage water will render the surface as firm at these points as at others and so prevent the formation of ditches." There was no agricultural obstacle that federal engineers could not overcome.[39]

After one year had elapsed, Elliott judged the experiment a mixed success. The tile system had prevented the onset of new depressions and gullies, but rainfall had been too deficient to warrant a firm conclusion about underdrainage's merits in arresting hillside soil erosion. The office's experiments also generally skirted the question of spacing distances between terraced embankments, a prevalent topic of inquiry and debate.[40]

Elliott's foray into soil erosion was limited, tentative, and deficient in scientific sophistication and precision. While leading the government's drainage activities, he never formulated a broader social critique about anthropogenic contributions to soil erosion. Moreover, the experimental work on Hardman's farm did not include a detailed survey of soil conditions or acknowledge that the USDA had already forged alliances with universities and other experiment stations to study the issue.

Elliott's dabbling in soil investigations put the OIDI and the Bureau of Soils on a collision course, especially as it related to studying irrigated tracts damaged by alkalinization and seepage. Contradictory congressional guidance fueled the rivalry. In the 1900 fiscal year appropriation, Congress instructed the Bureau of Soils to investigate "the cause and prevention of the rise of alkali in the soils of irrigated districts." Three years later, it directed the Office of Experiment Stations (e.g., the OIDI) to "investigate and report upon plans for the removal of seepage and surplus water by drainage." In November 1906, Whitney and Experiment Stations director A. C. True agreed that the division of labor produced confusion, delay, and duplicative efforts.

Anxious to stymie a contentious internal turf war, Secretary of Agriculture James Wilson approved an arrangement that harmonized the activities of the Bureau of Soils and the OIDI. It stipulated that the Bureau of Soils would investigate and reclaim alkali tracts that did not exceed twenty acres. Once the size of any investigation surpassed twenty acres, it would become the OIDI's prerogative. Furthermore, the OIDI's investigations into irrigated tracts damaged by seepage would continue until they encompassed "considerable areas of alkali land." At that point, the Bureau of Soils would step in and survey the extent and location of the poisoned tracts, the nature and character of the salts, and the salts' overall distribution. The agreement then required the Bureau of Soils to issue a report on the best methods for reclaiming those tracts.[41]

The Vanguard of Settler Colonialism: Federal Engineers, State Expertise, and Bureaucratic Rivalries

In 1907, the USDA reorganized the OIDI. First, Elwood Mead resigned and left civil service. Second, the OIDI's soaring popularity among congressmen and settlers prompted the USDA to split it into two entities: an OII and an Office of Drainage Investigations (ODI), both of which remained subordinate to the USDA's Office of Experiment Stations. In October 1907, Elliott became the ODI's new chief. The ODI also benefited from a surge in revenue. In 1907, Congress raised its annual appropriation to $150,000, more than double its $67,500 budget from 1904.[42]

From the moment he ascended to the ODI's leadership, Elliott strengthened his grip over federal drainage activities. He broadened the office's operations and professionalized its workforce, including recruiting and hiring rising stars from the drainage engineering profession. One of his most astute choices was Arthur E. Morgan. A self-taught engineer whom Franklin D. Roosevelt later tapped to serve on the Tennessee Valley Authority's first board of directors, Morgan ran a profitable civil engineering business in St. Cloud, Minnesota, that specialized in agricultural drainage assistance. In addition to running his private firm, Morgan had helped the Minnesota Engineers and Surveyors Society author a new drainage bill that the legislature ultimately passed (1905), and he represented the state at the 1906 Oklahoma City drainage congress. His business also thrived. By mid-1907, it possessed contracts totaling more than $300,000 across Minnesota. Impressed by a series of speeches Morgan delivered at state and national drainage conferences, Elliott encouraged him to take the civil service exam. After he aced the exam, Morgan was hired as an ODI supervisory drainage engineer with an annual salary of $2,000. Elliott immediately directed him to pack his bags and prepare for a hectic travel schedule. "The disagreeable feature of the profession of civil engineering is that it necessitates a rambling life," he explained. "Our work is widely scattered." Morgan spent much of the next two years on the road.[43]

C. G. Elliott is seated in the front row, second from right. Arthur E. Morgan is seated second from left, also in the front row. Office of Drainage Investigations personnel, 1909. Courtesy of Antioch College.

Morgan's brief stint with the ODI (1907–9) occurred as the office reached its pinnacle of influence. Officially, the office remained focused on farm drainage, but as the decade wound down it dedicated more resources to surveying larger undeveloped swamps, specifically Florida's Everglades and the St. Francis River drainage basin in Arkansas and Missouri. In his first year on the federal payroll, Morgan led examinations of overirrigated alkali tracts in Colorado's San Luis Valley before surveying and writing reports for proposed drainage projects in Arkansas, Louisiana, Mississippi, and North Carolina. In all, his investigations crisscrossed 1.1 million acres.[44]

As one of the ODI's three supervisory engineers, Morgan managed several field parties. Depending on the size and intricacy of a given assignment, the teams consisted of engineers, chainmen, axe men, rodmen, teamsters, and cooks. Although many of the seasonal laborers were plucked from local communities, the ODI actively sought out recent graduates from land grant institutions and technical schools as permanent staff members. Morgan demanded that his personnel adhere to strict standards of behavior, appearance, legible handwriting, and other protocols. Competence in the use of engineering instruments, including a reconnaissance transit, level rod, Y level, compass, and level with stadia, was a must. The office also mandated strict sartorial appearances and hygiene. The few surviving photographs of ODI field parties

Office of Drainage Investigations Encampment and Field Party in Arkansas, 1909. Courtesy of Antioch College.

depict well-dressed and freshly shaven men, even while operating in isolated, sparsely populated environs.[45]

Fieldwork was grueling, tedious, and lonely. Trudging through swamps and performing the same tasks day after day, ODI personnel often complained of physical and mental exhaustion. Malaria and other diseases strained their stamina. In a 1908 letter to his son, Morgan lamented that "traveling is demoralizing" and that local residents often bombarded him with dishonest work and land opportunities. Supervisory engineers like Morgan had to balance the process of building camaraderie and instilling an institutional culture with ensuring that the difficult and monotonous daily activities did not breed resentment, grumbling, and a toxic culture. Under Morgan's supervision, his Arkansas field party operated with a patriotic sense of duty, proudly unfurling American and ODI flags wherever they encamped.[46]

The immersive nature of the fieldwork—bodily, experientially, and intellectually—illuminates why the ODI always had a leg up on the Reclamation Service. As ODI field parties penetrated some of the last unmapped wetland ecosystems from 1907 onward, Elliott guided the ODI through its bitter rivalry with the Reclamation Service. At the Oklahoma City drainage congress in December 1906, delegates had rallied behind California Senator Frank P. Flint's drainage bill to nationalize drainage under the Reclamation Service. The Department of the Interior sent seven employees to Oklahoma

City and many stalwart allies of the Reclamation Service attended the proceedings, but Elliott was the USDA's lone representative. Afterwards, Mead and Elliott complained about the solidification of an alliance between the National Drainage Association (created at Oklahoma City) and the Reclamation Service. In February 1907, Mead alleged that the National Drainage Association's stealth purpose was to seize all federal drainage responsibilities. "There has been a great deal of log rolling and lobbying to get hold of our drainage work," Mead complained. "The National Drainage Association, which is [George Maxwell's] National Irrigation Association under new colors, ... [is] in close touch with the Reclamation Service but have never found time to call on us." Mead also scorned A. G. Bernard, a Minnesota member of the National Drainage Association's executive committee, as a "figurehead" who doubled as Maxwell's pawn. With so few reliable allies inside of government, the ODI's leaders felt besieged by a host of adversarial networks and politicians.[47]

Elliott dismissed the National Drainage Association as a "propaganda" organization. Initially, he did not take "an active part in forwarding the tenets of this association with respect to federal aid from the fact that we were not decided regarding the policy that should be pursued in forwarding the interests of the drainage work in general." The National Drainage Association's alignment with the Reclamation Service made Elliott reverse course. He fumed that the association billed the Reclamation Service as the best organization to lead a new federal drainage program, pointing out that the service possessed almost no in-house drainage engineering expertise, did not conduct drainage experiments, and possessed little understanding about drainage districts or state laws. "The work done so far by that association," he swiped at the service, "has been carried on largely by those who are not actively identified with practical drainage and who know little of the present needs or of the methods by which drainage reclamation has so far been achieved." Little wonder he derided the Reclamation Service's capacity to lead a federal drainage program.[48]

In the same letter, Elliott pointed to an "active and growing interest" in draining large swamps in Arkansas, Louisiana, and Mississippi. His own fieldwork, interactions with settlers and land developers, and correspondence convinced him that rural settlers favored his office's advisory consulting approach. "These people, as well as those in the Middle west," he taunted the Reclamation Service, "do not see any show for Federal assistance in the construction of drainage work[s], so that it is not probable that they will contend strongly for the provision of the Flint Bill." Earlier in the year and prior to his departure from civil service, Elwood Mead had construed the ODI's growing budget as evidence of "the growing circle of our acquaintance[s]" in Congress. Indeed, the ODI's workload continued to expand. Between July 1906 and June 1907, the office carried out drainage experiments, arranged demonstrational work, and assisted local communities in two-thirds of the states and territories. The ODI also catered to power brokers. In 1907, its leaders boasted about initiating a "great deal of work" in the district of Kansas Republican Rep. Charles Frederick Scott, the chairman of the House Committee on Agriculture.[49]

To fend off the Reclamation Service's frenzied pursuit of drainage, the ODI sought to "build some fences" around its work. In one such instance, Elliott aligned his office with Southern railroad corporations. In January 1907, D. E. King, the Missouri Pacific's industrial commissioner, struck an arrangement with the ODI and the USDA's Office of Public Roads to cosponsor an educational campaign for drainage and good roads across Arkansas and Louisiana. In areas traversed by the Missouri Pacific's trunk and feeder lines, such as the St. Francis and Washita basins and the Cypress Creek District, numerous marshy areas predominated. King hoped to awaken local settlers to the promise of drainage and good roads, promote dense settlement, increase property values, and enlarge the Missouri Pacific's carrying trade. In the autumn of 1906, the area had already emerged as a hotbed of drainage enthusiasm when Rep. Robert B. Macon, an Arkansas Democrat, unsuccessfully introduced legislation to divert $3 million from the reclamation fund to drain wetlands in the St. Francis basin.[50]

Beginning on January 28, 1907, the Missouri Pacific's demonstrational train embarked on its educational tour from Wynne, Arkansas. The train, carrying corporate representatives and ODI and Office of Public Roads personnel, visited twenty-one communities in Arkansas and Louisiana before reaching Little Rock two weeks later. At each stop, the railroad hosted a farmers' institute where Elliott and government engineer J. O. Wright detailed drainage methods, engineering aspects, and best practices for drainage district formation. The Missouri Pacific's educational campaign, which interfaced the ODI with Southerners and congressional Democrats, highlighted the office's bottom-up approach and penetration of the rural periphery with state expertise.[51]

The Missouri Pacific's first special train proved so successful that King and the ODI organized another one in the autumn. On October 28, 1907, Elliott, King, and Rep. Macon traveled on the Iron Mountain Railroad from Paragould, Arkansas, to Vidalia, Louisiana, armed with lists of large landowners in each community. At each stop, Elliott exhibited a chart depicting the steps involved in forming a drainage district under state law. The ODI chief also entertained the commissioner of the St. Francis Drainage District, founded two years earlier. By October 1907, the district encompassed 135,000 acres in Clay and Greene Counties. The district's major feature was a 38.5-mile-long ditch built parallel to the St. Francis River, draining a sprawling swath of marshland.[52]

As the ODI broadened its Southern presence, in February 1907 California Senator Flint queried the USDA on a handful of related subjects: the location and extent of "swamp and overflowed" lands; the quantity of wetlands that had already been cleared, drained, and cultivated; state laws authorizing the formation of drainage districts; European drainage laws; and the impact of drainage on public health. After the ODI hurriedly aggregated the available data, Secretary of Agriculture James Wilson responded that the country contained 75 million acres of "swamp and overflowed lands permanently unfit for cultivation." Although he deemed it "impossible" to determine the quantity of wetlands that had been converted into settler fields without further investigations, he estimated that one-half of the former wet prairie states of Illinois, Indiana, and Ohio had already been drained and that 58 million acres had been reclaimed in Illinois, Indiana, Ohio, Michigan, and Missouri. In general, drainage doubled annual crop yields and increased land values by $20 per acre.[53]

Wilson identified twenty-one states that had adopted "efficient and well-tried" drainage laws. The statutes all included two indispensable features. First, the initiation of a district originated with a petition signed by a baseline percentage of landowners whose property covered a specific proportion of the plot that was proposed to be drained. Second, the costs of the drainage system were distributed among landowners in proportion to their estimated future benefits. Wilson indicated that the scale of and expenditures for ongoing drainage projects varied considerably, respectively ranging from 30,000 to 100,000 acres and from $60,000 to $300,000.[54]

In terms of public health, Wilson proclaimed that etiological breakthroughs justified drainage investment. The conservation of settler bodies and "wastelands" always went hand in hand. In his report to Flint, the secretary crowed that due to extensive underdrainage in Indiana and Illinois, where malaria accounted for half of all deaths just fifty years earlier, "malarial fever" had disappeared."[55]

Wilson invoked a drainage project on James Island near Charleston to support his claims. Until the drainage of the island had commenced two decades

Unidentified man laying clay tile in North Carolina. Courtesy of the National Archives, College Park, MD.

earlier, the secretary reported, "white people were not able to remain upon them during the summer season." *Irrigation Age*, one of the nation's leading conservation periodicals, similarly weighed in on drainage's impact on preserving settler bodies in South Carolina. "The sanitary feature of drainage takes precedence over all others because the coast lands cannot be successfully cultivated by white laborers on account of their unhealthfulness," it argued in April 1907. "The fact is now recognized that these [South Carolina] lands must be drained before they will be made attractive to the better class of immigrant farmers." Preserving the health and longevity of settlers remained a powerful justification for the state's patronage of drainage.[56]

Circular 76 and Making Wetlands Legible

Shortly after Elliott seized the ODI's reins, the office took a major step in systemizing wetlands knowledge. With the exceptions of Nathaniel Southgate Shaler's 1885 and 1890 USGS reports, the federal government had done little to make the country's wetland resources known to its citizens despite hyping drainage as a means of achieving settler permanence, prosperity, and healthfulness.

In October 1907, the ODI took a step toward closing the knowledge gap by undertaking a thorough and comprehensive analysis of American wetlands. The USDA's Office of Experiment Stations published the findings in *Circular 76*, "Swamp and Overflowed Lands in the United States: Ownership and Reclamation." Authored by supervisory drainage engineer J. O. Wright, the circular inventoried national wetlands east of the Rocky Mountains. It lacked the scientific breadth and sophistication of Shaler's investigations, focusing solely on inventorying and mapping swamps to expedite their elimination. Over the course of the year, the office mailed blank questionnaires to officials in every county east of the 105th meridian, requesting estimates of the amount of "salt marsh," "swamp," and "overflowed" lands in their respective jurisdictions. The form also asked for an ownership breakdown of those tracts, including corporations, private parties, states, and the federal government.[57]

The questionnaire listed three possible wetland categories: salt marsh, swamp, or overflowed land. Although Shaler had pioneered one of the state's first wetland classification systems, Congress and the public continued to interchangeably refer to wet tracts as swamps, inundated land, overflowed land, marshes, or simply waste land. Government drainage engineers, however, understood the sheer diversity of lands characterized by an abundance of water. Consequently, the questionnaire asked respondents to designate wetlands using one of three categories. First, it defined *salt marsh* as "all lands along the coast that are covered or partly covered with salt water at high tide." Second, *swamp* encompassed "all land, whether open or timbered, above tide water that is too wet for cultivation." Third, *overflowed land* referred to "all bottom land along streams that cannot be cultivated safely because of overflow." In a final caveat, the questionnaire dictated that *shallow lakes* capable of being drained be classified as "swamp." Shaler's scientific categories were being translated, albeit haltingly and imperfectly, into lay knowledge.[58]

After collecting the questionnaires and aggregating the findings of ongoing fieldwork in Florida, Louisiana, Minnesota, and Wisconsin, Wright tabulated that 77 million acres of swamp and overflowed lands east of the 105th meridian were available for drainage, an area the combined size of New England, New York state, and the northern half of New Jersey. The per-acre cost of drainage depended on the scale and nature of the improvements. The cost of constructing levees and open ditches averaged from seven to sixteen cents per cubic yard while that of underground tiling ranged from $6 to $20 per acre. The amount of capital required to straighten, widen, or deepen

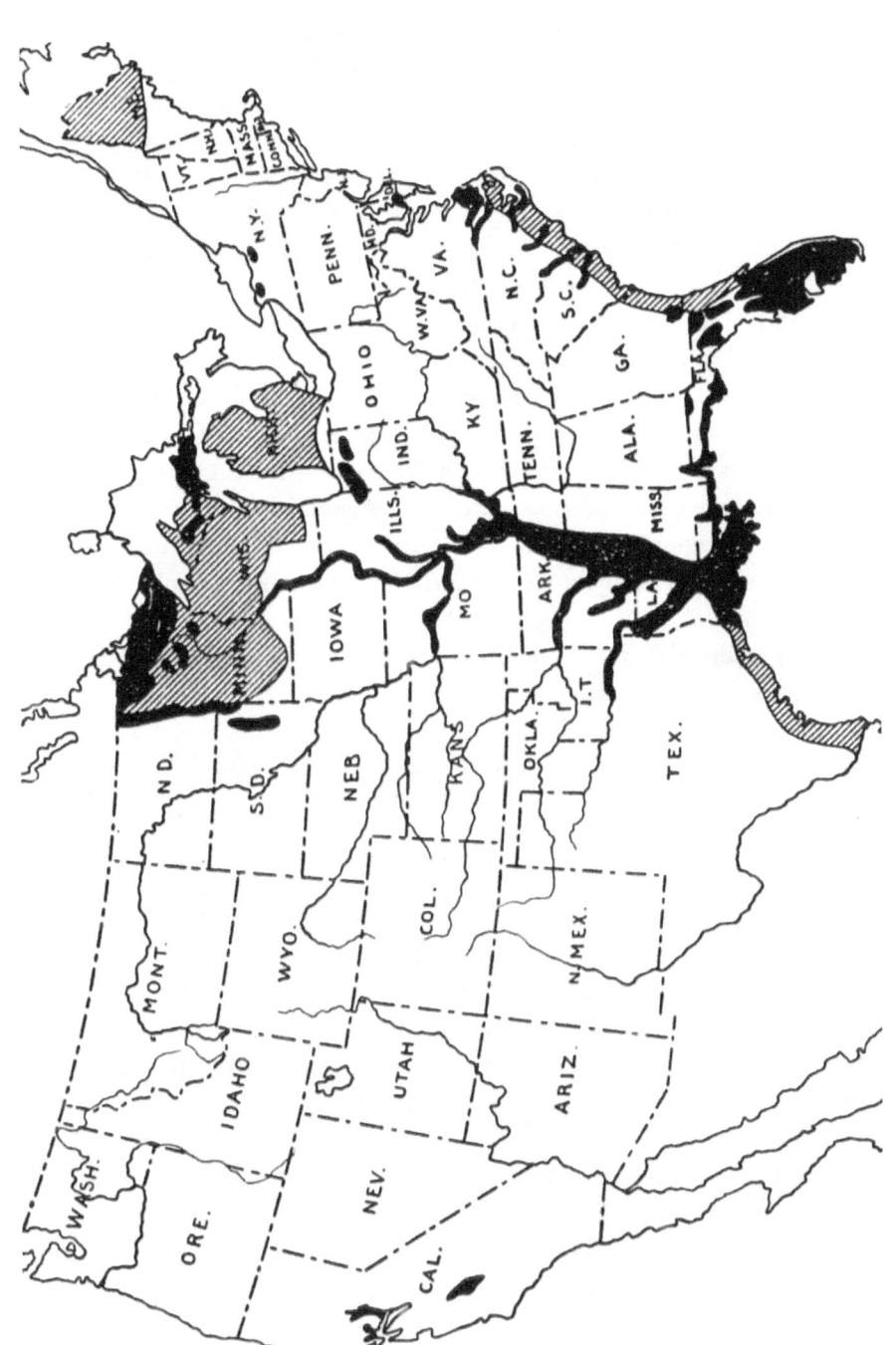

Map depicting wetland spaces available for drainage (based on 1907 USGS projections). The black areas represent solid wetlands while the ruled areas depict locales interspersed with wetlands. *National Geographic Magazine* 18 (May 1907): 298.

creeks or small watercourses also varied widely, especially if pumping stations or sluice gates were required. Ultimately, Wright put the average cost of reclamation at $15 per acre. According to the ODI's statistics, drained farmland across the country, bolstered by the era's thriving rural economy, fetched $60 to $100 per acre. Drained farms near cities commanded even bigger sums, sometimes selling for as much as $400 per acre. Wherever it took place, drainage proved a wise investment.[59]

Wright's circular meticulously addressed the subject of land ownership. According to federal records, as of June 30, 1906, Congress under the Swamp Land Acts had given away 82 million acres of swamp and overflowed lands to fifteen states, an expanse larger than New Mexico Territory. As a result, 95 percent of the 77 million acres of undrained wetlands were held in private ownership, leaving a trifling 3.9 million acres of swamp and overflowed lands in federal or state hands. Given these statistics, the future demand for drainage-related support to settlers would be centered on improving private property (farm drainage) rather than initiating a new generation of rural homes by draining and then privatizing public lands (swamp drainage).[60]

In large measure, this ownership breakdown helps explain why the ODI's advisory consulting approach triumphed over the Reclamation Service's construction strategy. In a December 1908 letter, Elliott took a victory lap: "[The] Flint Drainage Bill which was introduced and considered at the last Congress . . . was dropped because no one but Mr. Flint appeared to advocate its passage." The legislation's prospects looked no brighter in the next legislative session.[61]

Elliott could congratulate himself on the ODI's unprecedented expansion and creeping geographic reach. During the 1907–10 fiscal years, the office oversaw drainage investigations and surveys in thirty-two states and territories covering 9.2 million acres; the price tag for all of the investigations amounted to $291,000, or an average of approximately three cents per acre. In that time frame, the ODI spent most of its resources assisting riparian communities seeking to reclaim lands "subject to overflow as by floods" (5.9 million acres). The next-largest investment involved investigating "lands continually wet, swamps, etc." (2.5 million acres), followed by "lands requiring new outlets of improvement of water courses" (526,000 acres) and "irrigated lands needing drainage" (352,000 acres). Geographically, four out of the five states with the largest amount of land examined by the ODI were in the South: Florida (1.85 million acres), Louisiana (1.6 million acres), Arkansas (1.6 million acres), North Dakota (1.5 million acres), and Mississippi (638,000 acres).[62]

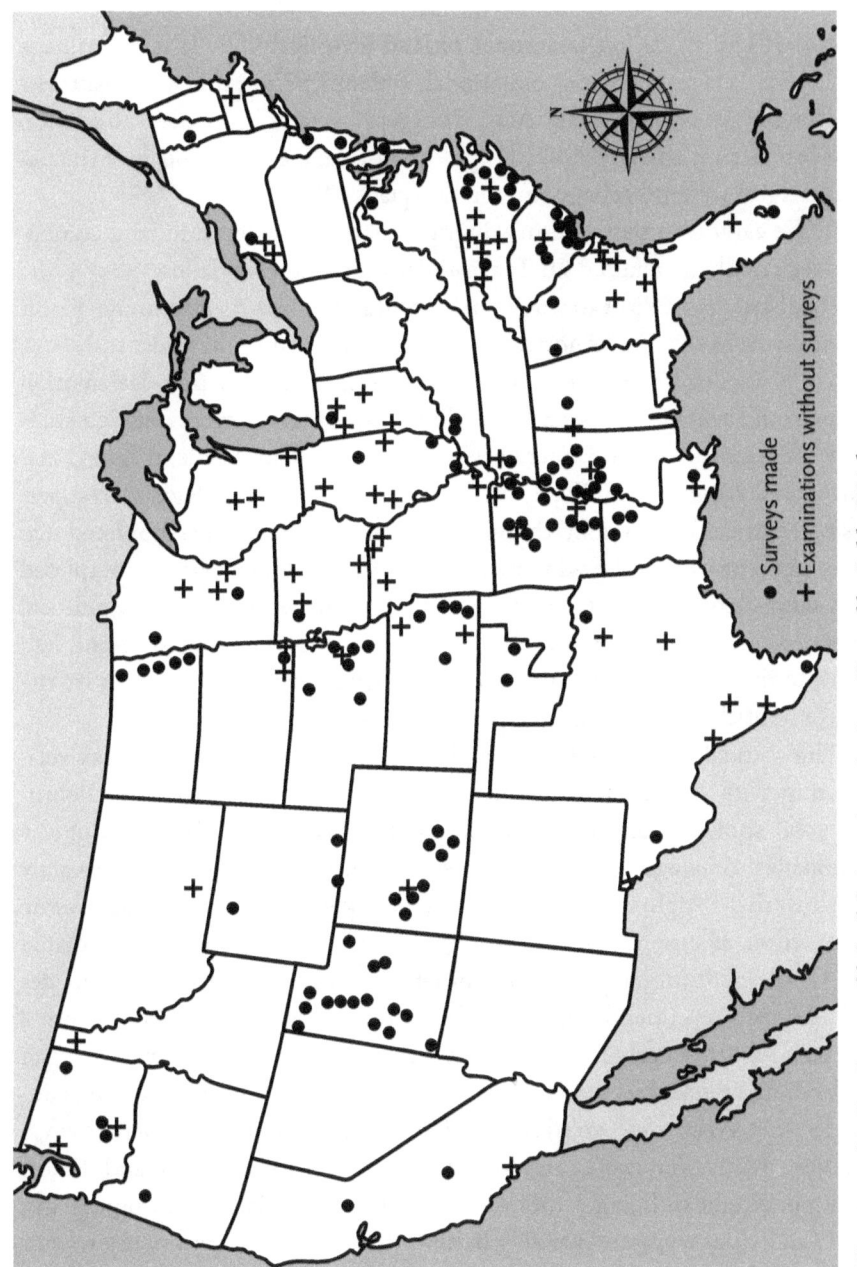

The geographic distribution of federal drainage investigations. Created by Bobby Wright.

The Everglades Scandal and the Illusion of Disinterested State Expertise

Florida led all states in the amount of land investigated by federal drainage engineers. Given the history of national wetlands policy, this was unsurprising. Under the Swamp Land Acts, Congress by 1907 had ceded 20 million acres of "swamp and overflowed" land to Florida, including much of the nation's largest wetland ecosystem: the Everglades.[63]

In the early twentieth century, Florida's political leaders and land companies repeatedly demanded ODI investigations of the Everglades. In 1904, the OIDI dispatched Elliott to carry out a preliminary survey, examining a tract spanning from sixty miles north of Miami to twenty-five miles west and south into the Everglades' interior. Although Elliott estimated that reclamation costs would average $50 per acre, he determined that far more detailed studies were required. In 1906, Secretary of Agriculture James Wilson agreed and earmarked $11,000 for an Everglades survey. The task was delegated to supervising engineer J. O. Wright, the author of *Circular 76*, and he completed two investigations during the winters of 1907 and 1908. Once Wright completed and submitted his Everglades report to Elliott in February 1909, a political firestorm erupted. The imbroglio tarnished the ODI's reputation and laid waste to the Progressive Era orthodoxy that federal engineers were above the fray of politics, partisanship, and personal greed.[64]

The scandal traced itself to Wright's second survey. Sharing a cozy relationship with Florida land companies and speculators, Wright in February 1908 spoke before 300 farmers and landowners in Miami about the Everglades' drainage prospects. The speech infuriated Elliott. He privately reprimanded Wright for prematurely releasing engineering data before the completion of surveys and the publication of a final report, which violated the ODI's institutional philosophy of letting experts guide and citizens decide. Elliott's reprimand portended future troubles. Once Wright submitted his final report in early 1909, ODI engineers, particularly Arthur E. Morgan and Elliott, attacked its rosy conclusions. After a thorough review, they concluded that Wright had erred in calculating evaporation rates, ditch flow velocities, excavation costs, and other critical factors. Morgan and Elliott accused Wright of being a tool of Florida land speculators and politicians. The ODI chief worked feverishly behind the scenes to suppress the report, update its findings, and warn investors about the magnitude of challenges incumbent in draining such an expansive ecosystem. For their part, Florida's politicians and land speculators were thrilled with Wright's conclusions and pressured Secretary of Agriculture Wilson to publish the report. In January 1910, they also lured Wright away from federal service and installed him as Florida's chief drainage engineer.[65]

Wright never forgave Elliott for attacking his professional competence and integrity. On January 20, 1912, he exacted payback by delivering a bombshell to the USDA's senior leadership, alleging that Elliott had committed "crooked" and "irregular" financial transactions in 1909 in connection with investigations in North Carolina. The accusations deepened the drama surrounding the tug-of-war over the release of Wright's Everglades report, prompted Elliott's firing, and led a Washington grand jury to indict Elliott and three others on April 1 of federal crimes. The charges carried a maximum fine of $5,000 and a five-year prison sentence. *The New York Times* summarized the allegations of submitting fraudulent expense forms: "In 1909 the appropriation for drainage work was exhausted. The work in hand was nearly finished, however, and the [ODI] officials borrowed money to complete it, paying it back out of the next year's appropriation by placing the lenders on the payrolls. It was not charged that any of the defendants profited by the transaction, nor that the Government suffered loss." In other words, Elliott and his lieutenants had cooked the books to benefit their mission, not themselves.[66]

Elliott's termination and indictment capped a stunning fall from grace. As one of the era's most accomplished federal engineers, known for unimpeachable character and honesty, he had matured the ODI into an organization with a national constituency, won bipartisan acclaim, professionalized the field of agricultural drainage engineering, and systemized drainage knowledge via publications and experimentation. He had also urged that settlers remain involved in water planning and policymaking. Seeking to give them a voice in their own political destinies, he used the ODI to tether them to state expertise and transformed his office into a storehouse of professionalized knowledge—knowledge gained through sweat, toil, dedication, and Indigenous dispossession.

In the end, Elliott would be vindicated. After months of testimony from Elliott, Wright, Morgan, Wilson, and others, the House Committee on Agricultural Expenditures released its final investigatory report on August 19, 1912. The voluminous report characterized Elliott's misapplication of funds as a minor indiscretion, accused Wright of self-dealing and of collusion with North Carolina land companies as a government employee, pointed to an assortment of inaccuracies in his Everglades reports, and concluded that Elliott's suppression of the report was not politically motivated. The congressional hearings captivated the national media; over 300 newspapers in almost every state offered breathless coverage. Only days after the report was published, President William Howard Taft ordered the Department of Justice to drop the indictment against Elliott. He was then reappointed to his old position with an enlarged salary. Elliott served a brief period longer before retiring from federal civil service and resuming private practice.[67]

The ODI quickly recovered. As the golden era of American agricultural drainage marched onward, the ODI remained as popular as ever. By 1913, the number of requests for ODI assistance across the country became "so numerous" that the USDA implemented a policy "requiring the interested parties [involved in] organiz[ing] a drainage district . . . to contribute approximately one-half of the cost of the survey, including the preparation of the plans and report." The policy also required local representatives to deposit half of the estimated costs of the survey and report in a local bank under the name of the ODI engineer in charge. The ODI's advisory consulting continued to kick-start, nurture, and sustain settler drainage ventures.[68]

Conclusion

At the beginning of the Progressive Era, settlers and local communities increasingly enlisted the state's assistance with drainage. The 1897 discovery of the mosquito vector for malaria, population growth, settler fears about a dwindling quantity of arable lands, and the desire to forge permanent and stable communities underwrote their demands for state support. A lingering, unresolved issue, however, was whether Washington should take control of drainage in the same manner as it had done with Western irrigation. Federalism, Supreme Court decisions, the Reclamation Service's financial woes, and the privatization of the majority of the nation's wetlands buffered drainage from the centralizing currents of the Progressive Era, as did the ODI's effective and widely popular associative approach.

The ODI's most consequential contribution to the conservation movement was fusing local autonomy and state expertise, unleashing intermediary organizations like drainage districts and state drainage agencies to achieve the state's paramount public policy objective of a drained and densely settled landscape. In doing so, it jump-started a silent revolution in American governance that witnessed an explosion of drainage districts, igniting the fastest and most sustained assault on wetlands in American history as settlers occupied unsettled watery tracts.

The triumph of the ODI's advisory consulting unleashed *both* centrifugal and centripetal forces. On the one hand, the state's dissemination of specialized knowledge and expertise empowered intermediary organizations that offered an alternative to centralization. Moreover, the management of drainage remained local and voluntary. Conversely, even though the federal government did not assume direct control of drainage, its presence in farmers' daily lives and reputation as a neutral dispenser of expertise grew in prestige and esteem, despite the scandalous conduct of J. O. Wright. While Elliott was the only federal employee involved in drainage in 1903, the ODI had grown to thirty employees by 1910. The Progressive Era state often proved far more adept at dispersing power than centralizing it, a fact that carried long-term adverse repercussions for one of the nation's most ecologically sensitive biomes and the Native communities that historically depended on their resources and raw materials.

Conclusion
We Opposed the Draining of Our So-Called Swamp Lands

Leave our swamps alone. Such was the message of Ed Prentice, Bay-mway-way-be-nais, and Nah-gon-nway-we-dung. During the spring of 1914, the three Anishinaabe men traveled from Minnesota to Washington to meet with Commissioner of Indian Affairs Cato Sells. After reaching Sell's office, they delivered a petition on behalf of the inhabitants of Red Lake Reservation and departed.[1]

The petition denounced several aspects of federal Indian administration, including new proposals to drain the Red Lake Reservation's swamps and peatlands. As chapter 6 described, in 1908 Reps. Halvor Steenerson and Andrew Volstead jammed a bill through Congress that authorized the state of Minnesota to impose drainage assessments on "ceded" Anishinaabe homelands to raise revenue for building drainage projects and stimulating dense white settlement. Although Minnesota's drainage agency had constructed over 500 miles of ditches and roads on the "ceded" tract over the past six years, Steenerson expressed disgust at the sluggish pace of agricultural development across northern Minnesota, especially on the diminished Red Lake Reservation. In 1914, he introduced legislation empowering the Bureau of Indian Affairs to sell timber from Red Lake to pay for drainage on the reservation. The congressman's newest scheme delighted the Bureau of Indian Affairs. "Unless drainage can be provided [on the reservation]," Assistant Commissioner E. B. Merritt concluded, "the plan of locating these Indians on tentative allotments in that section, upon which their industrial salvation largely depends, will be a practical failure."[2]

Ed Prentice, Bay-mway-way-be-nais, and Nah-gon-nway-we-dung rejected Merritt's rosy analysis. In their petition, they rebuked Steenerson's obsession with draining every swamp in Minnesota, as well as his ignorance about drainage's environmental consequences. They lashed out that drainage had already undermined the reservation's hydrology by drying out its peatlands, which magnified their combustibility. "We opposed the draining of our so-called swamp lands," the petition declared. "Our reason for opposing... drainage is that the great danger of losing much of the land thus drained [to] fire. These swamps contemplated to be drained readily catches fire and it is almost impossible to put such a fire out." Settler drainage activities jeopardized the integrity of Indigenous ecosystems.³

Wildfires were only one problem. As county and state officials steadily enlarged the system of state ditches and drains across northern Minnesota in the 1910s, Anishinaabe leaders condemned the resulting cascade of ecological and hydrological upheavals. Since 1908, drainage projects outside of the reservation had raised the water levels of Upper Red Lake and other upstream tributaries, just as Anishinaabe leaders had previously warned. The results were devastating. The influx of water from settler crop fields had elevated the volume of many tribal watercourses beyond their capacity and worsened seasonal flooding. The heightened water levels inundated riparian meadows and gardens throughout the reservation, disturbed fish spawning grounds, led to the sedimentation of tribal lakes and streams, and diminished the quality of wild rice beds that the Anishinaabeg had harvested for generations. In a very real sense, the consequences of settler drainage projects reverberated across Red Lake's ecosystems, eroding and then shattering centuries of Anishinaabe relationships with water, soils, peat, fish, and wild rice.⁴

In admonishing the expansion of drainage, the Anishinaabe delegation delivered a sharp censure to the long-standing narrative that drainage yielded no adverse consequences. Since the early republic, few settlers or government officials had considered that drainage delivered anything but material rewards, healthy bodies, and aesthetically pleasing landscapes. In part, this narrative proved attractive because of its assumption that swamps were anomalous landscapes divorced from lakes, streams, aquifers, and other bodies of water. They served no environmental purpose. Even Nathaniel Southgate Shaler's 1890 observation that swamps were integrated into nature's dynamic, interconnected webs failed to dampen enthusiasm for drainage. Federal investigations, surveys, and policies continued to aid states and local drainage districts in fashioning drained, domesticated, and dispossessed landscapes at an increasingly hurried pace and on an expanding scale.

The Anishinaabe opposition to expanded drainage, however, punctured this narrative, demonstrating that settler environmental values and assumptions were never universal, preordained, nor uncontested. Rather, changing social and cultural values entangled swamps, in addition to other landscapes, in webs of contingency that mediated how communities managed and interacted with them. *Perilous Waters* has aimed to contextualize these shifting interactions to showcase the emergence and endurance of drainage as a preeminent public policy objective in the United States. And whether settlers touted drained swamps as spaces to grow crops, raise sheep, establish eugenic sanctuaries, eradicate disease, or open new frontiers to stave off rural decline and preserve the nation's agrarian heritage, they held firm to the belief that watery tracts possessed no biological, botanical, or hydrological worth. The only useful swamp was one that had been drained and settled.

Hence by 1900 policymakers and conservationists had cemented drainage as one of the state's most indispensable conservation services. When paired with the arid West's federal irrigation program, drainage became a weapon in the state's arsenal for staking enduring settler claims to expropriated lands and filling them with homes. Gifford Pinchot, the head of the Forest Service and one of the Roosevelt administration's most enthusiastic cheerleaders for drainage and irrigation, summed up this sentiment in a 1909 speech to the National Irrigation Association in Spokane, Washington. Pinchot heralded the reclamation of internal *wastelands* (e.g., swamps and deserts) as a propellant for the United States' geopolitical ascendancy. And he raved about conservation's global ambitions: "The Nation that will lead the world will be a Nation of homes. The object of the great Conservation movement is just this, to make our country a permanent and prosperous home for ourselves and for our children, and for our children's children." Settler permanence and prosperity: Drainage had served these overriding social and economic end objectives since shortly after the start of colonization.[5]

The state additionally sponsored drainage to eliminate the refuges of *others* it deemed menacing and hostile: fugitive enslaved peoples and Indigenous combatants during colonial wars. As the state diverted away murky swamp waters, thinned out dense vegetation, and then filled those tracts with settler amenities and infrastructure—homes, roads, fences, crop fields, ditches, culverts, telephone poles, and railroads—conservationists hyped drainage as the coup de grâce for extinguishing any and all opposing claims to *reclaimed* lands. But as Ed Prentice, Bay-mway-way-be-nais, and Nah-gon-nway-we-dung intimately understood, drainage inaugurated new settler homes at the expense and erasure of centuries of Indigenous interactions with the water, land, and soil. Precolonization Indigenous communities in the Ohio River Valley, California's Great Central Valley, Chicago's muddy portage, Red Lake Reservation, and numerous other places prized biodiverse wetlands as nutrient- and protein-rich "supermarkets." Rather than shunning America's wetlands, Native peoples incorporated them into their subsistence practices, extracting edible and medicinal plants, dyes, fish, fur-bearing small mammals, waterfowl, fruits, wild rice, and other raw materials.[6]

Drainage blotted out these rich and diverse Indigenous connections with wetlands, fortifying the narrative that soggy spaces were uninhabitable and unproductive until they were reclaimed and distributed to settlers. As an insidious feature of this process, the state's drainage infrastructure and technologies—most of them unobtrusively hidden beneath the surface and administered by a kaleidoscope of inscrutable local institutions—normalized colonial landscapes bereft of surface water, aquatic vegetation, and wildlife associated with wetlands. The normalization process reinforced the stubborn belief that swamps, wet prairies, marshes, and other watery tracts were irregular and consigned to a singular fate of drainage. The combined effect of locally administered and inconspicuous infrastructure tipped the balance of power in favor of settlers in the rural periphery following decades of colonial violence, epidemic disease, deceitful diplomacy, and dispossession.

As World War I approached, settlers increasingly deployed the term "warfare" to communicate the violence that drainage inflicted upon wetlands. In one such example, *The Minneapolis Star Tribune* extolled the conquest of state and "ceded" swamps. "Carrying on open warfare on nature is a stupendous task," it editorialized in 1912, "yet for nearly 20 years the state of Minnesota has been defying natural causes in wresting bog lands and swamps from the grasp of Dame Nature." In allocating $11 million since 1893 to drain 9,000 square miles, Minnesota—bolstered by generous federal resources, subsidies, and expertise—had waged "warfare" against swamps that enabled it to "reclaim" thousands of acres before "restor[ing]" them to settlers.[7]

The elimination of swamps and other undesirable landforms went hand in hand with the state's violence against inhabitants deemed unworthy of possessing them. Ever since Tench Coxe's 1814 report on arts and manufactures, settlers mythologized the alleged vacancy of swamps despite the longstanding Indigenous exploitation and occupation of them. This mythmaking crucially hinged on the epistemological policing and gatekeeping of federal scientists and engineers. Nathaniel Southgate Shaler spearheaded this policing. When he conducted the federal government's first two investigations of national wetland resources, he omitted Indigenous and Black experiences, knowledge, and history in swamps and Southern rice fields. Later, land grant universities dedicated to scientific agriculture discounted Indigenous agrarian practices in wetlands and never incorporated them into their courses and curricula. Progressive Era natural resource agencies broadened these erasures by relying exclusively on European and American sources in their internal catalogues of swampland resources, uses, and policies.[8]

The epistemological reductionism allowed agents of the state to cast swamps as empty and unused. It also undergirded a presumption that settlers were the only rightful inheritors of the land, which had been underdeveloped from the beginning. The very logic and rhetoric of land reclamation legitimized settler claims to liminal spaces, where the act of *reclaiming* them from their hydrological fluidity and imbalances conferred full ownership—or "restoration"—to the next generation of settlers. Drainage, especially following the Civil War, hardened into one of the political and social structures that buttressed settler colonialism.

The act of agricultural drainage may have helped forge settler permanence and prosperity, but Ed Prentice, Bay-mway-way-be-nais, and Nah-gon-nway-we-dung knew that it also triggered a torrent of environmental disruptions. Going forward, the state's unrelenting assault on wetlands only intensified over the course of the twentieth century and has especially accelerated during the last decade and a half. Ecologists and wetland scientists now estimate that agricultural drainage, suburban sprawl, and other forms of development have eliminated more than half of the conterminous United States' wetlands, an area the approximate size of California. Even in the face of softening public attitudes and overwhelming scientific evidence about wetlands' beneficial biological and hydrological functions, the decade between 2009 and 2019 witnessed a 50 percent increase in wetland losses across the US mainland. In that period, approximately 670,000 acres of vegetated wetlands—a geographic area exceeding the size of Rhode Island—was drained, filled, plowed, or paved over. As the US Fish and Wildlife Service's sixth annual wetlands status and trends report (2024) bluntly declared, "We are failing as a Nation to sufficiently protect out wetlands." Little wonder that one of the most consequential, destructive, and ongoing acts of water engineering and land modification in American history shows no sign of slowing. Worse yet, the US Supreme Court recently paved the way for even more wetlands losses by gutting long-standing, modest federal wetlands protections under the Clean Water Act in a series of decisions that whitewashed the assorted and undeniable public benefits of swamps and wetlands.[9]

Far from being disease-ridden wastelands, wetlands have always provided an assortment of "ecosystem goods and services." Wetlands enhance biodiversity, sequester carbon dioxide, store floodwaters, filtrate groundwater, combat river and stream sedimentation, shield shorelines from erosion and hurricane-related damage, and offer habitat for migratory waterfowl, numerous other species, and aquatic vegetation. In 2024, the US Fish and Wildlife Service tallied up these benefits, calculating that America's dwindling wetlands provide $7.7 billion in annual ecological services related to flood control, commercial fishing, recreation, and water quality and supply. Outside of their monetary value, wetlands possess intrinsic value and remain among nature's most biologically and hydrologically productive ecosystems.[10]

The historic erasure of American wetlands, which gathered steam in the decades before the Civil War and accelerated during the Gilded Age and Progressive Era, has exacerbated the Anthropocene's most daunting challenge: the global climate crisis. Wetlands are vital in reducing atmospheric carbon levels. Scientific studies estimate that global wetlands, which cover 5–8 percent of the earth's surface, disproportionately store 20–30 percent of global soil carbon. Drainage projects around the globe thus result in a "net transfer of carbon from the soil to the atmosphere," eliminating one of nature's most "important sink[s] for atmospheric carbon dioxide" and aggravating the anthropogenic climate crisis. Moreover, freshwater inland wetlands sequester ten times more carbon than coastal saltwater marshes. Unfortunately, the geographic distribution of the most heavily drained agricultural spaces in the United States—the former wet prairies of southern and western Minnesota, northern Iowa, Wisconsin, Illinois, Indiana, and Ohio and the prairie pothole region stretching from the eastern Dakotas and into southern Manitoba, Saskatchewan, and Alberta—has eliminated the bulk of North America's natural capacity for soil carbon storage.[11]

In addition to forfeited carbon storage, agricultural drainage inflicts tremendous damage on hemispheric water bodies. Researchers at the University of Illinois and Cornell University have identified intensive subsurface tile drainage systems in the corn belt arcing from southern Minnesota across Iowa, Wisconsin, Illinois, Indiana, and Iowa as the number-one contributor to hypoxia in the Gulf of Mexico. In their calculations, the 53 million acres of farmland drained by buried tile systems forms the basis of a "leaky" agricultural system. Mimicking subterranean gutters, the tiles rapidly transport fertilizer-based nutrients such as nitrogen into the Mississippi River's watershed following precipitation events. The seasonal gushes of these nutrients down the Mississippi and into the Gulf of Mexico have created a sprawling, oxygen-depleted "dead zone" that kills off fish and marine life, triggers species migrations, and undermines the reproductive capacity of many forms of marine life. In 2023, the hypoxic zone measured 3,058 square miles, an area nearly as large as Yellowstone National Park.[12]

The future of American wetlands has seldom looked bleaker. When it comes to continental swamps, myth, misperception, and erasure still reign supreme—a testament to the resilience of entrenched colonial ideas, values, and discourse.

If any glimmers of hope or optimism remain, they predominantly reside with Indigenous efforts over the past several generations to reclaim and revitalize heritage, economic practices, cultural knowledge, and ceremonies related to wetlands. In recent decades, a growing number of federally recognized tribes have inaugurated programs to protect, inventory, monitor, and restore wetlands. In March 2023, for instance, Gaa-Miskwaabikaang, the Red Cliff Band of Lake Superior Chippewa, unveiled a pilot program to monitor and study its interior wetlands and coastal estuaries at the northernmost tip of Wisconsin's Bayfield Peninsula. The Red Cliff also hired a wetland scientist to help coordinate a conservation plan. Comprising 7 percent of the 15,000-acre reservation, wetlands provide the Red Cliff with blueberries, cranberries, wild rice, fish, waterfowl, and ceremonial medicinal plants. According to the tribe's wetland program plan, the reservation's twenty-two-mile Lake Superior shoreline contains "several coastal wetland complexes where reservation streams flow into Lake Superior. The coastal estuary areas are revered for their cultural significance and are exceptionally critical for fish and aquatic life habitat, migratory bird and other wildlife habitat, floodwater storage, plant communities, and shoreline protection." The program plan aspires to monitor, regulate, and restore tribal wetlands to reclaim and preserve historical cultural and environmental practices.[13]

The Red Cliff wetland program envisions a combination of ancestral knowledge and contemporary science guiding its managerial approach. Mark Duffy, the Red Cliff Band of Lake Superior Chippewa's conservation officer and a tribal member, explained in 2023 that the band hired a scientist who "can get out there and see and make recommendations about how to protect our wetland areas. It is very important for us to have that because in the world of today that expert holds a weight, along with our traditional knowledge.... Together, it creates a better buffer zone for us to make sure that we maintain, protect, and enhance the biodiversity that wetlands bring." Carolyn L. C. Gougé-Powless, a fellow tribal member, associates the Red Cliff's rediscovery of wetlands with other aspects of social and cultural renewal. "One thing that has really happened is the revitalization ... of our language [along with] the gathering, hunting, the fishing," she observes. "The same with the water. You will hear that there are water blessings. We bless our water. If we are disrespectful to our water, the water can change just like that on us. We can never take things for granted because we are never guaranteed that it will be there." Other Wisconsin tribes have embraced a renewed consciousness about the significance of wetlands. Beginning in the early 2000s, the Oneida Nation started restoring 3,000 acres of agricultural fields to native wetlands, forests, prairies, and grasslands on its reservation near Green Bay. In addition, the Oneida partners with conservation societies and universities to monitor the return of rare bird species to restored wetlands.[14]

Around the Great Lakes basin, Indigenous communities are likewise lining up to restore northern wild rice (*Zizania palustris*) stands. An annual aquatic plant native to the region's wetlands, shallow lakes, watery bottomlands, and shallow-moving streams, northern wild rice was a dietary staple of the Anishinaabeg for centuries. It also attracted nonmigratory and migratory waterfowl and served as nurseries for spawning fish. The arrival of railroads, logging, and agriculture in the second half of the nineteenth century, however, decimated the sprawling, dense rice beds. Dams, drainage projects, and road construction reengineered the wet landscape, choking out the dense rice stands and depriving Native communities of a long-standing component of cultural and environmental identity.

In 2017, the Lac du Flambeau Band of Lake Superior Chippewa launched a program to reintroduce northern wild rice to their reservation's waters. After seeding thousands of pounds of wild rice in lakebeds and wetlands, the restoration effort finally triumphed in the summer of 2023 when wild rice stands germinated in the Lac du Flambeau's lakes and wetlands. Since that time, the band has recommenced harvest ceremonies surrounding the revered grain and inspired other tribes to mobilize resources and join external stakeholders in enlarging the restoration program. In Michigan, all twelve federally recognized tribes have formed a collective—the Michigan Wild Rice Initiative—to share knowledge, pursue resources, and cooperate to reintroduce the wetland plant across the Great Lakes basin. Moreover, the initiative also welcomes participation from state government departments, federal agencies, conservation organizations, and universities. Roger Labine, the tribal water resource technician for the Lac Vieux Desert Band of Lake Superior Chippewa in Michigan's Upper Peninsula, proudly captured the collective enthusiasm in March 2024: "There's a big movement [to restore northern wild rice beds], just like reviving our language." Later, he emphasized the high stakes of the restoration effort. "If manoomin [e.g., northern wild rice] disappears, the Anishinaabe will disappear. The manoomin spirit is returning to these tribal communities." Wetland resources remain just as integral to Indigenous identity as language, heritage, culture, and geography.[15]

Despite these renaissances, formidable obstacles to conserving and restoring wetlands across the United States remain, especially drainage's enduring legacy of federalism and localism. Since the Swamp Land Acts privatized much of the nation's wetlands and the states subsequently inaugurated drainage districts and other local drainage bodies, the radical diffusion of authority for drainage has fortified an institutional legacy of local control that has seldom been overcome or contested. The proliferation of thousands of drainage districts and county drainage boards across dozens of states served national political goals, reengineering soggy landscapes to preserve settler bodies and expand agricultural production. Today, however, this untidy maze of sovereign institutions stifles the ability of federal and state agencies to monitor agricultural runoff and water quality or implement uniform conservation and restoration plans. Iowa and Illinois alone, for instance, respectively contain 3,700 and 1,700 self-governing drainage districts, demonstrating why one environmental historian has characterized the expansion of district governments and "their relationship to county, state, and federal institutions" as the twentieth century's biggest "untold story."[16]

Combined, federalism and localism have wildly succeeded at underwriting drainage's short-term wealth creation goals rather than affording opportunities for long-term, collective stewardship once shifting environmental values and scientific knowledge dictated new approaches to wetland management. Small wonder that drainage's outsized historical role and unobtrusive rural presence continue to reverberate so profoundly in the present.

Notes

Abbreviations

CCDCR	Clay County District Court Records
CG	*Congressional Globe*
CR	*Congressional Record*
HF	T. J. Hyman File
Hill Papers	James J. Hill Papers
LDR	Law Department Records
LOC	Library of Congress
MHS	Minnesota Historical Society
NARA	National Archives and Records Administration
PR	President's Records
PSF	President's Subject Files
RG	Records Group
SPM&M	St. Paul, Minneapolis and Manitoba Railroad
SPM&M LB	SPM&M Letter Book

Introduction

1. "The GREAT Drainage Convention!," *Crookston (MN) Times*, July 3, 1886.
2. "The GREAT Drainage Convention!"
3. "Call for a Drainage Convention for the Red River Valley," *Crookston Times*, June 5, 1886; "The GREAT Drainage Convention!"
4. "The GREAT Drainage Convention!"
5. "Drainage Convention," *Crookston Times*, December 11, 1886.
6. "What Drainage Has Done for the State," *St. Paul Daily Globe*, November 15, 1903.
7. "Irrigation and Drainage," *Fourteenth Census*, 7:14, 365, 371.
8. Dahl, *Wetlands Losses*, 1.
9. US House of Representatives Select Committee, *Swamp Lands in Missouri and Arkansas*, 30th Cong., 2nd Sess., February 28, 1849, H. Rep. 130, 1. The US Geological Survey (USGS) now characterizes *wetland* as "a generic term for all the different kinds of wet habitats—implying that it is land that is wet for some period of time, but not necessarily permanently wet. Wetlands have numerous definitions and classifications in the United States as a result of their diversity, the need for their inventory, and the regulation of their uses." In the nineteenth and early twentieth centuries, however, settlers and politicians overwhelmingly balked at any utilization for wetlands that did not involve agricultural exploitation. As a result, they interchangeably used "swamp," "marsh," "bog," "lowland," "overflowed land," and other terms that pointed to a tract's partial, seasonal, or annual wetness. Tiner, "Technical Aspects of Wetlands," 1.

10. Byrd, *Description of the Dismal Swamp*, 21, 23. Pervasive fears about the deadly impact of swamps on settler bodies—particularly the miasmatic diseases they emitted—is covered in many publications and will be explored in chapter 1. See, for instance, Judd, *Untilled Garden*, 235–40; Nash, *Inescapable Ecologies*, 64–68, 70–74, 115; Valenčius, *Health of the Country*, 79–84, 109–10, 114–32, 137–52; Carlson, "'Vast Factories of Febrile Poison'" and Dillon, "Civilizing Swamps in California." In a similar vein, English writers and political authorities complained that residents of the Fenlands—the country's largest wetland—were sickly, deformed, misshapen, and ashen due to residing inside of the watery and miasmatic tract. See Mulry, *Empire Transformed*, chap. 2; and Di Palma, *Wasteland*, 93–102.

11. Warfare against Indigenous communities helped forge settlers' antipathy towards swamps. During King Philip's War, New Englanders scorned swamps for affording their Algonquian adversaries with cover, concealment, and hideouts impervious to assault—similar to how the Crown criticized inaccessible English fenlands and Irish bogs for housing rebellious, unruly subjects following the Restoration. For Puritan colonists, the Algonquians blended into wooded swamps with perfect ease, becoming indistinguishable with nature. So well did they conceal and camouflage themselves that colonists adopted "swamp" as a verb to describe their adeptness at disengaging from pitched battles and withdrawing into the interior of swamps. As the historian Jill Lepore concludes, the violence of King Philip's War hardened attitudes that swamps were "hideous and dangerous places, the most foreign and un-English land in all the New World." Lepore, *Name of War*, 85–88, 186, quote at 85; and Mulry, *Empire Transformed*, 128–29. This critique of swamps found renewed expression in the nineteenth century when the US government orchestrated three wars against the Seminoles (1817–18, 1835–42, and 1855–58). Swamps' contributions to the army's indecisive military operations against the Seminole Indians, particularly during the Second Seminole War, intensified the disparagement of them as uncivilized spaces, a theme touched on in Monaco, *Second Seminole War*.

12. "Some Account of the Great Dismal Swamp," 170. For recent studies rebutting the settler myth that Black bodies proved immunologically superior for work in tropical climates or Southern swamp and cane fields, see Espinosa, "Question of Racial Immunity"; Willoughby, "'His Native, Hot Country'"; and Hogarth, "Myth of Innate Racial Differences."

13. Hening, *Statutes at Large*, 460. On North Carolina's adoption of the 1705 law, see Nevius, *City of Refuge*, 17. For a sampling of historical studies that evaluate the dynamic connection between swamps and maroon communities, see Nelson, "Hidden Away in the Woods and Swamps"; Lockley and Doddington, "Maroon and Slave Communities"; Diouf, *Slavery's Exiles*; Nevius, *City of Refuge*, 15–19; and Morris, *Dismal Freedom*. For an overview of the blossoming historical field of marronage studies, see Nevius, "New Histories of Marronage."

14. "Some Account of the Great Dismal Swamp," 170; and Ruffin, *Nature's Management*, 208.

15. Emphases in original. *Instructions to the Surveyors General*, 16. See also White, *History of the Rectangular Survey System*, vii, 88, 112, www.blm.gov/sites/blm.gov/files/histrect.pdf.

16. C. G. Elliott, "Road Improvement," *Prairie Farmer*, March 20, 1886, 180.

17. Hill, *Highways of Progress*, 185–87. Quote at 185.

18. Quoted in Allen, "Deep History," 17.

19. The three Swamp Land Acts eventually donated sixty-five million acres of public "swamp and overflowed" lands to fifteen public land states. There is a voluminous literature

that examines the laws. For two of the newest interpretations, see Leshy, *Our Common Ground*, 58–60; and Allen, "Deep History."

20. Meyer, *Human Impact*, 73; "Irrigation and Drainage," *Fourteenth Census*, 7:14, 365, 371. For overviews of drainage districts, see Herget, "Taming the Environment"; McCorvie and Lant, "Drainage District Formation"; and Thompson, *Wetlands Drainage*.

21. Baeten, "Making Wet Places Drier."

22. Agricultural land drainage is almost completely omitted in environmental histories of the United States: Melosi, *Water in North American Environmental History*; Opie, *Nature's Nation*; Fiege, *Republic of Nature*; Kline, *First Along the River*; Steinberg, *Down to Earth*; Crane, *Environment in American History*. Published in 2014, *Oxford Handbook of Environmental History* includes no references to wetlands or drainage in its index: Isenberg, *Oxford Handbook of Environmental History*.

23. On the statist symbolism of dams and irrigation infrastructure, see Cullather, "Damming Afghanistan," esp. 520–22; and Cullather, *Hungry World*, 110. In one of the most influential, controversial interpretations of irrigation, Karl Wittfogel, an American scholar of Chinese history, argued that irrigation's rise in ancient societies led to the concentration of political sovereignty in the hands of a technocratic elite, inevitably contributing to the emergence of despotic "hydraulic societies." See Wittfogel, *Oriental Despotism*. Although Wittfogel's influence on American irrigation historians, most notably Donald Worster, would span decades, his thesis was not applied to drainage until 2012: Imlay and Carter, "Drainage on the Grand Prairie."

24. Hays, *Conservation and the Gospel of Efficiency*. There is a mammoth literature on the American conservation movement, most of which focuses on irrigation, flood control, national parks, and mineral, forest, and range policies. See, for example, Van Hise, *Conservation of Natural Resources*; Burton and Kates, *Readings in Resource Management*; Swain, *Federal Conservation Policy*; Penick, "Progressives and the Environment"; Fox, *American Conservation Movement*; Koppes, "Efficiency/Equity/Esthetics"; Phillips, *This Land, This Nation*; Tyrrell, *Crisis of the Wasteful Nation*; Taylor, *Rise of the American Conservation Movement*; Johnson, *Escaping the Dark, Gray City*; and Wellock, *Preserving the Nation*. A rare study of conservation that integrates drainage into the movement's history is Meyer, *Progressive Environmental Prometheans*.

25. There is a vast literature on irrigation and the federal reclamation program in the American West. For a sampling of some of the landmark works that set the historiographical agenda, see Hundley, *Dividing the Waters*; Hundley, *Water and the West*; Maass and Anderson, *. . . and the Desert Shall Rejoice*; Dunbar, *Forging New Rights*; Pisani, *From the Family Farm*; Worster, *Rivers of Empire*; Reisner, *Cadillac Desert*; Pisani, *To Reclaim A Divided West*; and Pisani, *Water and American Government*. For a discussion of the irrigation literature's key themes, see Rowley, "Introduction." As Paul S. Sutter reminds us, the study of Western water development exerted a tremendous impact on overall US environmental historiography, especially in its demolishing of the "conservation-preservation obsession" and its "sophisticated early grappling with the environmental-management state." See Sutter, "World with Us," 101n11.

26. Vileisis, *Discovering the Unknown Landscape*; Prince, *Wetlands of the American Midwest*, quote at 1; Pisani, "Beyond the Hundredth Meridian"; Thompson, *Wetlands Drainage*; Langston, *Where Land and Water Meet*; Stine, *America's Forested Wetlands*; Wilson, *Seeking Refuge*; and Garone, *Fall and Rise*.

27. The USGS studies were Shaler, "Preliminary Report on Sea-Coast Swamps"; and Shaler, "General Account." On the Office of Drainage Investigations' personnel, structure, and responsibilities, see "Organization, Work, and Publications of Drainage Investigations."

28. Balogh, *Government Out of Sight*, 77, 19; and Balogh, *Associational State*.

29. Warren, *Hunter's Game*; Spence, *Dispossessing the Wilderness*; Jacoby, *Crimes against Nature*; Kantor, "Ethnic Cleansing"; Krakoff, "Settler Colonialism and Reclamation"; Tyrrell, *Crisis of the Wasteful Nation*; and Nelson, *Saving Yellowstone*. For the relationship between conservation and eugenics, see Allen, "'Culling the Herd'"; Brechin, "Conserving the Race"; Spiro, *Defending the Master Race*; Stern, *Eugenic Nation*, 7, 22, 25, 139–72; Leonard, *Illiberal Reformers*, 68–69; Powell, *Vanishing America*; Rosenberg, "No Scrubs"; Goode, *Agrotopias*, 10–11, 152–63; and McNally, *Cast Out of Eden*, 175–83.

30. Halvor Steenerson to Charles D. Walcott, January 3, 1906, RG 115, Records of the Bureau of Reclamation, General Administrative and Project Records, 1902–1919, entry 3, box 97, folder 110-G, "General Correspondence re. Federal Legislation for Drainage of Swamp and Overflow Lands," NARA, Denver, CO; and Frymer, *Building an American Empire*, 24.

31. Frederick H. Newell, "The Undrained Empire of the South," March 2, 1911, box 6, Frederick Haynes Newell Papers, LOC, Washington, DC.

32. Gifford Pinchot, "The Conservation of Natural Resources," September 3, 1907, box 771, Gifford Pinchot Papers, LOC. For an incisive, persuasive, and original analysis of how colonial projects such as land drainage, harbor upgrades, canals, railways, telegraphs, and other *internal improvements* served as the handmaiden of dispossession, see Nelson, *Muddy Ground*, 3–5, 7, 13, 166–70, 174–75, 182.

33. Whyte, "Settler Colonialism," 136. Several works have sharpened my understanding of *settler colonialism*—the structure by which settler societies expropriated Indigenous territory, violently removed Native people onto reservations, settled their homelands, and then erased any Indigenous connections with the land, water, and soil: Wolfe, "Settler Colonialism"; Veracini, *Settler Colonialism*; Veracini, "Introducing: Settler Colonial Studies"; Tuck and Yang, "Decolonization Is Not a Metaphor"; and Hixon, *American Settler Colonialism*.

34. Wolfe, "Settler Colonialism," 388. On the mutually reinforcing elements of reclamation and wilderness preservation, see Rozum, *Grasslands Grown*, 279, 292–93, and Nelson, *Saving Yellowstone*, 99. When Congress established Yellowstone National Park in 1872, it did so to inaugurate "a public park or pleasuring-ground for the benefit and enjoyment of the people." See "Origin of the National Park Idea," National Park Service, n.d., accessed December 22, 2022, www.nps.gov/articles/npshistory-origins.htm.

Chapter One

1. Camporesi, *Fear of Hell*, 15–16; and Rigby, *Topographies of the Sacred*, 202–3.

2. Historical surveys of American wetlands do not fully probe how the early modern Hippocratic agenda, as well as fears about the erratic American climate, shaped social discourse about swamps: see Prince, *Wetlands of the American Midwest*; Schmid, "Wetlands as Conserved Landscapes"; Stine, *America's Forested Wetlands*; and Vileisis, *Discovering the Unknown Landscape*. As William B. Meyer argues, by "far and away the most important rea-

son for the widespread dislike and fear of wetlands in 17th-, 18th-, and 19th-century North America was their association with disease." Meyer, "From Past to Present," 91.

3. Lindemann, *Medicine and Society*, 179–80. On the "New Hippocratism," see Cassedy, "Meteorology and Medicine"; Hannaway, "Environment and Miasmata"; Pelling, "Contagion/Germ Theory/Specificity"; Rusnock, "Hippocrates, Bacon, and Medical Meteorology"; Wear, "Place, Health, and Disease"; and Brown, "From Foetid Air to Filth," 517–18.

4. There is a voluminous literature on Hippocrates and the Hippocratic Corpus. I have relied on Glacken, *Traces on the Rhodian Shore*, 80–88; Hannaway, "Environment and Miasmata"; and Porter, *Greatest Benefit to Mankind*, 55–62. For a broader historical perspective on the elusive phenomenon of miasma, see Parker, *Miasma*.

5. On the "fatalism" of Hippocratic medicine, see Riley, *Eighteenth-Century Campaign*, ix–x. On Hippocratic treatments, see Porter, *Greatest Benefit to Mankind*, 60–61.

6. Borca, "Towns and Marshes," 74–84.

7. Borca, "*Palus Omni Modo Vitanda*," esp. 7; and Borca, "Towns and Marshes," 74, 78, and 81. Early modern English writers also argued that the circulation of water impeded the formation of miasma. See Wear, "Making Sense of Health," 143.

8. Golinski, *British Weather*; Hannaway, "Environment and Miasmata"; and Rusnock, "Hippocrates, Bacon, and Medical Meteorology."

9. Levere, "Measuring Gases and Measuring Goodness," 105; Golinski, *Science as Public Culture*, 86, 93, 117–20; Hankins, *Science and the Enlightenment*, 94–95; Ihde, *Development of Modern Chemistry*, 29–30; Porter, *Greatest Benefit to Mankind*, 254; and Bowler and Morus, *Making Modern Science*, 61–71.

10. Dobson, *Contours of Death and Disease*, 16–17; Fleming, *Historical Perspectives*, 11–32; Golinski, *British Weather*, 139–140, 155, 158–59, 168, 185; and Riley, *Eighteenth-Century Campaign*, ix–xiii, 89–112.

11. Golinski, *British Weather*, 138–39, 152–57. On the British association of fevers and epidemics with the urban poor, see Pickstone, "Dearth, Dirt and Fever Epidemics."

12. As Lydia Barnett argues, "The premodern term *climate*, which designated a local natural economy of air, water, and soil and their combined effects on living things, more closely approximates the modern definitions of "environment" or "ecosystem" than the modern term *climate*." Emphases in original. Barnett, "Theology of Climate Change," 220–21. See also Bewell, *Romanticism and Colonial Disease*; Charters, *Disease, War, and the Imperial State*; Golinski, *British Weather*, 170–202; Golinski, "American Climate"; and Weidenhammer, "Patronage and Enlightened Medicine," esp. 32.

13. On ancient and Enlightenment climate theories, see Glacken, *Traces on the Rhodian Shore*; Grove, *Green Imperialism*; and Fleming, *Historical Perspectives*. For climatic discourse resulting from early American colonization, see Jankovic, "Climates as Commodities"; Vogel, "Letter from Dublin"; and White, "Unpuzzling American Climate." On the "puzzling" early American climate, see Kupperman, "Puzzle of the American Climate"; Kupperman, "Fear of Hot Climates"; Meyer, *Americans and their Weather*, 17–42; and Wear, "Place, Health, and Disease." On the New World's atmospheric wonders, see Delbourgo, *Most Amazing Scene of Wonders*, 50–51, 237–38. For the Little Ice Age, see Brooke, *Climate Change*, 370–72; and Brian Fagan, *Little Ice Age*. For the provocative argument that the demographic collapse of Indigenous communities across the hemisphere after 1492 may have fueled the Little Ice Age, see Koch, Brierley, Maslin, and Lewis, "Earth System Impacts."

14. Suazo, "Translating Dumont de Montigny's Frog," 115; and Gerbi, *Dispute of the New World*, 7–8. See also Glacken, *Traces on the Rhodian Shore*, 680.

15. Quoted in Roger, *American Enemy*, 7; quoted in Miller, "American Nationalism," 79; and Robertson, *History of America*, 1:258.

16. Emphasis in original. Coxe, *Statement*, xiii. For a sample of the literature analyzing the British belief in the economic and epidemiological supremacy of European land exploitation, see Vogel, "Letter from Dublin"; Armitage, *Ideological Origins*, chap. 3; Bewell, *Romanticism and Colonial Disease*; Cronon, *Changes in the Land*; Merchant, *Ecological Revolutions*; and Zilberstein, *Temperate Empire*. More broadly on the English "cult of improvement," see Drayton, *Nature's Government*, esp. 52–54.

17. Colden, "Observations on the Fever," 324, 320; and Colden, "Account," 310.

18. Antill, "Essay," 120; and Williamson, "Change of Climate," 272–80, quotes at 278, 279, and 280.

19. Jefferson, *Notes on the State of Virginia*, 47, 80. On Jefferson's rebuttal of Buffon's theory of degeneration, see Colden, *Fate of the Mammoth*, chap. 5; Dugatkin, *Mr. Jefferson*; Gerbi, *Dispute of the New World*, 252–58; Miller, *Jefferson and Nature*, 61–63; and Wilson, *Shadow and Shelter*, 7–8.

20. Wood, *Empire of Liberty*, 394; and Fleming, *Historical Perspectives*, 21–32.

21. Genesis 1:28, English Standard Version.

22. Isenberg, *White Trash*, 48, 53, 56, 93, 312, 320, 343n35, quote at 56.

23. Wilson, *Swamp*, 124–32; Mulry, *Empire Transformed*, 129–30.

24. For an excellent analysis of the belief that European forms of land exploitation were making colonial climates more genial and temperature, see Zilberstein, *Temperate Empire*.

25. Patterson, "Yellow Fever Epidemics"; and Grob, *Deadly Truth*, 102.

26. On the sugar industry's impact on *A. aegypti*, see McNeill, *Mosquito Empires*, especially chap. 2.

27. On this dispute, see, for instance, Golinski, "Debating the Atmospheric Constitution," 149–65; and Apel, *Feverish Bodies*.

28. Currie, "An Enquiry," 128, 135, 138–39. On Van Breda's eudiometrical experimentation, see Zuidervaart, "Eighteenth-Century Medical-Meteorological Society," 55–58.

29. Currie, "Enquiry," 141–42.

30. Caldwell, *Oration*, 5–10, quotes at 5, 6.

31. Emphasis in original. Caldwell, *Oration*, 13–19, quotes at 14.

32. Caldwell, *Oration*, 20, 21, 22.

33. Caldwell, *Oration*, 21.

34. Nelson, "Landscape of Disease," 557–59, quotes at 557–58.

35. Patterson, "Yellow Fever Epidemics," 858.

36. Troves of historical and archaeological evidence from outside of North America belie the persistent settler accusation that swamps constituted vacant wastelands and were inimical to community formation. As the political scientist James C. Scott argues, global wetlands historically gave rise to some of the earliest sites of ancient sedentism. In Mesopotamia, the lower Nile Valley, eastern China's Hangzhou Bay, the Indus River Valley, Southeast Asia's Hoabinhian sites, Peru's Lake Titicaca, and Mexico's Teotihuacan, nutrient-dense wetland ecosystems fueled the emergence of sedentary villages and urbanism. Far from being the "mirror image of civilization," wetlands supplied sedges and reeds

(building materials); an assortment of edible plants, fruits, and nuts; and small mammals, waterfowl, birds, migrating mammals, fish, mollusks, crustaceans, and other protein sources. "The density and diversity of resources that are lower in the food chain," Scott contends, "make sedentism more feasible ... [and explain] the *wetland origins* of early sedentary villages and early urbanism." Emphasis added. Scott, *Against the Grain*, 43–57, 127–28 (quotes at 50 and 55).

37. Brooks, "'Every Swamp Is a Castle,'" 50, 62.

38. Sleeper-Smith, *Indigenous Prosperity*, 6, 22, 30–31, 37, 39, 41–42.

39. Sleeper-Smith, *Indigenous Prosperity*, 43. Sleeper-Smith also eloquently likens the rich, productive, and biodiverse Ohio River valley wetlands to Indigenous "breadbaskets." See 39.

40. Garone, *Fall and Rise*, 48, 50.

41. Garone, *Fall and Rise*, 50.

42. Nelson, *Muddy Ground*, 91–92, 169.

43. John Ferdinand Smyth, quoted in Lockley and Doddington, "Maroon and Slave Communities," 127–28.

44. For a good overview of the technological limitations that impeded early drainage progress, see Baugher, "What Is It?"

Chapter Two

1. Ruffin, *Nature's Management*, 194, 201.

2. On Ruffin's career as an agricultural reformer and "protoconservationist," see Kirby, "Introduction," xi–xxvii; and Stoll, *Larding the Lean Earth*, 120–22, 150–60. On Ruffin's political vision of drainage, see Allmendinger, *Ruffin*, 115.

3. W. P., "Drainage and Irrigation," *Cultivator*, June 1851, 201; J. R. S., "Importance of Underdraining," *Cultivator*, September 1865, 274. On the origins, evolution, and growth of the antebellum reform movement, see Ron, *Grassroots Leviathan*; Pawley, *Nature of the Future*; and Phillips, "Antebellum Reform." For the rural press, see Douglas, "'To Improve the Soil and the Mind'"; Marti, "Agricultural Journalism"; and Ron, *Grassroots Leviathan*, 44–46.

4. Ruffin, *Nature's Management*, 208.

5. Ruffin, *Nature's Management*, 182. For thoughtful discussions of the conditions that rendered drainage a collective matter, see Meyer, "From Past to Present," 91–92; and Cumbler, *Northeast and Midwest United States*, 134.

6. While historians once argued that the United States was "stateless" and that national political authority was weak or hollow, Brian Balogh argues that the federal government "did not govern *less*. Americans *did*, however, govern *less visibly*." Rather than forging a bureaucratic state, nineteenth-century policymakers relied on the law, subsidies, the courts, trade policies, and third parties (local and state governments) to govern through an associational arrangement. "Intermediary institutions" served as the arrangement's linchpin. As long as midlevel institutions delivered results, the federal government remained free to serve as a coordinator and facilitator. As Balogh concludes, "Where local and state government was up to the task, or where voluntary and private groups might fulfill public purposes, Americans preferred that the national government enable rather than command." The state's approach to drainage, which relied heavily on intermediary institutions

(drainage districts, county drainage boards, state drainage agencies, and so forth), epitomized this associational governance model. Balogh, *Associational State*, 23 (emphasis in original); and Balogh, *Government Out of Sight*, 3.

7. In 1844, the *Cincinnati Weekly Herald and Philanthropist* underscored this evolving terminology. Malaria, it explained, "consists in certain invisible effluvia or emanations from the surface of the earth, which were formerly called Marsh Miasmata, but to which it has of late years become fashionable to apply the foreign term Malaria." See "An Important Article to Sufferers of Fever and Ague," *Cincinnati Weekly Herald and Philanthropist*, November 27, 1844.

8. For studies of Western settlers' understanding of miasmatic diseases and their relation to marshes and other stagnant waters, see Valenčius, *Health of the Country*, 84, 109–10, 114–32, 137–52; Nash, *Inescapable Ecologies*, 64–68, 70–74, 115; Carlson, "'Vast Factories of Febrile Poison'"; and Dillon, "Civilizing Swamps." Also useful are Urban, "An Uninhabited Waste"; Baldwin, "How Night Air Became Good Air'"; and Krueger, "Motherhood on the Wisconsin Frontier." For the population statistics, see Howe, *What Hath God Wrought*, 41.

9. Quoted in Beier, *Health Culture in the Heartland*, 4; and Meinig, *The Shaping of America*, 240.

10. R. B. J****, "Remarks on Marsh Effluvia," *Farmer's Register*, July 1, 1837, 142.

11. "Health," *Illinois Farmer*, January 1, 1856; and A. J. Murray, "Miasma," *Western Rural*, March 10, 1870, 78.

12. Rusticus, "Inland Swamps," *American Farmer*, June 20, 1823, 100; and A Medical Man, "Medical Tropics," *New England Farmer*, July 1870, 338.

13. "Drainage of Land," *Cultivator*, June 1849, 175; and E. Woolverton, "Draining Prairie Land," *Genesee Farmer*, May 1856, 141.

14. "Drainage of Land," 175; "On Draining," *American Farmer*, May 7, 1819, 43; and D. A. A. Nichols, "Wool Growing in Missouri," *Valley Farmer*, April 1860, 112.

15. N. T. Sorsby, "Wet Soils and Their Drainage. No. II," *Southern Cultivator*, May 1849, 65; and T. Kerr, "Malaria," *Prairie Farmer*, July 1, 1856, 212.

16. "Drainage of Land," 175; and "On Draining Lands," *Ohio Cultivator*, January 1, 1847, 2.

17. Pawley, *Nature of the Future*, 18; and Ruffin, *Nature's Management*, 227.

18. "Drainage of Land," 175; John Johnston, "The Effects of Drainage on Tillage," *Cultivator*, February 1854, 66; and "Draining Deepens the Soil," *Cultivator*, June 1858, 179.

19. Emphasis in original. Henry Cowles, D. B. Kinney, and Henry Shipherd, "Report on Drainage," *Ohio Cultivator*, September 15, 1849, 278.

20. A. D. G., "Advantages of Draining," *Cultivator*, July 1858, 207; and R., "Advantages of Under-Drainage," *Cultivator*, June 1859, 188.

21. "On Draining as a Means of Improving Lands—Ammonia in Rain Water—Drain Pipe Machines," *Ohio Cultivator*, December 15, 1852, 369; and J. H., "Underdraining," *Genesee Farmer*, August 1853, 235.

22. "Drainage of Land," *Cultivator*, June 1849, 175; and R., "Advantages of Under-Drainage," *Cultivator*, June 1859, 188.

23. "Reclaiming Bogs and Swamps," *Boston Cultivator*, August 5, 1843, 242; and Rusticus, "Inland Swamps," 100. Government scientists kept this idea alive and active well into the twentieth century. See, for instance, Mitchell, "To Farm America's Swamps."

24. John Foster, "Extensive and Profitable Improvements," *Ohio Cultivator*, March 15, 1850, 81.

25. Biebighauser, *Wetland Drainage*, 4, 73–74; Bidwell and Falconer, *History of Agriculture*, 318; and Weaver, *History of Tile Drainage*, 11–17, 20.

26. John Johnston, "Valuable Letter on Draining," *Cultivator*, September 1856, 268; "The Great Tile-Drainer," *New-York Daily Tribune*, October 29, 1859; Biebighauser, *Wetland Drainage*, 73–74; Vileisis, *Discovering the Unknown Landscape*, 122–24; and Weaver, *History of Tile Drainage*, 57–89.

27. Weaver, *History of Tile Drainage*, 222; and John Johnston, "Draining—Its Importance and Results," *Cultivator*, July 1860, 222.

28. Thos. Messenger, "Draining of Swamp Lands on Long Island," *Cultivator*, February 1864, 52; W. W. Stark, "Drainage of the Savannah Bottoms," *Southern Cultivator*, September 1849, 129; and H. W. Lester, "Thorough Underdraining with Tile," *New England Farmer*, May 1861, 231.

29. Edgar Sanders, "Effects of Drainage on Swamp Land," *Prairie Farmer*, July 9, 1864, 17.

30. "Drainage of Wet Lands," *Ohio Cultivator*, April 1, 1851, 103.

31. Sub-Soil, "Effects of Drainage on Fruit," *Genesee Farmer*, August 1855, 250.

32. "Draining," *Cultivator*, November 1837, 151; Messenger, "Draining of Swamp Lands," 52; and "Underdraining with Stone," *Cultivator*, July 1857, 209.

33. "Benefits of Underdraining," *Cultivator*, January 1863, 19; "Drainage of Wet Lands," 103; and John Johnston, "A Little More about Draining," *Cultivator*, November 1856, 333 (emphasis in original).

34. "Draining of Land—What Hinders More Frequent Trials of It?," *Cultivator*, January 1858, 11.

35. "Drainage of Wet Lands," 103.

36. W. G. Edmundson, "Agricultural Resources of the Great West," *Cultivator*, October 1852, 345.

37. See, for instance, the discussion in Steinberg, *Slide Mountain*, 11–15, 180n17.

38. Ruffin, *Nature's Management*, 174.

39. Ruffin, *Nature's Management*, 175; and Fiege, *Irrigated Eden*, 35. In a later essay, Fiege extended the metaphor to the mobility of weeds in the American West: Fiege, "The Weedy West." Shannon Stunden Bower, in her book on land drainage in southern Manitoba, builds on Fiege's idea of a hydrological commons by conceptualizing the relationship between private property and surface water as an "ecological commons, in which the private property landscape is overlaid by elements of the natural world that are of common concern to all landowners. In southern Manitoba, surface water was a particularly significant variety of mobile nature, creating an ecological commons among those touched by its flow patterns." See Bower, *Wet Prairie*, 13.

40. "Drainage of Wet Lands," 103; and "Reclamation of Southern Swamps," *DeBow's Review*, November 1854, 525.

41. "South Carolina," December 8, 1849, *North Carolinian*.

42. The best overview of colonial commissions of sewers is Hart, "Colonial Land Use Law." In his history of the development of settler agriculture in Concord, Massachusetts, Brian Donahue argues that the English "legal custom" of commissions of sewers was "transferred to New England" as early as 1644. New England's system of mixed husbandry, Donahue explains, hinged on the routine harvesting of meadow hay, as well as controlling the flow of water into and out of meadows and removing obstructions from watercourses. In

1644, the Massachusetts General Court instructed four local settlers to form a commission "to set some order which may conduce to the better surveying, improving, and draining of the meadows, and saving and preserving of the hay there gotten, either by draining the same, or otherwise, and to proportion the charges layed [sic] out about it equally and justly." As Donahue puts it, commissions of sewers were "an institutional means to compel all those meadow owners who might benefit (rather than ratepayers in general) to contribute to improving their soggy property," a distinguishing feature of later drainage districts. See Donahue, *Great Meadow*, 63, 94.

43. For drainage districts and other local institutions dedicated to surface water removal, see Herget, "Taming the Environment"; McCorvie and Lant, "Drainage District Formation"; Imlay and Carter, "Drainage on the Grand Prairie"; and Thompson, *Wetlands Drainage*, 13–14.

44. This is my own tabulation.

45. There is a lengthy secondary literature on the Swamp Land Acts. See Prince, *Wetlands of the American Midwest*, 140–48; Vileisis, *Discovering the Unknown Landscape*, 90–91; Leshy, *Our Common Ground*, 58–60; Reuss, *Designing the Bayous*, 40–47; Palmer, "Swamp Land Drainage," 17–31; Harrison and Kollmorgen, "Land Reclamation in Arkansas"; Bogue, "Swamp Land Act"; Gates, *History of Public Land Law Development*, 321–35; Peterson, "Failure to Reclaim"; and Strausberg, "Indiana and the Swamp Lands Act." The newest historical interpretation of the Swamp Land Acts is Allen's "Deep History." Allen argues that, "Among members of America's ruling class, privatization [of swamps] was understood to be the most effective way to secure the country against the threats posed by unimproved people and to extend the practical reach of the state. . . . [T]he Swamp Land Acts facilitated the enclosure of the de facto multiracial wetland commons that formed in the United States, and helped to ensure that a formal commons was never allowed to take shape." Allen's analysis, with the notable exception of downplaying pervasive settler fears about swamps' deleterious impact on settler bodies, supports my argument that settlers enlisted the state to drain and improve swamps to build stable and resilient communities on expropriated Indigenous homelands. Davis, "Deep History," 28–29.

46. See, for instance, the floor debate in early 1849: *CG*, 30th Cong., 2nd Sess., February 26, 1849, 594.

47. Dahl and Allord, "Technical Aspects of Wetlands," 21.

48. W. B., "The Muck Bed, and its Future Prospects," *New England Farmer*, December 11, 1858, 1.

Chapter Three

1. Details about Andrew Lommeland's life, farm, and family are gleaned from the 1880 US Census and testimony "Paper Books" from subsequent court cases involving Clay County farmers and the SPM&M. The legal records of the SPM&M, the predecessor to the Great Northern Railway, are part of the Great Northern Railway Corporate Records (hereafter GN Records). See *State of Minnesota. Supreme Court, April Term, 1886. Andrew A. Lommeland vs. St. Paul, Minneapolis and Manitoba Railway Company*, vol. 31 (St. Paul: H. M. Smyth Printing Company, n. d.), 3, 47–49, LDR, 1877–1970, GN Records, MHS, St. Paul, MN (hereafter Lommeland Paper Book). See also US Census Office, *Nonpopulation Cen-*

sus Schedule, Minnesota, 1860–1880, microfilm reel 4, "Agriculture Schedules, 1880, Aitkin–Crow Wing Counties," (St. Paul: Minnesota Historical Society, 1977), frame 677 (hereafter *Nonpopulation Census Schedule*).

2. Lommeland Paper Book, LDR, GN Records, 46–51, quote at 48.

3. Lommeland Paper Book, LDR, GN Records, 29–32, 49–51, quote at 51.

4. Prince, *Wetlands of the American Midwest*, 228–29, 231.

5. Pielou, *After the Ice Age*, 193, 294; Chapman, Fischer, and Ziegenhagen, *Valley of Grass*, 3; Severson and Sieg, *Nature of Eastern North Dakota*, 8–9; Almquist, "Farm Drainage," 15; and Tester, *Minnesota's Natural Heritage*, 10–14.

6. Quoted in Bower, "Watersheds," 799–800. On wetland prairie ecosystems, see Hewes, "Northern Wet Prairie"; Hewes and Frandson, "Occupying the Wet Prairie"; and Prince, *Wetlands of the American Midwest*, 41, 53, 57. See also Pemble, *Natural History*, 13.

7. Murray, *Valley Comes of Age*, 127–29.

8. Angevine, *Railroad and the State*, 121–23.

9. Mercer, *Railroads and Land Grant Policy*, 56; Hidy et al., *Great Northern Railway*, 2–4; Veenendaal, *Saint Paul and Pacific*, 24, 26–27.

10. Veenendaal, *Saint Paul and Pacific*, 35–41.

11. Lubetkin, *Jay Cooke's Gamble*; White, *Railroaded*, chap. 2; Malone, *James J. Hill*, 35; and Mercer, *Railroads and Land Grant Policy*, 52–53.

12. Lansing, "From Wheat to Wheaties"; Cronon, *Nature's Metropolis*, 376–77; Robinson, *History of North Dakota*, 135–36; and Kane, *Waterfall That Built a City*.

13. Drache, *Day of the Bonanza*, 34–35, 38, 42–45, 51–52, 71–82, 207. On the emergence of the bonanzas, see also Briggs, "Early Bonanza Farming"; Murray, *Valley Comes of Age*, 105–9, 121, 131–33; and Robinson, *History of North Dakota*, 137–40. President Hayes's visit to Dakota Territory is described in "President Hayes," *Cincinnati Daily Gazette*, September 7, 1878.

14. Drache, *Day of the Bonanza*, 41–42 (quotation), 68, 71–72; Engelhardt, *Gateway to the Northern Plains*, 25; Murray, *Valley Comes of Age*, 67, 117n32, 121, 134 (Table V); and Murray, "Agricultural Development," 63.

15. Murray, *Valley Comes of Age*, 110, 120, 127–29, 136–37, 141.

16. Murray, *Valley Comes of Age*, 96, 140, 158, 204; and Drache, *Day of the Bonanza*, 29.

17. Murray, *Valley Comes of Age*, 131.

18. Malone, *James J. Hill*, 4–5, 10–14, 23–24, 28–30; and Martin, *James J. Hill*, pt. 1.

19. Though the St. Paul and Pacific's original land grant—which encompassed approximately 3 million acres—appears diminutive when compared to the NP's 40-million-acre grant, it was the seventh largest of the seventy-five federal railroad land grants authorized by Congress and belies the "considerably less than a half truth" that the GN, the SPM&M's successor, reached the Pacific in 1893 without the aid of a land grant. Malone, *James J. Hill*, 3–63, quote at 33; Rae, "The Great Northern's Land Grant"; Hidy et al., *Great Northern Railway*, 36; Murray, *Valley Comes of Age*, 65; and Veenendaal, *Saint Paul and Pacific*, 135.

20. Murray, *Valley Comes of Age*, 124.

21. As Benjamin Palmer observed, "From the time of the acceptance by the legislature of the lands granted by Congress in 1860 no attention seems to have been given to the clause in the act by which the grant was made." Palmer, "Swamp Land Drainage," 88–94, quote at 94.

22. On Western railroads as "surrogate governments," see Pisani, "George Maxwell," 201, and Orsi, *Sunset Limited*, xiv–xv, 170, 180, 377.

23. Hill, *Highways of Progress*, 44, 185. During the 1880s, Hill touted drainage as a strictly a financial mechanism for boosting the SPM&M's carrying trade, selling off its land grant, and reducing maintenance costs associated with track washouts. His personal correspondence and public speeches in the decade do not describe drainage as a tool to preserve the nation's agrarian heritage, avert a Malthusian calamity, or thwart the alleged mobocracy that accompanied urbanization. These social and cultural apprehensions did not influence his motivations for conservation and water management until the 1890s and early 1900s. Pisani, *To Reclaim a Divided West*, 288–94; and Carlson, "'There May Be Bloodshed,'" 402–3. The most thorough study of Hill's promotion of settler agriculture and land reclamation is Strom, *Profiting from the Plains*.

24. Drache, *Day of the Bonanza*, 95–96.

25. George C. Reis to Hill, September 13, 1879, PR, 1872–1970, GN Records.

26. Reis to Hill, September 13, 1879, May 25, 1880, and May 26, 1880, PR, GN Records; Hill to Reis, September 15, 1879, SPM&M LB No. 1, "June 25–October 15, 1879"; and Hill to Reis, May 27, 1880, SPM&M LB No. 3, "March 5–October 7, 1880," Hill Papers, James J. Hill Reference Library, St. Paul, Minnesota. The Hill Papers are now housed at the Minnesota Historical Society. Since the bulk of the archival research for this chapter occurred at the Hill Reference Library, I have maintained the former citations.

27. "The Opening," *Red River Valley News* (Glyndon, MN), April 1, 1880; "Railway News," *Red River Valley News*, May 27, 1880; "Observations," *Red River Valley News*, July 22, 1880; and Editorial, *Moorhead (MN) Weekly News*, December 28, 1882. The heavy precipitation pattern that accompanied the Red River Boom is described in Murray, *Valley Comes of Age*, 120.

28. Reis to Hill, May 25, 1880, and May 26, 1880, PR, GN Records; and Hill to Reis, May 27, 1880, SPM&M LB No. 3, "March 5–October 7, 1880," Hill Papers.

29. In 1888, attorneys for Brevik explained that their client, sometime in October 1880, permitted railroad engineers to excavate a ditch across a sixteen-foot strip of his land near the section line. See "Ole O. Brevik v. SPM&M," CCDCR, case 1822, MHS.

30. Hill to R. B. Angus, July 9, 1880, SPM&M LB No. 3, "March 5–October 7, 1880," Hill Papers; "Railway News" and "Glyndon's Fame Abroad," *Red River Valley News*, June 24, 1880; "Big Crops in the Red River Valley," *Minneapolis Star Tribune*, June 21, 1881; "State Drains: Which Have Been Constructed in the Minnesota Half of the Red River Valley," *Grand Forks Herald*, April 16, 1899; and "What Drainage has Done for the State," *St. Paul Daily Globe*, November 15, 1903.

31. "Land Drainage. II," *Grand Forks Herald*, December 4, 1881; and "Crop Prospects," *Red River Valley News*, June 17, 1880.

32. "The Lands of the Red River Valley—The Best System for Draining Them," *Fisher (MN) Bulletin*, November 19, 1881; and Editorial, *Moorhead Weekly News*, December 28, 1882.

33. "Municipal Court Decision," *Moorhead Weekly News*, August 23, 1883; and "Gabriel Wilson v. SPM&M," case 1489, CCDCR.

34. "Lommeland Paper Book," LDR, GN Records, 46, 48.

35. Holcombe and Bingham, *Compendium of History and Biography*, 262.

36. "Lommeland Paper Book," LDR, GN Records, 15–20, quote at 20; and Hill to H. M. Hogenson and others, May 5, 1881, SPM&M LB No. 5, "April 12, 1881–March 23, 1882," Hill Papers. Hill delegated responsibility for dealing with the Moland flooding to Allen Manvel, the SPM&M's assistant general manager. Manvel asked the railroad's chief engineer, C. C. Smith, to evaluate the situation, but it is unclear if Smith ever visited Moland or surveyed the ditch. See Manvel to Smith, May 5, 1881, SPM&M LB for the Assistant General Manager's Office, "May 3–July 24, 1881," Hill Papers.

37. James B. Power to Hill, June 14, 1882, PR, GN Records.

38. Power to Hill, June 14, 1882, PR, GN Records.

39. St. Paul, Minneapolis, and Manitoba Railroad Company, *Letters from Golden Latitudes*, 15–16; and Murray, *Valley Comes of Age*, 111–12, 135–36.

40. "Drainage," *Fisher Bulletin*, September 24, 1881, and March 4, 1882; and Power to Joseph Dilworth, February 21, 1882, letterpress book, "December 1880—April 1882," box 2, James B. Power Papers (hereafter Power Papers), North Dakota Institute for Regional Studies, Fargo.

41. Power to Hill, June 14, 1882, PR, GN Records; Power to A. J. Norrish, July 29, 1882, book, "April 1882–March 1885," Box 3, Power Papers; Untitled, *Fisher Bulletin*, July 21, 1883; "The Land of Plenty," *Fisher Bulletin*, July 28, 1883; Untitled, *Fisher Bulletin*, August 11, 1883; "The Draining Boom," *Moorhead Weekly News*, October 12, 1882.

42. *State of Minnesota. County of Clay. District Court, 11th Judicial District. Calendar of Causes. November General Term, 1882.* (Moorhead: n.p., 1882), cases 26–41. Copy available at Minnesota State University Archives, Moorhead, Minnesota. The above source wrongly listed October 14 (instead of November 14) as the date when the farmers' filed suit. Official court documents list November 14 as the filing date. *Clay County Naturalization Records, 1872–1954*, vol. 1, Final Papers, 1874–1900 (St. Paul: Minnesota Historical Society, 1996), frames 276–279, 283, 287, 291, CCDCR. Ole Matisen, Halvor Rasmusson, and William Lloyd followed the first group of farmers and took the oath of citizenship in mid-January 1883.

43. The affidavits for the transferred cases are housed in RG 21, Records of the District Courts of the United States, Third Division (St. Paul), Law "C" Cases (1879–84), boxes 25 and 26, cases 398–412, NARA, Kansas City, Missouri (hereafter KC). Untitled, *Red River Valley News*, April 19, 1883; Untitled, *Red River Valley News*, August 30, 1883; Untitled, *Red River Valley News*, October 25, 1883; Untitled, *Red River Valley News*; November 30, 1883; "Hogen M. Hogenson, v. SPM&M," CCDCR, case 3611.

44. Gibbs v. Williams, 25 Kan. 149, 153 (1881); Davis, "The Law of Diffused Surface Water," 227n1.

45. Graham, "Reasonable Use Rule," 225. Also see Kinyon and McClure, "Interferences with Surface Waters"; and Dobbins, "Surface Water Drainage."

46. Fraley, "Water, Water, Everywhere," 93–97, quotes at 93; and Graham, "Reasonable Use Rule," 227. On wetlands as ecological nuisances in late nineteenth- and early twentieth-century jurisprudence, see Nagle, "From Swamp Drainage to Wetlands Regulation." In addition to the civil law and common enemy doctrines, two states by World War II adopted a "reasonable use" framework, which made "reasonableness" the standard for determining liability in diffused surface water disputes. See Graham, "Reasonable Use Rule;" Kinyon and McClure, "Interferences with Surface Waters"; and Fraley, "Water, Water, Everywhere," 98–99.

47. "State of Minnesota, Supreme Court. October Term, A. D. 1883. Hogen M. Hogenson, Appellant, vs. The St. Paul, Minneapolis and Manitoba Railroad Company, Respondent. Appellant's Brief," LDR, GN Records.

48. "State of Minnesota, Supreme Court. October Term, A. D. 1883. Hogen M. Hogenson, Appellant, vs. The St. Paul, Minneapolis and Manitoba Railroad Company, Respondent. Appellant's Brief," LDR, GN Records.

49. Hogen M. Hogenson vs. St. Paul, Minneapolis & Manitoba, 31 Minn. 224 (1883), quote at 226.

50. Allen Manvel to Solomon G. Comstock, November 30, 1885, box 8, file "November–December, 1885," Solomon G. Comstock Papers (hereafter Comstock Papers), Minnesota State University–Moorhead Archives, Moorhead. Also see Manvel to Comstock, September 22, 1886, box 9, "September–October, 1886," in the same collection.

51. For the SPM&M's settlements with farmers, see Clay County Deed Record Books, vol. 5, 324, 486; vol. 9, 565, 591–92, 594–98; vol. 11, 124, 144, 153–54; vol. 13, 54; and Clay County Miscellaneous Record Books, vol. B, 619–20, 623–26, 637–40; vol. C, 1–6, 8, 100–101, 291–95. Both volumes are available at the Clay County Recorder's Office, Clay County Courthouse, Moorhead, Minnesota. On the board of arbitration, see "Arbitration," *Moorhead Weekly News*, January 13, 1887; "The Flowage Suits," *Moorhead Weekly News*, March 3, 1887; "Judgments," *Moorhead Weekly News*, June 2, 1887; and "Flowage and Arbitration," *Red River Valley News*, May 19, 1887. For Hill's $100,000 estimate, see Hill to H. W. Donaldson, June 19, 1893, HF, Hill Papers.

52. Lommeland Paper Book, LDR, GN Records, 107–10; and Andrew A. Lommeland vs. St. Paul, Minneapolis & Manitoba Railway Company, 35 Minn. 412 (1886).

53. Hill to Henry H. Oberg, July 28, 1886, SPM&M LB No. 9, "June 15–November 26, 1886," Hill Papers. Later in 1906, Hill reflected at length on his 1880s drainage experiences during a public speech. See "A Drainage Talk," *Grand Forks Herald*, January 12, 1906.

Chapter Four

1. Elias Steenerson, "Memoirs of Pioneer Days," unpaginated mimeograph, Elias Steenerson Papers, Norwegian-American Historical Association Library, St. Olaf College, Northfield, Minnesota. For the amount of land owned by the SPM&M on the Minnesota side of the valley in 1886, see "The GREAT Drainage Convention!," *Crookston Times*, July 3, 1886.

2. Holcombe and Bingham, *Compendium of History and Biography*, 167–69.

3. Steenerson, "Memoirs."

4. Christopher Steenerson to Hill, February 22, 1886, PR, GN Records.

5. Malone, *James J. Hill*, 102–50; and Martin, *James J. Hill*, 360–78, 388–98.

6. Hill to C. Steenerson, March 17, 1886, SPM&M LB No. 8, "February 14, 1885–June 15, 1886," Hill Papers.

7. Hill to E. D. Childs, May 13, 1886, SPM&M LB No. 8, "February 14, 1885–June 15, 1886"; Childs to Hill, May 12, 1886, General Correspondence, Hill Papers; Steenerson, "Memoirs."

8. "Call for a Drainage Convention for the Red River Valley," *Crookston Times*, June 5, 1886; C. E. Page to Hill, July 8, 1886, PR, GN Records; Hill to Page, July 9, 1886, Hill to

Childs, July 23, 1886, SPM&M LB No. 9, "June 15–November 26, 1886," Hill Papers; and W. S. Alexander to Comstock, June 28, 1886, box 9, file "June 1886," Comstock Papers.

9. "The GREAT Drainage Convention!"; "The Red River Drainage Convention at Crookston, Minn.," *Kittson County Enterprise*, July 10, 1886; and Elliott, *Practical Farm Drainage; Why, When and How to Tile Drain*, unpaginated preface. On Elliott's life and career, see Hager, *Hydraulicians in the USA*, 2010. On Elliott's 1901 takeover of the *Drainage Journal*, see *Drainage Journal* 23 (May 1901): 145–46.

10. "The GREAT Drainage Convention!"
11. "The GREAT Drainage Convention!"
12. "The GREAT Drainage Convention!"
13. "The GREAT Drainage Convention!"
14. "The GREAT Drainage Convention!"; Murray, *Valley Comes of Age*, 93, 134.
15. "The GREAT Drainage Convention!" On the vilification of drainage opponents, see Prince, *Wetlands of the American Midwest*, 208.
16. "The GREAT Drainage Convention!"
17. "The Drainage Committee," *Bismarck Tribune*, July 30, 1886; and "Red River Drainage," *Manitoba Weekly Free Press*, August 12, 1886.
18. *Red River Valley Drainage*, 9, 22–23, quotes at 23. The report is available at the Minnesota Historical Society in St. Paul and the Polk County Historical Society in Crookston.
19. *Red River Valley Drainage*, 24.
20. "Drainage Convention," *Crookston Times*, December 11, 1886.
21. "Drainage Convention."
22. "Drainage Convention."
23. "Drainage Convention."
24. "Drainage Convention"; and "Drainage of the Red River Valley," 236.
25. *Red River Valley Drainage*, 6, 9, 13, 21–22, 24, quote at 21; and "Drainage Convention." Minnesota received the following amounts of "swamp and overflowed" lands in the six valley counties: Clay (11,564 acres), Kittson (49,952), Marshall, (73,386), Norman (39,774), Polk (69,912), and Wilkin (5,000).
26. *Minnesota Farmer*, January 21, 1887; "Red River Valley Drainage," *St. Paul Daily Globe*, February 26, 1887; and "Drainage Law," *Moorhead Weekly News*, January 20, 1887.
27. *Executive Documents*, 30; and "Drainage," *St. Paul Daily Globe*, January 6, 1887.
28. "Proceedings," *Red River Valley News*, June 16, 1887; and "Commissioners' Proceedings," *Warren (MN) Sheaf*, July 19, 1894.
29. "Red River Valley Drainage," *St. Paul Daily Globe*, February 26, 1887; "Draining the Red River Valley," *St. Paul Daily Globe*, February 27, 1887; and "The Red River Valley," *St. Paul Daily Globe*, March 4, 1887.
30. "The Drainage Committee," *St. Paul Daily Globe*, September 21, 1887; "Biennial Message of Gov. A. R. McGill, of Minnesota," January 11, 1889, *People's Press* (Owatonna, MN); and "Minnesota Legislative Doings," *Bismarck Tribune*, March 26, 1891.
31. House Committee on Rivers and Harbors, *Red River of the North, Minnesota*, 52nd Cong., 1st Sess., February 10, 1892, Ex. Doc. 127, 2, 22, quote at 2. See also House Committee on Rivers and Harbors, *Red River of the North*, 51st Cong., 2nd Sess., March 3, 1891, Ex. Doc. 292.
32. [Ezra G. Valentine], *Statement and Statistics*, 3–4, HF, Hill Papers. On Valentine's authorship of the pamphlet, see Ezra G. Valentine to Hill, May 11, 1893, in the same file. On the

repeated failures of the Red River Valley Drainage Commission to secure a state appropriation before 1893, see Valentine to Hill, May 4, 1893, HF, and Hill to Levi S. Myers, March 9, 1887, SPM&M LB No. 10, "November 26, 1886–June 6, 1887," Hill Papers; and *Legislative Manual*, 287–88.

33. S. F. 182, "AN ACT to appropriate Moneys for the purpose of opening of closed Water-courses leading into the Red River and its Tributaries, and for opening existing streams in the Red River Valley in the counties of Wilkin, Clay, Norman, Polk, Marshall, Kittson, Grant, and Traverse, in this State," and Valentine to Hill, May 4 and May 6, 1893, HF, Hill Papers.

34. *General Laws of Minnesota for 1893*, chap. 221, 371, and "Elevator Bill a Go," *Duluth News Tribune*, March 31, 1893.

35. On the GN's 999-year lease, see Malone, *James J. Hill*, 129.

36. Hill to Valentine, May 8, 1893, Hill to Donaldson, June 19, 1893, Hill to Nelson, June 19, 1893, and Hill to Valentine, June 19, 1893, HF, Hill Papers. W. P. Clough, the GN's vice president, remarked that the 1893 bill "is in the same form as that in which it appeared on previous years, except the blunder of the [railroad] name." See Clough to Hill, April 28, 1893, in the same file.

37. Emphasis in original. Donaldson to Hill, June 16, 1893, HF, Hill Papers.

38. Hill to Valentine, November 26, 1895, PSF, GN Records; *Report of the State Drainage Commission of Minnesota to the Governor and State Legislature on the Condition of the State Ditches Located in the Red River Valley for the Biennial Period Ending Feb. 1st, 1899* (Minneapolis: University Press of Minnesota, 1899), 15; *Report of the Board of Drainage Commissioners of the State of Minnesota to the Governor of Minnesota on the Condition of the State Ditches Located in the Red River Valley for the Year Ending December 31st, 1901* (Crookston, MN: Crookston Journal, 1902), 12, 16. On the board of audit's initial composition and organization, see Nelson to Hill, July 15, 1893, HF, Hill Papers.

39. Nelson to Hill, July 15, 1893, N. D. Miller to Hill, December 31, 1894, HF, Hill Papers; Hill to Valentine, November 26, 1895, PSF, file 2922, GN Records; *General Laws of Minnesota for 1895*, chap. 164, 372; *Report of the Board of Drainage Commissioners*, 12, 16; and "What Drainage has Done for the State," *St. Paul Daily Globe*, November 15, 1903.

40. *Report of the State Drainage Commission*, 3–4, 13, 17, 23, 47–48, 57; *Report of the Board of Drainage Commissioners*, 14; and Valentine to Hill, PSF, file 3776, GN Records.

41. Hanson, "Damming Agricultural Drainage," 142.

42. Prince, *Wetlands of the American Midwest*, 242.

43. Pisani, "George Maxwell"; and White, *Railroaded*, 394–95.

44. Steenerson, "Memoirs."

Chapter Five

1. Shaler, "Swamps of the United States," 232.

2. Pisani, *To Reclaim A Divided West*, 139; and Hurt, *American Agriculture*, 179.

3. Shaler, "Swamps of the United States," 232. For the postfrontier anxiety of the 1880s, see Wrobel, *End of American Exceptionalism*, 13–25. Fears about the impact of the frontier's closing on national identity and culture reached a crescendo after Frederick Jackson Turner, a young Wisconsin historian, delivered his "frontier thesis" at Chicago's 1893

World's Fair. Ultimately, anxieties about a landless future were more cultural than real as the pace of homesteading and, after 1909, enlarged homestead entries dramatically accelerated. On this "second phase of homesteading," see Gregg, "Imagining Opportunity."

4. Daniels, "Immigration in the Gilded Age," 21. Shaler's most important critique of the new immigration was "European Peasants as Immigrants," which will be discussed in depth later in the chapter. On Shaler's racial views, see Livingstone, "Science and Society"; Baker, *From Savage to Negro*, 45–48; and Haller, "Nathaniel Southgate Shaler," 173–74.

5. Shaler, "Swamps of the United States," 232–33; and Wrobel, *End of American Exceptionalism*, 53–68.

6. For an articulation of the concept of "legibility," see Scott, *Seeing Like a State*.

7. Shaler, "Preliminary Report"; Shaler, "General Account." The quotation regarding the Atlantic Coast Division's initial purpose is filched from Shaler, "Swamps of the United States," 232.

8. Livingstone, *Nathaniel Southgate Shaler*, 206. The concept of *watersheds* will be presented later in the chapter.

9. There is a recent, growing body of scholarship that evaluates the overlapping motivations, rhetoric, aims, and crossover appeal of conservationists and eugenicists in the late nineteenth and early twentieth centuries: Allen, "'Culling the Herd'"; Brechin, "Conserving the Race"; Spiro, *Defending the Master Race*; Stern, *Eugenic Nation*, 7, 22, 25, 139–72; Powell, *Vanishing America*; Rosenberg, "No Scrubs"; Goode, *Agrotopias*, 10–11, 152–63; and McNally, *Cast Out of Eden*, 175–83.

10. Livingstone, *Nathaniel Southgate Shaler*, 12–14.

11. Livingstone, *Nathaniel Southgate Shaler*, 14–15.

12. Numbers, *Creationists*, 19–20. On Agassiz's scientific career, see Lurie, *Louis Agassiz*; and Irmscher, *Louis Agassiz*. Agassiz's role in inaugurating scientific racism in the United States is surveyed in Louis Menand, "Morton, Agassiz, and the Origins of Scientific Racism."

13. Livingstone, *Nathaniel Southgate Shaler*, 20–29.

14. Shaler and Shaler, *Autobiography*, 57, 434; Numbers, *The Creationists*, 16–19.

15. Stern, *Eugenic Nation*, 14–16. On Shaler's neo-Lamarckism, see Haller, "Nathaniel Southgate Shaler," 173–74; and Livingstone, *Nathaniel Southgate Shaler*, 55–57, 62–63, 169, 171, 188, 191, 207–8.

16. Livingstone, *Nathaniel Southgate Shaler*, 29–31.

17. Livingstone, "Geologist by Profession," 147–53, quote at 153. For a year-by-year list of Shaler's publications, see Livingstone, *Nathaniel Southgate Shaler*, 341–55.

18. Shaler and Shaler, *Autobiography*, 261; Livingstone, *Nathaniel Southgate Shaler*, 33. Comprising some 700,000 acres in eastern England, the Fens (or fenlands) range for over seventy miles from the outskirts of Cambridge in the south to Lincoln in the north. In the seventeenth century, an assortment of English and Dutch "projectors," with the Crown's hearty assistance, began reclaiming hundreds of thousands of submerged acres into farms. The drainage of the fens became enmeshed in the broader process of English early modern state formation and, according to one scholar, constituted "an expression of changing... attitudes toward the proper management and exploitation of the natural environment and the state's role in facilitating such." Ash, *Draining of the Fens*, 2. See also Darby, *The Changing Fenland*.

19. On Shaler's tenure with the Kentucky Geological Survey, see Zabilka, "Nathaniel Southgate Shaler," 408–31; and Livingstone, *Nathaniel Southgate Shaler*, 33–35.

20. Lowenthal, *George Perkins Marsh*, esp. 267–312. On Marsh's pioneering of the watershed concept, see Worster, *Shrinking the Earth*, 77–89, 235n1. As Worster explains, Marsh's concept of watershed referred to "a bounded, limited area of land within which living things were linked and interdependent." In his tenure as the USGS director (1881 to 1895), John Wesley Powell would further refine the concept of watersheds, describing them as "hydrological basins." For Shaler's affinity for Marsh's ideas and message, see, for instance, Livingstone, "Nature and Man in America," 371; Livingstone, *Nathaniel Southgate Shaler*, 195–96; and Koelsch, "Legendary 'Rediscovery,'" 510–11.

21. Shaler, *Geological Survey of Kentucky*, 241.

22. Shaler, *Geological Survey of Kentucky*, 66–67, 79–80. While Marsh, Shaler, and many other opponents of clear-cutting subscribed to the relationship between forest cover and stream flow, some scientists by the beginning of the Progressive Era rejected the idea. Dodds, "Stream-Flow Controversy."

23. Worster, *Shrinking the Earth*, 85; and Shaler, *Geological Survey of Kentucky*, 66–68, quote at 68. In 1883, Shaler expounded on his views about flooding and flood control in "The Floods of the Mississippi Valley." Ellet's report was one of two War Department studies commissioned by Congress in 1850 after devastating floods submerged the Mississippi River Valley for a second consecutive year, inundating 27 million acres. His report championed French water management theories, especially the construction of reservoirs on the Mississippi's and Ohio's upstream tributaries to minimize downstream flooding. A rival study, coauthored by US Army topographers Andrew Atkinson Humphreys and Henry L. Abbot, appeared in 1861. Humphreys and Abbot recoiled at Ellet's reservoir proposal; in its place, they promoted the continuous construction of levees alongside the Mississippi, which would intensify the river's velocity and thereby "scour" and deepen its channel, lessening seasonal overflows. In tying the issue of flooding to nothing more than water volume, Humphreys and Abbot helped birth the Corps of Engineers' "levees only" policy and severed the relationship of riparian wetlands to streamflow. See, for instance, Shallat, *Structures in the Stream*, 104–5, 174–76; Morris, *Big Muddy*, 141–47; and Doyle, *The Source*, 69–77.

24. Shaler, *Geological Survey of Kentucky*, 58.

25. Shaler, *Geological Survey of Kentucky*, 69, 72.

26. Shaler, *Geological Survey of Kentucky*, 78, 102, 306–7, 399–400.

27. Rabbitt, *Brief History*, 5–6; White, *"It's Your Misfortune,"* 128–135; Goetzmann, *Army Exploration*, 432.

28. Rabbitt, *Brief History*, 3, 6–7; Worster, *River Running West*, 360–70.

29. Powell's life and military and scientific career have attracted significant scholarly attention: see Darrah, *Powell of the Colorado*; Stegner, *Beyond the Hundredth Meridian*; Worster, *River Running West*; and Ross, *Promise of the Grand Canyon*.

30. White, *"It's Your Misfortune,"* 58; Worster, *River Running West*, 416; Ross, *Promise of the Grand Canyon*, 281–82; Stegner, *Beyond the Hundredth Meridian*, 272–73, 272 (quote).

31. John Wesley Powell to Nathaniel Southgate Shaler, March 8, 1884, "Papers Relating Chiefly to Work for US Geological Survey, 1884–98," box 1, Nathaniel Southgate Shaler Papers (hereafter Shaler Papers), Harvard University Archives, Cambridge, MA; and Livingstone, *Nathaniel Southgate Shaler*, 38. Shaler had also recently published on the characteristics,

habitat, and geographic distribution of the American swamp cypress: "The American Swamp Cypress." For an overview of Shaler's USGS tenure, consult Berg, "Nathaniel Southgate Shaler," 83–143.

32. Shaler, "Preliminary Report," 362.

33. Shaler, "Preliminary Report," 361. One wetland ecologist argues that Shaler's 1890 USGS report ("General Account") represented "the nation's first national wetland classification system." Although the more detailed and sophisticated 1890 system had its roots in the report described in the above text, the 1885 three-tier system marked the government's first true system for classifying wetlands. Tiner, *Wetland Indicators*, 3.

34. Shaler, "Preliminary Report," 380–81, 381 (quote); and Shaler, "Report of Prof. N. S. Shaler," in *Sixth Annual Report*, 21.

35. Shaler, "Preliminary Report," 378, 380, 390–98.

36. Shaler, "Fluviatile Swamps of New England"; *Selma (CA) Enterprise*, October 19, 1899. In 1886, newspapers across the country praised Shaler's USGS investigations for disclosing that 50,000 square miles of swamps and inundated lands east of the Mississippi were available for drainage. See *Daily Republican* (Monongahela, PA), May 21, 1886.

37. Shaler to Powell, June 17, 1885, RG 57, Records of the Geological Survey, National Archives Microfilm Publications microcopy no. 590, Letters Received by the United States Geological Survey, 1879–1901, "May 27, 1885–July 31, 1885," reel 28, NARA, College Park, Maryland (hereafter CP); and Worster, *River Running West*, 362. Despite his celebrated and much-studied involvement with Western irrigation, Powell was an early and enthusiastic convert to drainage's economic promise. In 1878, three years before he assumed the USGS directorship, he touted the "millions of acres" that could be "redeemed" by draining Florida's wetlands, tidelands on the Gulf Coast, alluvial lands in Southern river valleys, and swamps and small lakes in the Great Lakes region. House of Representatives, *Surveys of the Territories*, 45th Cong., 3rd Sess., December 3, 1878, H. Mis. Doc. 5, esp. 20.

38. For a description of Shaler's team, see Shaler, "Inundated Lands," RG 57, microcopy no. 590, Letters Received by the United States Geological Survey, 1879–1901, "January 2, 1886–February 23, 1886," reel 31.

39. Powell to Shaler, August 6, 1886, July 2, 1888, box 1, Shaler Papers. The surviving letters from Cobb's correspondence offer a goldmine of firsthand settler impressions about swamps. RG 57, Data Collected by Collier Cobb concerning Swampland, Timber and Drainage, 1887, entry 162, box 1, folder "Collier Cobb Entry 69."

40. On the Powell irrigation survey and its political consequences, see Manning, *Government in Science*, 168–203; Pisani, *To Reclaim a Divided West*, 127–68; Worster, *Unsettled Country*, 1–30; and Ross, *Promise of the Grand Canyon*, 300–325.

41. Shaler, "General Account," 255–339.

42. Shaler, "General Account," 262–63, quote at 262.

43. Shaler, "General Account," 263.

44. Shaler, "General Account," 263–64; and Shaler, "Swamps of the United States," 233.

45. Shaler, "General Account," 303.

46. Shaler, "General Account," 303; and "Woods and Character," *Boston Globe*, March 25, 1894.

47. Shaler, "General Account," 303–4; and Soper and Osbon, "Occurrence and Uses of Peat," 3.

48. Shaler, "General Account," 304–5.
49. Shaler, "General Account," 305–6.
50. Shaler, "General Account," 307.
51. Shaler, "General Account," 308. On George Perkins Marsh's eager support for swamp drainage, especially in Italy's Maremma coastal region, see Hall, "Provincial Nature," 196; and Hall, "Restoring the Countryside," 96.
52. Shaler, "General Account," 310.
53. Livingstone, "Geologist by Profession," 151. Although in the 1890s Shaler still gave an occasional paper on geomorphological subjects to the Geological Society of America, he could not keep pace with new research in the field of geology during the last fifteen years of his life as he shouldered increasing administrative duties at Harvard. Livingstone, *Nathaniel Southgate Shaler*, 44–45, 350–53.
54. Adams, "World Conquerors," 191. For the contours and broader development of Anglo-Saxonism, see Horsman, *Race and Manifest Destiny*; Bederman, *Manliness and Civilization*; Jacobson, *Barbarian Virtues*; Kramer, "Empires, Exceptions, and Anglo-Saxons"; and Tuffnell, *Made in Britain*.
55. Shaler, "Summing Up of the Story," 613–14.
56. Cadle, *Mediating Nation*, 135; Livingstone, *Nathaniel Southgate Shaler*, 122–23; and Shaler, "European Peasants as Immigrants," 647.
57. Shaler, "European Peasants as Immigrants," 647.
58. Shaler, "European Peasants as Immigrants," 647; Haller, "Nathaniel Southgate Shaler," 186; and Livingstone, "Science and Society," 192.
59. Shaler, "European Peasants as Immigrants," 649, 650.
60. Shaler, "European Peasants as Immigrants," 650, 651, 652.
61. Shaler, "European Peasants as Immigrants," 652–53.
62. LeMay, *Guarding the Gates*, 97–98; Higham, *Strangers in the Land*, 141; Adams, "World Conquerors," 209–10; and Livingstone, *Nathaniel Southgate Shaler*, 153–55.
63. Shaler, *Nature and Man in America*, 229.
64. Shaler, *Nature and Man in America*, 227, 229.
65. Shaler, "Future of the Negro," 150. For a racial and epidemiological contextualization of Shaler's proposal, see Anderson, "Immunities of Empire," 94, 118.

Chapter Six

1. Steenerson to Charles D. Walcott, January 3, 1906, RG 115, entry 3, box 97, folder 110-G; Meyer, "Red Lake Ojibwe"; and Meyer, *White Earth Tragedy*, 51–57, 198. On the federal government's allotment policy, see Hoxie, *Final Promise*; and Carlson, *Indians, Bureaucrats, and Land*. Throughout the remainder of the book, I parenthesize the term "ceded," which permeated the correspondence of federal policymakers and conservationists involved in the drainage movement. The parentheses highlight that many Indigenous land cessions were made as part of an unequal power relationship between Indigenous communities and the federal government. Even if tribes voluntarily signed formal cession agreements, they seldom constituted a legitimate agreement between equals with the federal government negotiating from a position of good faith and concern for the future welfare of tribal communities.

2. House Subcommittee of the Committee on Indian Affairs, *Drainage of Certain Lands*, 4, 6–7, quote at 6.

3. Steenerson to Walcott, January 3, 1906, RG 115, entry 3, box 97, folder 110-G; and "Drainage Bill," *New Ulm (MN) Review*, January 10, 1906. For biographical details on Steenerson, see *Compendium of History and Biography of Central and Northern Minnesota*, 143–44; Holcombe and Bingham, *Compendium of History and Biography*, 263–66, 344–45; and "Steenerson, Halvor," *Biographical Directory of the United States Congress*, n.d., accessed March 4, 2016, http://bioguide.congress.gov/scripts/biodisplay.pl?index=S000842. On the origins and early growth of the federal reclamation program, see Worster, *Rivers of Empire*; Reisner, *Cadillac Desert*; Pisani, *To Reclaim A Divided West*; Pisani, *Water and American Government*; Rowley, *Bureau of Reclamation*; and Billington and Jackson, *Big Dams*.

4. The historical literature on the origins, political dimensions, and statist implications of the Progressive Era conservation movement is vast. The classic study remains Hays, *Conservation and the Gospel of Efficiency*. See also Richardson, *Politics of Conservation*; Swain, *Federal Conservation Policy*; Pisani, "Many Faces of Conservation"; Schulman, "Governing Nature"; Tyrrell, *Crisis of the Wasteful Nation*; Taylor, *American Conservation Movement*; and Johnson, *Escaping the Dark, Gray City*.

5. On Newell's and the Reclamation Service's bureaucratic aggrandizement, a form of "water imperialism" that sought the consolidation of all federal water resources responsibilities in the Reclamation Service, see Carlson, "Forging Transcontinental Alliances," 141.

6. Despite the many useful historical and political studies of wetlands, the turbulent rivalry between the Reclamation Service and the USDA over drainage remains uncharted scholarly waters. See Vileisis, *Discovering the Unknown Landscape*; Prince, *Wetlands of the American Midwest*; Gaddie and Regens, *Regulating Wetlands Protection*; Stine, *America's Forested Wetlands*; and Garone, *Fall and Rise*. Three works that situate drainage within the framework of state-sponsored conservation in the early 1900s include Carlson, "Other Kind of Reclamation"; Meyer, *Progressive Environmental Prometheans*, particularly chap. 2; and Nygren, *State of Conservation*.

7. Quoted in Fite, *The Farmers' Frontier*, 135; and Pisani, *To Reclaim a Divided West*, 273–325.

8. Pisani, *To Reclaim a Divided West*, 289–90; Pisani, "George Maxwell"; and Bennett and Kohl, *Settling the Canadian-American West*, 20.

9. Pisani, *Water and American Government*, 1.

10. Pisani, "Tale of Two Commissioners," 637–43. Another excellent study of Newell's federal career is Jackson, "Engineering in the Progressive Era."

11. Pisani, *Water and American Government*, 2–4, 30. For a list of the initial reclamation projects, see Rowley, *Bureau of Reclamation*, 130.

12. Emphasis added. Pisani, "Many Faces of Conservation," 125; and Pisani, "Tale of Two Commissioners," 639.

13. House of Representatives, *Hearings Before the Committee on Irrigation of Arid Lands*, 58th Cong., 3rd Sess., March 2, 1905, H. Doc. 381, 44.

14. "Newell Visits Sacramento," *Gridley (CA) Herald*, February 9, 1906.

15. A Bill Appropriating the Receipts from the Sale of Public Lands in the State of Minnesota to the Construction of Drainage Works for the Reclamation of Swamp and Overflowed

Lands, H.R. 10062, 59th Cong., 1st Sess., January 4, 1906; and "Bill for Swamp Land Drainage," *Duluth News-Tribune*, January 5, 1906.

16. Walcott to Steenerson, February 23, 1906, RG 115, entry 3, box 97, folder 110-G. Walcott estimated that Minnesota had received only 4.5 million of the 70 million acres ceded under the Swamp Land Acts.

17. "Hansbrough Roasts Newell," *Grand Forks Herald*, September 25, 1904; Webster Ballinger, "Funds for Drainage," *Grand Forks Herald*, February 11, 1905; "A Drainage Convention," *Grand Forks Herald*, November 21, 1905; and "A Drainage Talk," *Grand Forks Herald*, January 12, 1906.

18. Louis W. Hill to J. W. Blabon, August 22, 1904, PSF, file 4013, GN Records; "A Drainage Convention"; "A Drainage Talk," *Grand Forks Herald*, January 12, 1906; Stewart, *Report on the Drainage*, 7; Providing for the Segregation of One Million Dollars from the Reclamation Fund Created by the Act of June Seventeenth, Nineteen Hundred and Two, and for other Purposes, S. 3687, 59th Cong., 1st Sess., January 25, 1906; Providing for the Segregation of One Million Dollars from the Reclamation Fund Created by the Act of June Seventeenth, Nineteen Hundred and Two, and for other Purposes, H. R. 13196, 59th Cong., 1st Sess., January 26, 1906. For the Reclamation Service's pessimism on North Dakota's irrigation potential, see Hafermehl, "To Make the Desert Bloom," 20. On James J. and Louis Hill's rocky relationship with the Reclamation Service, see Carlson, "'There May Be Bloodshed,'" 402–3.

19. A Bill Providing for the Use of Three Million Dollars of the Money that would Otherwise become a Part of the Reclamation Fund for the Drainage of Certain Lands in North Carolina and Virginia, and for other Purposes, H. R. 16804, 59th Cong., 1st Sess., March 15, 1906; "To Drain Dismal Swamp," *Washington Post*, March 16, 1906; and "To Redeem Rice Lands," *State (Columbia, SC)*, May 30, 1906. On the overwhelming Southern support for the Reclamation Act, see Pisani, *To Reclaim a Divided West*, 319.

20. "Irrigation May Get Blow," *Evening Statesman* (Walla Walla, WA), February 26, 1906; "Funds to Drain Dakota Swamps," *Idaho Statesman* (Boise), March 23, 1906; and "Reclamation Fund Threatened," *Forestry and Irrigation*, 112–13. On Westerners' varied responses to federal conservation measures, see Hays, *Conservation and the Gospel of Efficiency*, 256–57; White, *"It's Your Misfortune,"* 407–9; and Smith, *Conservation Constitution*, 106-17.

21. "Reclamation Fund Exhausted," *Irrigation Age*, 180–81; Walcott to Hitchcock, February 25, 1906, RG 115, entry 3, box 148, folder 286-3, "House Bills—59th Congress—1st Session"; and Pisani, *Water and American Government*, 3–4.

22. Steenerson introduced two similar bills within twelve days: A Bill Appropriating the Receipts from the Sale and Disposal of Public Lands in Certain States to the Construction of Works for the Drainage or Reclamation of Swamp and Overflowed Lands, H. R. 16007, 59th Cong., 1st Sess., March 1, 1906; and A Bill Appropriating the Receipts from the Sale and Disposal of Public Lands in Certain States to the Construction of Works for the Drainage or Reclamation of Swamp and Overflowed Lands, H. R. 16550, 59th Cong., 1st Sess., March 12, 1906. The legislation stated that public land sales in Alabama, Arkansas, Florida, Illinois, Indiana, Iowa, Louisiana, Michigan, Minnesota, Mississippi, Missouri, Ohio, and Wisconsin would finance the drainage fund.

23. Guy Elliott Mitchell to George H. Maxwell, March 9, 1906, Maxwell to Walcott, March 13, 1906, and Walcott to Maxwell, March 19, 1906, box 2, Charles D. Walcott Collec-

tion, Smithsonian Institution Archives, Washington, DC; and Mitchell, "Land Reclamation by Drainage."

24. Trachtenberg, *Incorporation of America*, 87–88; and Kline, *First Along the River*, 60.

25. Frederick H. Newell, "Drainage," undated memorandum, RG 115, entry 3, box 207, folder 676-2, "National Drainage Assn. Meetings."

26. Guy Elliott Mitchell, "To Farm the Swamps," March 25, 1906, *Minneapolis Star Tribune*.

27. Newell, "Drainage," undated memorandum, RG 115, entry 3, box 207, folder 676-2. On the gendered, social, and cultural underpinnings of Progressive Era land reclamation, see Lovett, *Conceiving the Future*, 67. See also Autobee, "Every Child in a Garden"; Lovett, "Land Reclamation as Family Reclamation"; and Lovett, "'Rooted in the Soil.'" The symbolic and political significance of the *home* in the Gilded Age and beyond is a consistent theme in White, *Republic for Which It Stands*, esp. 5, 31, 136–71.

28. House Committee on the Public Lands, *Drainage in North Dakota*, 59th Cong., 1st Sess., June 14, 1906, H. Rep. 4929, 1–2; and House Committee on the Public Lands, *Drainage of Dismal Swamp in North Carolina and Virginia*, 59th Cong., 1st Sess., June 22, 1906, H. Rep. 4994, 1–4. For Cannon's opposition to federalized drainage, see "Conversion of Speaker Cannon," *Idaho Statesman*, July 10, 1906.

29. "Drainage in the South," *Watchman and Southron* (Sumter, SC), June 27, 1906; *CR*, Senate, 59th Cong., 1st sess., June 26, 1906, 9247–49. Lever's joint resolution is described in Elwood Mead to A. F. Lever, January 12, 1907, RG 8, entry 6, box 19, folder "July 1906–July 1907," NARA, KC.

30. *CR*, Senate, 59th Cong., 1st Sess., June 26, 1906, 9248.

31. *CR*, Senate, 59th Cong., 1st Sess., June 26, 1906, 9248. Latimer's resolution was ultimately referred to the Committee on Agriculture and Forestry and not acted on.

32. Thoburn, *First Biennial Report*, 29.

33. "For a Drainage Act," *Dallas Morning News*, August 28, 1906; and "About the Drainage Question," *Chandler (OK) News*, October 11, 1906.

34. Circular, "First Annual Session, National Drainage Congress," RG 115, entry 3, box 207, folder 676-2, "National Drainage Assn. Meetings."

35. Arthur Powell Davis to Newell, December 11, 1906, RG 115, entry 3, box 207, file 676-2, "National Drainage Assn. Meetings"; B. Campbell to Louis W. Hill, December 15, 1906, PSF, file 4013, GN Records; "New Drainage Association," *Duluth News-Tribune*, December 12, 1906; and "First National Drainage Congress Will Convene in Oklahoma City This Morning," *Daily Oklahoman* (Oklahoma City), December 5, 1906.

36. "Flint, Frank Putnam," *Biographical Directory of the United States Congress*, n.d., accessed March 4, 2016, http://bioguide.congress.gov/scripts/biodisplay.pl?index=F000207.

37. Kelley, *Battling the Inland Sea*, 4–7; and Thompson, "Historic Flooding in the Sacramento Valley." On the Sacramento River Valley's role in Progressive Era water politics, see Carlson, "Forging Transcontinental Alliances"; and O'Neill, *Rivers by Design*, chap. 8.

38. On the multipurpose concept, see Pisani, "Conservation Myth." For the USGS's and Reclamation Service's identification of the Sacramento Valley as a laboratory for multipurpose conservation, see Newell to George Pardee, December 30, 1904, January 23, 1905, February 8, 1905, Newell folder, George Pardee Collection, Bancroft Library, Berkeley, CA; Lippincott, "General Outlook for Reclamation Work"; "Reclamation Scheme," *Woodland*

(CA) Daily Democrat, January 21, 1905; Carlson, "Forging Transcontinental Alliances"; and Pisani, *From the Family Farm*, 325–34.

39. A Bill Providing for the reclamation of lands in the Sacramento and San Joaquin valleys in the State of California, S. 5376, 59th Cong., 1st Sess., March 27, 1906. In addition to his public remarks, Flint privately groused to Secretary of Agriculture James Wilson and Mead about the USDA paying far too little attention to agricultural conditions in California and requested an accounting of the department's in-state activities. Wilson to Flint, July 7, 1906, RG 16, Records of the Office of the Secretary of Agriculture, microcopy no. 440, vol. 1, "June 1–November 21, 1906," reel 58, NARA, CP.

40. C. E. Grunsky, "State Aid to Irrigation, Drainage and Flood Control Enterprises," June 1918, box 18, folder "Reports on Agriculture, Irrigation," 5, Elwood Mead Papers, Bancroft Library; and "Memorandum By C. E. Grunsky. Senate Bill No. 5376 introduced by Mr. Flint," April 1906, RG 115, entry 3, box 153, folder 287-4, "Senate Bills, 59th Congress, 1st Session." Grunsky served with the Reclamation Service until 1907. On his career and legacy, see Detwiler, *Who's Who in California*, 282; and Kelley, *Battling the Inland Sea*, 190–91, 238–41.

41. Flint to Walcott, April 4, 1906, box 1, Walcott Collection; and Hitchcock to Walcott, April 16, 1906, RG 115, entry 3, box 153, folder 287-4, "Senate Bills, 59th Congress, 1st Session." Despite President Roosevelt persuading Congress to create a temporary Inland Waterways Commission in May 1907, lawmakers and other natural resources agencies had cooled to the multipurpose concept by that time. In the turf-conscious, zero-sum culture of Progressive Era water politics, the Reclamation Service and Forest Service, to quote historian Donald J. Pisani, branded the IWC as a "threat" and only lent "lukewarm support." From the perspective of the Department of the Interior, Pisani further argues that the multipurpose concept resembled little more than an "advertising technique" as well as a "weapon" to ostracize the US Army Corps of Engineers in federal water planning. Pisani, "Water Planning in the Progressive Era," 391, 403–4, 410–11; and Pisani, "Conservation Myth," quotes at 155.

42. H. C. Rizer to Flint, November 28, 1906, Flint to Rizer, November 28, 1906, Newell to Flint, December 1, 1906, RG 115, entry 3, box 97, folder 110-G; and A Bill For the establishment of a drainage fund and the construction of works for the reclamation of swamp and overflowed lands, S. 6626, 59th Cong., 2nd Sess., December 5, 1906. The public land states not covered under the 1902 law included Alabama, Arkansas, Florida, Illinois, Indiana, Iowa, Louisiana, Michigan, Minnesota, Mississippi, Missouri, Ohio, and Wisconsin.

43. A. P. Davis to Newell, December 11, 1906, RG 115, box 207, file 676-2, "National Drainage Assn. Meetings"; and "Drainage Problems being Considered at Conference," December 6, 1906, and "Ask Help of U.S.," December 7, 1906, *Daily Oklahoman*.

44. Kansas v. Colorado, 206 U.S. 46 (1907).

45. For the case's background and assortment of legal implications, see Smith, *Conservation Constitution*, 141–42, 144–45, 222, 234–35, 251–52; Brodhead, *David J. Brewer*, 162–64; Littlefield, *Conflict on the Rio Grande*, 73–74, 97–99; Pisani, "State vs. Nation"; and Sherow, *Watering the Valley*, 6, 103–19.

46. Kansas v. Colorado, 206 U.S. 46 (1907), quotes at 87, 89, and 92; United States v. Rio Grande Dam and Irrigation Company, 174 U.S. 690 (1899).

47. Wright, "Swamp and Overflowed Lands," 8; Bien, "Memorandum," February 1, 1908, RG 115, entry 3, box 97, folder 110-G.

48. "The Swamp Lands Question," *Washington Post*, September 2, 1907; and "Drainage of Swamp Lands," undated memorandum, RG 57, Topographical Division, General Administrative Files, 1890–1948, entry 146, box 5, folder 20.

49. W. A. Richards to Hitchcock, March 8, 1906, RG 115, entry 3, box 148, folder 286–3, "House Bills, 59th Congress, 1st Session"; and Richards to Hitchcock, January 28, 1907, RG 49, Records of the Bureau of Land Management, entry 673, "Press Copies of Letters Sent Concerning Swamplands," Letter Press Books, book 213 (January 2, 1907 to April 30, 1907), NARA, Washington, DC.

50. Wilson, "Reclaiming the Swamp Lands"; Mitchell, "The National Drainage Problem"; Wilson, "Swamp Lands and Their Reclamation"; and Mitchell, "To Farm America's Swamps."

51. Mitchell, "To Farm America's Swamps," 438.

52. R. P. Teele to Fortier, December 6, 1907, RG 8, series "Correspondence between Berkeley and Washington, D.C.," box 33, folder "November–December, 1907," NARA, San Francisco, CA (hereafter SF); "Marylanders Honored," *Baltimore Sun*, November 27, 1907; and "Drainage Legislation," *Galveston Daily News*, December 2, 1907.

53. "Washington Gossip," *Charlotte Observer*, December 24, 1907; and Napoleon B. Broward to J. A. Dapray, box 6, folder "January 1908," Napoleon Bonaparte Broward Papers, George A. Smathers Library, Gainesville, FL (hereafter Broward Papers). Elected as governor in 1905, Broward convinced the Florida legislature to create a special "Everglades drainage district" with the power to impose acreage assessments on landowners. The legislature passed Broward's plan, but it quickly came under assault. As the Everglades drainage district's largest ratepayers, Florida railroad corporations challenged its constitutionality. On April 6, 1907, federal judge James W. Locke, siding with the railroads, ruled that the district's taxing power was unconstitutional. Locke's ruling forced Broward to promote federalized drainage as the next best alternative for draining the Everglades, which remained the country's largest undrained wetlands ecosystem. See McCally, *The Everglades*, 92.

54. This is my own individual count.

55. In addition to Garfield, Newell, and Morris Bien, the attendees included Senators Moses E. Clapp (R-MN), Flint, Latimer, Newlands, and Furnifold M. Simmons (D-NC). The House members included Steenerson, John Humphrey Small (D-NC), Hannibal L. Godwin (D-NC), Robert B. Macon (D-AR), Joseph J. Russell (D-MO), Charles R. Thomas (D-NC), and Robert Wallace (D-AR). Newell to James R. Garfield, January 21, 1908, RG 48, Records of the Office of the Secretary of the Interior, Central Classified Files, 1907–36, entry 749, box 734, file 2–9, NARA, CP; and Newell to Latimer, January 21, 1908, "Memorandum of meeting held in the office of the Secretary of the Interior at 10 a.m., January 25, 1908," "Memorandum on conference held in the office of the Secretary of the Interior, 10 a.m., Saturday, February 1, 1908," and "Memorandum," February 1, 1908, RG 115, entry 3, box 97, folder 110-G. Though Morris Bien publicly told Flint that his bill met "the various difficult conditions in which this subject [drainage] is involved," he harbored substantial doubts about federal drainage's constitutionality. Bien to Flint, February 4, 1908, in folder 110-G. Senate Bill 4855 was Flint's legislation.

56. All of the quotations are taken from a memorandum dated January 25, 1908. The memorandum begins with the phrase, "In accordance with a generally expressed..." RG 115, entry 3, box 97, folder 110-G.

57. Senate, *Drainage or Reclamation of Swamp or Overflowed Lands*, 60th Cong., 1st Sess., February 24, 1908, S. Rep. 289, 1–2.

58. Asbury Latimer to Broward, January 30, 1908, Box 6, folder "January 1908"; and Latimer to Broward, February 4, 1908, box 6, folder "February 1908," Broward Papers.

59. The debates, quotes, and vote on Flint's bill are in *CR*, Senate, 60th Congress, 1st Sess., April 15, 1908, 4769–74, April 17, 1908, 4859–66, and April 20, 1908, 4970–4971.

60. John Sharp Williams to Walter L. Fisher, Juley 24, 1912, RG 48, Legislation Files, 1905–36, entry 753, box 83, folder, "General Land Office Swamp Lands Legislation."

61. House Committee on Public Lands, *Reclamation of Certain Lands in Minnesota*, 59th Cong., 2nd Sess., January 28, 1907, H. Doc. 607, 1–2, quote on 1.

62. "Got Favorable Action from Interior Dept.," December 26, 1907, *Bemidji (MN) Daily Pioneer*; and House Committee on Public Lands, *Drainage Survey of Certain Lands in Minnesota*, 61st Cong., 1st Sess., May 13, 1909, H. Doc. 27, 2.

63. Quoted in Righter, *Battle Over Hetch Hetchy*, 122.

64. Steenerson to Bien, January 9, 1908, RG 115, entry 3, box 97, folder 110-G; and A Bill to Authorize the Drainage of Certain Lands in the State of Minnesota, H.R. 19541, 60th Cong., 1st Sess., March 19, 1908.

65. The debates, quotes, and votes on Volstead's bill are in *CR*, House, 60th Cong., 1st Sess., April 20, 1908, 4988–92, April 21, 1908, 5023–24, and May 11, 1908, 6104–08; and Senate, 60th Cong., 1st Sess., April 25, 1908, 5236–37, May 12, 1908, 6123–24.

66. House of Representatives, *Drainage of Certain Lands in Minnesota*, 64th Cong., 1st Sess., July 25, 1916, H. Rep. 1039, 1.

Chapter Seven

1. Elliott to A. D. Crooks, November 13, 1909; and Elliott to P. P. Campbell, November 19, 1909, both reprinted in "The Drainage Proposition," *Neodesha (KS) Daily Sun*, December 6, 1909. The contours of the 1905 Kansas law authorizing the formation of drainage districts were described in "The Drainage District Law," *Fredonia (KS) Daily Herald*, December 22, 1908.

2. F. M. Robertson, "Overflow and Drainage," *Fredonia Daily Herald*, January 14, 1909; and "Drainage Work Becomes Imperative," *Fredonia Daily Herald*, January 6, 1909.

3. "Petition for Benedict Drainage District Filed," *Fredonia Daily Herald*, January 16, 1909; "Publication Notice," *Wilson County Citizen (Fredonia, KS)*, January 22, 1909; "Drainage Petition Did Not Conform," *Neodesha Daily Sun*, February 2, 1909; and "South-west Webster," *Fredonia Daily Herald*, March 11, 1909.

4. "South-west Webster."

5. For a concise description of the ODI's personnel, structure, and slate of publications, see "Organization, Work, and Publications of Drainage Investigations." On the descriptive title of "consulting advisors," see Elwood Mead to Stephen Milancthon Sparkman, June 2, 1906, reprinted in *The Sun* (Jacksonville, FL), June 30, 1906.

6. Elliott to Campbell, November 19, 1909; "To Investigate River Drainage," *Neodesha Daily Sun*, January 8, 1910; "Drainage an Easy Proposition," *Neodesha Daily Sun*, January 10, 1910; "Drainage Question Up," *Fredonia Daily Herald*, January 13, 1910; and "Commissioners Act on Drainage," *Neodesha Daily Sun*, January 14, 1910.

7. "Drainage Investigation," *Neodesha Register*, March 18, 1910.
8. *Wilson County Citizen*, July 14, 1911.
9. Hamilton, "Building the Associative State," 215, 218.
10. Elliott to R. W. McAlister, June 12, 1908, RG 8, entry 6, box 21, folder "Jan 1908–July 1908," NARA, KC.
11. "To Straighten the Neosho," *Fredonia Daily Herald*, April 17, 1906.
12. "Irrigation and Drainage," *Fourteenth Census*, 365, 371.
13. Zavodnyik, *Rise of the Federal Colossus*, 30.
14. Zavodnyik, *Rise of the Federal Colossus*, 145; and Carpenter, *Forging of Bureaucratic Autonomy*, 188, 201–13, 217, quotes at 212.
15. Kluger, *Turning On Water*, 27–40; and Pisani, *To Reclaim a Divided West*, 304–10.
16. Maxwell to Lippincott, February 8, 1903, RG 41, George H. Maxwell Papers, Louisiana State Museum Historical Center, New Orleans, Louisiana (hereafter Maxwell Papers); Kluger, *Turning On Water*, 29–30; Pisani, *To Reclaim a Divided West*, 306–10; and Pisani, *From the Family Farm*, 335–50.
17. Maxwell to Lippincott, February 8, 1903, RG 41, box 8, Maxwell Papers; Pisani, *From the Family Farm*, 345–50; and Dunbar, *Forging New Rights*, 126–28. On the USDA–Department of the Interior rivalry, see Pisani, *Water and American Government*, 237–38; and Soffar, "Differing Views."
18. "Annual Report of Irrigation and Drainage Investigations, 1904," 19; and Mead to R. P. Teele, April 5, 1904, RG 8, Correspondence of R. P. Teele, 1904–9, entry 3, box 19, folder "Untitled," NARA, KC. On the irrigation origins of drainage as a federal activity, see Epperson, *Draining the Swamp*, 31.
19. Thompson, *Wetlands Drainage*, 17, 47; Hurt, *American Agriculture*, 221–22; and Macleod, "Food Prices, Politics, and Policy," 383–84.
20. "C. G. Elliott '77, Drainage Expert, Is Dead in East," *Daily Illini* (Urbana), September 28, 1926. For brief biographical sketches of Elliott, see Thompson, *Wetlands Drainage*, 46–47; and Hager, *Hydraulicians in the USA*, 2010.
21. C. G. Elliott, "Vacation Journal, 1875," September 15, 1875, Vacation Journals, 1876–78, 11/5/31, 10, University of Illinois Archives, Urbana, Illinois.
22. Elliott, "Vacation Journal," 3–5, 12.
23. Elliott, *Practical Farm Drainage*; and Windsor, "Artificial Drainage," 107, 270. Media accounts of Elliott's speeches routinely praised them as "very able" and "plainly" delivered. For one such example, see "Engineering Notes," *Daily Illini*, February 15, 1892.
24. Elliott, "Farm Drainage"; and Elliott, "Drainage of Farm Lands." After publishing the *Drainage Journal* for only one year, Elliott sold it to *Irrigation Age* and fixed his attention on making the transition from private practice to government service. See "A Change of Ownership," 145; and Elliott to P. Lombillo Clark, March 25, 1907, RG 8, entry 6, box 6, folder "July 1906–May 1909," NARA, KC.
25. I am indebted to Bob Morrissey for assisting me in navigating the University of Illinois's Special Collections and digital databases. For the 1880s and 1890s, I identified around a handful of theses dedicated to agricultural drainage. On the civil engineering profession and land reclamation more broadly during the Progressive Era, see Meindl, Alderman, and Waylen, "Environmental Claims-Making"; Teisch, *Engineering Nature*; and Jackson, "Engineer as Lobbyist."

26. For a list of qualifications for an agricultural drainage engineer, see Elliott, *Engineering for Land Drainage*, 12–14, quote at 45.

27. Elliott, *Engineering for Land Drainage*, 14–15.

28. Elliott to Louis C. Bulkeley, March 4, 1910, RG 8, entry 6, box 4, folder "Oct 16, 1909–Dec 22, 1910," NARA, KC; and "Elliott is Swamp Doctor," *Washington Herald*, September 15, 2010. During the Progressive Era and beyond, government entomologists, especially John Bernhard Smith, implored state health departments and local communities to eradicate mosquito breeding habitats by draining swamps, introducing mosquito-devouring fish, or oiling wetlands. See Patterson, *Mosquito Crusades*, 19–57; and McWilliams, *American Pests*, 121–25. For case studies of anti-mosquito hysteria and control in Florida, see Patterson, *Mosquito Wars*; and Patterson, "Trials and Tribulations of Amos Quito." For the extension of mosquito suppression into the Panama Canal zone, see Sutter, "Nature's Agents?"; and Sutter, "'The First Mountain.'"

29. "Drainage Investigations," no date, RG 8, entry 1, Orders and Circulars, 1903–13, box 1, NARA, KC; For the OIDI's first annual drainage report, see Elliott, "Report on Drainage Investigations, 1903."

30. Elliott, "Report on Drainage Investigations, 1903," 5.

31. Elliott, "Report on Drainage Investigations, 1903," 5–6.

32. Elliott to F. Woodard, July 10, 1906, RG 8, entry 6, box 34, folder "July 1906–Sept. 1907," NARA, KC; and Mead to Sparkman, June 2, 1906, reprinted in *The Sun* (Jacksonville, FL), June 30, 1906. See also House of Representatives, *Annual Report of the Office of Experiment Stations for the Year Ended June 30, 1905*, 59th Cong., 1st Sess., H. Doc. 924, 35.

33. *Annual Report of the Office of Experiment Stations for the Year Ended June 30, 1905*, 35.

34. L. G. Hardman to Mead, July 2, 1906, Mead to Hardman, July 5, 1906, and Elliott to Hardman July 7, 1906, RG 8, entry 6, box 13, folder "July 1906–June 1907," NARA, KC.

35. "Northwest Experiment Farm," 3–4.

36. Mead to William Robertson, June 26, 1906, and Robertson to Mead, July 25, 1906, RG 8, entry 12, Agreements and Contracts, 1906–12, box 1, folder "Cooperative Agreements and Contracts, 1906–1912," NARA, KC; and "Northwest Experiment Farm," 4, 6, 81–82.

37. Sutter, *Let Us Now Praise*, 38, 40–44, 46, 53–54; and Stalcup, "Public Interest, Private Lands," 53–54, 69–72.

38. Elliott, "Report on Drainage Investigations, 1903," 60.

39. Elliott, "Report on Drainage Investigations, 1903," 61.

40. Elliott, "Report on Drainage Investigations, 1903," 61.

41. Milton Whitney and A. C. True to Wilson, November 26, 1906, RG 8, series "Correspondence between Berkeley and Washington, D.C.," box 32, folder "July–December, 1906," NARA, SF.

42. True to [anonymous], October 9, 1907, Teele to Samuel Fortier, October 17, 1907, RG 8, series "Correspondence between Berkeley and Washington, D.C.," box 33, folder "July–October, 1907," NARA, SF; and House of Representatives, *Annual Report of the Office of Experiment Stations for the Year Ended June 30, 1907*, 60th Cong., 1st Sess., H. Doc. 984, 33. On the growth of federal drainage expenditures, see Purcell, "Plumb Lines, Politics, and Projections," 167. For Mead's rich post-OIDI career, see Kluger, *Turning On Water*, 57–149; Rodgers, *Atlantic Crossings*, 345–53; and Rook, "An American in Palestine."

43. Arthur E. Morgan to Elliott, May 1, 1907, Series IV (Engineering), subseries A (Early Career), box 6, folder "Engineering Correspondence 1907," Elliott to Morgan, May 6, 1907, Series IV (Engineering), subseries A (Early Career), box 7, folder "U.S.D.A. Investigations 1907," Arthur E. Morgan Papers, Antioch College Special Collections, Yellow Springs, Ohio (hereafter Morgan Papers); and Purcell, *Arthur Morgan*, 45, 47–53.

44. Purcell, *Arthur Morgan*, 53, 55.

45. See, for instance, the discussion in Morgan to Elliott, October 28, 1908, Series IV (Engineering), subseries A (Early Career), box 7, folder "U.S.D.A. Investigations 1908"; and Morgan to L. J. Ervin, January 19, 1909, Series IV (Engineering), subseries A (Early Career), box 7, folder "U.S.D.A. Investigations 1909," Morgan Papers.

46. Morgan to Arthur Ernest Morgan Jr., May 31, 1908, in *The Morgan Heritage*, Morgan Papers.

47. "The National Drainage Association," *Irrigation Age* 22 (February 1907): 110–12; and Mead to Fortier, February 18, 1907, RG 8, series "Correspondence between Berkeley and Washington, D.C.," box 33, folder "July–October, 1907," NARA, SF.

48. Elliott to James Cosgrove, October 15, 1907, RG 8, entry 6, box 6, folder "July 1906–May 1909," NARA, KC.

49. Elliott to Cosgrove, October 15, 1907, RG 8, entry 6, box 6, folder "July 1906–May 1909," NARA, KC; Mead to Fortier, February 12, February 18, February 23, and March 7, 1907; Teele to Fortier, December 23, 1907, RG 8, series "Correspondence between Berkeley and Washington, D.C.," box 33, folder "July–October, 1907," NARA, SF; and House of Representatives, *Annual Report of the Office of Experiment Stations for the Year Ended June 30, 1907*, 60th Congress, 1st Sess., H. Doc. 984, 39.

50. Mead to S. Fortier, March 7, 1907, RG 8, series "Correspondence between Berkeley and Washington, D.C.," box 33, folder "July–October, 1907," NARA, SF; "For Reclamation of Lands in Arkansas and Missouri," 18; "The Growing Interest in Drainage," 136; "News of the Railroads," *Daily Arkansas Democrat*, January 16, 1907; and "Louisiana Affairs," *Times-Democrat* (New Orleans, LA), January 18, 1907.

51. "The Growing Interest in Drainage," 136; "News of the Railroads."

52. Elliott to S. M. Woodward, October 28, 1907, "Stopping Points on Drainage Education Trip," n.d., RG 8, "Correspondence Regarding Field Trips, 1907," entry 10, box 1, folder "Drainage Educational Trip," NARA, KC; and "Talking Drainage of the St. Francis," *Paragould (AR) Daily Soliphone*, October 29, 1907.

53. Wilson to Flint, March 2, 1907, RG 16, microcopy no. 440, reel 279.

54. Wilson to Flint, March 2, 1907, RG 16, microcopy no. 440, reel 279.

55. Wilson to Flint, March 2, 1907, RG 16, microcopy no. 440, reel 279. Wilson's assessment was largely accurate. As the medical historian Margaret Humphreys argues, malaria by 1900 "was in retreat everywhere in the United States" with the exceptions of small pockets in the South and California. Humphreys explains that in the South a complex web of geographic, racial, and social factors kept malaria alive and active. The region's subtropical climate, poor rural housing clustered around marshy areas, insufficient tax revenue to support government programs, limited access to pesticides, substandard medical care, and significant population of African American bodies created a stable supply of *Plasmodium falciparum* malaria parasites to maintain the disease's presence beneath the Mason–Dixon line. Humphreys, *Malaria*, 38–68, quote at 49. See also Grob, *Deadly Truth*, 185–86.

56. Wilson to Flint, March 2, 1907, RG 16, microcopy no. 440, reel 279; and "Drainage and Irrigation Legislation," 167.

57. Wright, "Swamp and Overflowed Lands," 7–8.

58. Wright, "Swamp and Overflowed Lands," 7.

59. Wright, "Swamp and Overflowed Lands," 8–10.

60. Wright, "Swamp and Overflowed Lands," 6, 8. Wright's statistics about swampland cessions were inaccurate. In total, the states selected over 80 million acres under the Swamp Land Acts, but Congress ultimately ended up awarding title to only 65 million acres. Leshy, *Our Common Ground*, 59.

61. Elliott to Paul Clagstone, December 16, 1908, RG 8, entry 6, box 6, folder "July 1906–May 1909," NARA, KC.

62. House of Representatives, *Expenditures for Drainage Investigations*, 61st Cong., 3rd Sess., H. Doc. 1180, 2, 5–8.

63. Prince, *Wetlands of the American Midwest*, 145.

64. Purcell, *Arthur Morgan*, 57–58; and Purcell, "Plumb Lines," 168–71.

65. Purcell, *Arthur Morgan*, 58–60; and Purcell, "Plumb Lines," 172–82.

66. "Four Indicted in Everglades Inquiry," *New York Times*, April 2, 1912; Purcell, *Arthur Morgan*, 62–64; and Purcell, "Plumb Lines," 185–91.

67. Purcell, "Plumb Lines," 191–95.

68. E. W. Allen, "Memorandum for the Assistant Secretary," June 20, 1913, RG 16, General Correspondence of the Secretary, 1906–70, entry 17, box 54. In 1915, the Agricultural Appropriation Act transferred the ODI to the newly created Office of Public Roads and Rural Engineering.

Conclusion

1. Bay-mway-way-be-nais, Nah-gon-nway-we-dung, and Ed Prentice to Cato Sells [1914], RG 75, Records of the Bureau of Indian Affairs, Central Classified Files, 1907–1939, Red Lake, box 93, 6491–1939–339 to 37626–1914–343, NARA, Washington, DC (hereafter 1914 Petition).

2. E. B. Merritt to Charles A. Lindbergh, December 5, 1914, RG 75, Central Classified Files, box 93.

3. 1914 Petition, RG 75, Central Classified Files, box 93.

4. Meyer, "The Red Lake Ojibwe," 257.

5. Pinchot, "Homebuilding for the Nation," August 10, 1909, box 517, Pinchot Papers, LOC.

6. Brooks, "'Every Swamp Is a Castle,'" 50.

7. "State Reclaims Thousands of Acres of Swamp Lands; Vast Fertile Areas are being Restored to the Settlers," *Minneapolis Star Tribune*, December 1, 1912.

8. See, for instance, "List of Books on Drainage" [1911], RG 115, entry 3, box 347, folder 1363, "Reclamation Expansion"; and Sleeper-Smith, *Indigenous Prosperity*, 65.

9. Lang, Ingebritsen, and Griffin, *Status and Trends of Wetlands*, 6 (quote), 8–9; and Tiner, *In Search of Swampland*.

10. Johnson et al., "Prairie Wetland Complexes," 128; McKenna et al., "Synergistic Interaction," 1–2; and Lang, Ingebritsen, and Griffin, *Status and Trends of Wetlands*, 10.

11. Nahlik and Fennessy, "Carbon Storage in US Wetlands," 2.

12. David, Drinkwater, and McIsaac, "Sources of Nitrate Yields in the Mississippi River Basin," 1658. Published in early 2024, the 2022 *Census of Agriculture* counted 199,617 American farms covering 53.1 million acres that were drained by tile. US Department of Agriculture, *2022 Census of Agriculture*, 100.

13. Red Cliff Band of Lake Superior Chippewa Environmental Department and Environmental Protection Agency, "Red Cliff Band of Lake Superior Chippewa Wetland Program Plan: 2023–2027," March 2023, www.epa.gov/system/files/documents/2023-07/Red%20Cliff%20Wetland%20Program%20Plan%202023–27.pdf.

14. "Wisconsin Tribes: Leading the Way in Protecting and Restoring Wetlands and Watersheds," Wisconsin Wetlands Association, April 21, 2023, https://www.wisconsinwetlands.org/updates/wisconsin-tribes-leading-the-way-in-protecting-and-restoring-wetlands-and-watersheds/; and Tony Kuchma, "Part I: Restoring an Altered Landscape," Wisconsin Wetlands Association, May 12, 2022, www.wisconsinwetlands.org/updates/wetland-restoration-and-bird-monitoring-on-the-oneida-nation/.

15. Spoorthy Raman, "Culture and Conservation Thrive as Great Lakes Tribes Bring Back Native Wild Rice," *Mongabay*, March 4, 2024, https://news.mongabay.com/2024/03/culture-and-conservation-thrive-as-great-lakes-tribes-bring-back-native-wild-rice/; and Sheri McWhirter, "Michigan Tribal Experts Want to Save Wild Rice. Here's How They'll Do It," *MLive*, January 30, 2025, www.mlive.com/environment/2025/01/michigans-tribal-experts-want-to-save-wild-rice-heres-how-theyll-do-it.html.

16. Pisani, *Water and American Government*, 292. On the entrenched legacy of federalism and localism in water resources management, see Carlson, "Other Kind of Reclamation"; Kelley, "Context and the Process"; Pisani, *To Reclaim a Divided West*; Pisani, "Federalism and the American West"; and Taylor, *Federalism of Wetlands*. The drainage district statistics are gleaned from "Production and Environmental Benefits of Drainage," n.d., Monona County, IA, https://mononacountyiowa.gov/drainage/about_drainage_districts/; and "Union Drainage Districts," n.d., Village of Tinley Park, IL, www.tinleypark.org/government/departments/finance_department/union_drainage_districts.php.

Bibliography

Primary Sources

ARCHIVES AND MANUSCRIPT COLLECTIONS
Berkeley, CA
   Bancroft Library, University of California, Berkeley
      Elwood Mead Papers
      George C. Pardee Papers
Cambridge, MA
   Harvard University Archives
      Nathaniel Southgate Shaler Papers
Cleveland, OH
   Western Reserve Historical Society Library
      Theodore E. Burton Papers
College Park, MD
   National Archives and Records Administration
      Records Group 16, Records of the Office of the Secretary of Agriculture
      Records Group 48, Records of the Office of the Secretary of the Interior
      Records Group 57, Records of the United States Geological Survey
Denver, CO
   National Archives and Records Administration
      Records Group 115, Records of the Bureau of Reclamation
Fargo, ND
   North Dakota Institute for Regional Studies
      James B. Power Papers
Ft. Worth, TX
   National Archives and Records Administration
      Records Group 8, Records of the Bureau of Agricultural Engineering
Gainesville, FL
   George A. Smathers Library
      Napoleon Bonaparte Broward Papers
Kansas City, MO
   National Archives and Records Administration
      Records Group 8, Records of the Bureau of Agricultural Engineering
      Records Group 21, Records of the District Courts of the United States
Moorhead, MN
   Minnesota State University Archives
      Solomon G. Comstock Papers
New Orleans, LA
   Louisiana State Museum Historical Society
      George H. Maxwell Papers

Northfield, MN
  Norwegian-American Historical Association Archives
    Elias Steenerson Papers
San Francisco, CA
  National Archives and Records Administration
    Records Group 8, Records of the Bureau of Agricultural Engineering
St. Paul, MN
  James J. Hill Reference Library
    James J. Hill Papers
  Minnesota Historical Society
    Clay County (MN.) Court Civil Case Files Records
    Great Northern Railway Records
Urbana, IL
  University of Illinois Archives
    Vacation Journals, 1876–1878
Washington, DC
  Library of Congress
    Frederick Haynes Newell Papers
    Gifford Pinchot Papers
    James Rudolph Garfield Papers
  National Archives and Records Administration
    Records Group 49, Records of the Bureau of Land Management
    Records Group, 75, Records of the Bureau of Indian Affairs
  Smithsonian Institution Archives
    Charles D. Walcott Collection
Yellow Springs, OH
  Antioch College Special Collections
    Arthur E. Morgan Papers

GOVERNMENT DOCUMENTS

"A Change of Ownership." *Drainage Journal* 23 (May 1901): 145.
"Annual Report of Irrigation and Drainage Investigations, 1904." United States Department of Agriculture, Office of Experiment Stations, *Bulletin* 158. Washington, DC: Government Printing Office, 1905.
Dahl, T. E. *Wetlands Losses in the United States 1780's to 1980's*. Washington, DC: US Department of the Interior, Fish and Wildlife Service, 1990. https://www.fws.gov/sites/default/files/documents/Wetlands-Losses-in-the-United-States-1780s-to-1980s.pdf.
Dahl, T. E., and Gregory J. Allord. "Technical Aspects of Wetlands: History of Wetlands in the Conterminous United States." *National Water Summary on Wetland Resources*, United States Geological Survey Water Supply Paper 2425, 1996. https://water.usgs.gov/nwsum/WSP2425/history.html.
Elliott, C. G. "Drainage of Farm Lands." United States Department of Agriculture, *Farmers' Bulletin* 187. Washington, DC: Government Printing Office, 1904.

———. "Farm Drainage." United States Department of Agriculture, *Farmers' Bulletin* 40. Washington, DC: Government Printing Office, 1899.

———. "Report on Drainage Investigations, 1903." United States Department of Agriculture, Office of Experiment Stations, *Bulletin* 147. Washington, DC: Government Printing Office, 1904.

Executive Documents of the State of Minnesota for the Fiscal Year Ending July 31, 1886. Vol. 1. St. Paul, MN: Pioneer Press Company, 1887.

Fellows, A. L. *Second Biennial Report of the State Engineer to the Governor of North Dakota for the Years of 1905 and 1906*. Bismarck: Tribune, State Printers and Binders, 1906.

House Subcommittee of the Committee on Indian Affairs. *Drainage of Certain Lands Held in Trust for the Chippewa Indians in Minnesota: Hearing on H.R. 15666*. Washington, DC: Government Printing Office, 1908.

Instructions to the Surveyors General of Public Lands of the United States, for those Surveying Districts Established in and Since the Year 1850; Containing, also; A Manual of Instructions to Regulate the Field Operations of Deputy Surveyors. Washington, DC: Government Printing Office, 1871.

"Irrigation and Drainage." *Fourteenth Census of the United States Taken in the Year 1920*. Vol. 7. Washington, DC: Government Printing Office, 1922.

The Legislative Manual of the State of Minnesota. St. Paul, MN, 1907.

"Northwest Experiment Farm, Crookston, Minn.: Installation of an Experimental Drainage System." University of Minnesota, *Bulletin* 110. Delano, MN: Eagle Printing Co., 1908.

"Organization, Work, and Publications of Drainage Investigations." United States Department of Agriculture, *Circular* 88. Washington, DC: Government Printing Office, 1910.

Report of the Board of Drainage Commissioners of the State of Minnesota to the Governor of Minnesota on the Condition of the State Ditches Located in the Red River Valley for the Year Ending December 31st, 1901. Crookston, MN: Crookston Journal, 1902.

Report of the State Drainage Commission of Minnesota to the Governor and State Legislature on the Condition of the State Ditches Located in the Red River Valley for the Biennial Period Ending Feb. 1st, 1899. Minneapolis: University Press of Minnesota, 1899.

Shaler, Nathaniel Southgate. "General Account of the Fresh-Water Morasses of the United States, with a Description of the Dismal Swamp District of Virginia and North Carolina." In *Tenth Annual Report of the United States Geological Survey to the Secretary of the Interior 1888-'89*, 255–339. Washington, DC: Government Printing Office, 1890.

———. *Geological Survey of Kentucky: Reports of Progress*. Vol. 3. Frankfort, KY: Yeoman Press, 1877.

———. "Preliminary Report on Sea-Coast Swamps of the Eastern United States." In *Sixth Annual Report of the United States Geological Survey to the Secretary of the Interior*, 353–98. Washington, DC: Government Printing Office, 1885.

Smith, A. G. "Tile Drainage on the Farm." United States Department of Agriculture, *Farmers' Bulletin* 524. Washington, DC: Government Printing Office, 1913.

Soper, E. K., and C. C. Osbon. "The Occurrence and Uses of Peat in the United States." United States Geological Survey, *Bulletin* 728. Washington, DC: Government Printing Office, 1922.

Stewart, John T. *Report on the Drainage of the Eastern Parts of Cass, Traill, Grand Forks, Walsh, and Pembina Counties, North Dakota.* United States Department of Agriculture, Office of Experiment Stations, *Bulletin* 189. Washington, DC: Government Printing Office, 1907.

Teele, R. P. "Land Reclamation Policies in the United States." United States Department of Agriculture, *Bulletin* 1257. Washington, DC: Government Printing Office, 1924.

Thoburn, J. B. *First Biennial Report of the Oklahoma Territorial Board of Agriculture, 1903–1904.* Guthrie, OK: State Capital Co., 1905.

Tiner, Ralph W. "Technical Aspects of Wetlands: Wetland Definitions and Classifications in the United States." *USGS Water Supply Paper* 2425, 1997. https://water.usgs.gov/nwsum/WSP2425/definitions.html.

US Department of Agriculture. *2022 Census of Agriculture: United States Summary and State Data.* Vol. 1, Geographic Area Series, pt. 51. Washington, DC: US Department of Agriculture, 2024. https://www.nass.usda.gov/Publications/AgCensus/2022/Full_Report/Volume_1,_Chapter_1_US/usv1.pdf.

US House of Representatives. *Annual Reports of the Department of Agriculture for the Fiscal Year Ended June 30, 1906.* 59th Cong, 2nd Sess., November 24, 1906. H. Doc. 6.

———. *Annual Report of the Office of Experiment Stations for the Year Ended June 30, 1905.* 59th Cong., 1st Sess., March 12, 1906. H. Doc. 924.

———. *Annual Report of the Office of Experiment Stations for the Year Ended June 30, 1906.* 59th Cong., 2nd Sess., March 30, 1907. H. Doc. 820.

———. *Annual Report of the Office of Experiment Stations for the Year Ended June 30, 1907.* 60th Cong., 1st Sess., April 29, 1908. H. Doc. 984.

———. *Annual Report of the Office of Experiment Stations for the Year Ended June 30, 1908.* 60th Cong., 2nd Sess., April 28, 1908. H. Doc. 1561.

———. *Annual Report of the Office of Experiment Stations for the Year Ended June 30, 1909.* 61st Cong., 2nd Sess., April 30, 1910. H. Doc. 128.

———. *Annual Report of the Office of Experiment Stations for the Year Ended June 30, 1910.* 61st Cong., 3rd Sess., March 8, 1911. H. Doc. 1458.

———. *Annual Report of the Office of Experiment Stations for the Year Ended June 30, 1911.* 62nd Cong., 2nd Sess., July 1, 1912. H. Doc. 136.

———. *Appropriating the Proceeds from Sale of Public Lands to Minnesota for Certain Purposes.* 59th Cong., 1st Sess., June 15, 1906. H. Rep. 4940.

———. *Disposal of Unallotted Lands of the Omaha Indian Reservation.* 60th Cong., 2nd Sess., February 24, 1909. H. Doc. 1479.

———. *Drainage in North Dakota.* 59th Cong., 1st Sess., June 14, 1906. H. Rep. 4929.

———. *Drainage in North Dakota.* 59th Cong., 1st Sess., June 23, 1906. H. Rep. 4929, pt. 2.

———. *Drainage of Certain Lands in Minnesota.* 60th Cong., 1st Sess., April 2, 1908. H. Rep. 1376.

———. *Drainage of Certain Lands in Minnesota.* 64th Cong., 1st Sess., July 25, 1916. H. Rep. 1039.

———. *Drainage of Dismal Swamp in North Carolina and Virginia.* 59th Cong., 1st Sess., June 22, 1906. H. Rep. 4994.

———. *Drainage Survey of Certain Lands in Minnesota.* 61st Cong., 1st Sess., May 13, 1909. H. Doc. 27.

———. *Expenditures in the Department of Agriculture*. 62nd Cong., 2nd Sess., August 19, 1912. H. Rep. 1207.

———. *Providing for Payment of Cost of Drainage of Omaha Indian Allotted Lands*. 60th Cong., 2nd Sess., January 25, 1909. H. Rep. 1924.

———. *Reclamation of Certain Lands in Minnesota*. 59th Cong., 2nd Sess., January 28, 1907. H. Doc. 607.

———. *Red River of the North*. 51st Cong., 2nd Sess., March 3, 1891. H. Ex. Doc. 292.

———. *Red River of the North, Minnesota*. 52nd Cong., 1st Sess., February 16, 1892. H. Ex. Doc. 127.

———. *Surveys of the Territories*. 45th Cong., 3rd Sess., December 3, 1878. H. Mis. Doc. 5.

———. *Swamp Lands in Missouri and Arkansas*. 30th Cong., 2nd Sess., February 28, 1849. H. Rep. 130.

———. *Swamp Lands in Missouri and Arkansas*. 31st Cong., 1st Sess., February 20, 1850. H. Rep. 108.

———. *Thirteenth Annual Report of the Reclamation Service, 1913–1914*. 63rd Cong., 3rd Sess., 1915. H. Doc. 1255.

———. *To Enable Certain Indians to Protect their Lands from Overflow, Etc*. 59th Cong., 1st Sess., June 4, 1906. H. Rep. 4677.

US Senate. *Drainage of Certain Lands in Minnesota*. 60th Cong., 1st Sess., April 25, 1908. S. Rep. 571.

———. *Drainage of Certain Lands in North Dakota, Etc*. 59th Cong., 1st Sess., February 5, 1906. S. Rep. 689.

———. *Drainage of Winnebago Lands*. 60th Cong., 2nd Sess., February 5, 1909. S. Doc. 922.

———. *Drainage or Reclamation of Swamp or Overflowed Lands*. 60th Cong., 1st Sess., February 24, 1908. S. Rep. 289.

———. *Drainage or Reclamation of Swamp or Overflowed Lands*. 60th Cong., 1st Sess., April 14, 1908. S. Rep. 506.

———. *Drainage or Reclamation of Swamp or Overflowed Lands*. 60th Cong., 1st Sess., April 18, 1908. S. Rep. 542.

———. *Establishment of a Drainage Fund, Etc*. 59th Cong., 2nd Sess., March 2, 1907. S. Rep. 7342.

———. *Everglades of Florida*. 62nd Cong., 1st Sess., August 7, 1911. S. Doc. 89.

———. *Examinations for the Drainage of Lands Made by the Department of Interior*. 60th Cong., 1st Sess., January 7, 1908. S. Doc. 151.

———. *Swamplands of the United States*. 60th Cong., 1st Sess., April 21, 1908. S. Doc. 443.

White, C. Albert. *A History of the Rectangular Survey System*. Washington, DC: Government Printing Office, 1983. https://www.blm.gov/sites/blm.gov/files/histrect.pdf.

Wright, J. O. "Swamp and Overflowed Lands in the United States: Ownership and Reclamation." United States Department of Agriculture, Office of Experiment Stations, *Circular 76*. Washington, DC: Government Printing Office, 1907.

NEWSPAPERS AND FARM JOURNALS

Albuquerque Journal
American Farmer
Anaconda (MT) Standard
Baltimore Sun
Bellingham (WA) Herald
Bemidji (MN) Daily Pioneer
Bismarck Tribune
Boston Cultivator
Chandler (OK) News
Charlotte Observer
Chicago Daily Tribune
Cincinnati Daily Gazette
Cincinnati Weekly Herald and Philanthropist
Country Gentleman
Crookston (MN) Northern Tier
Cultivator
Daily Arkansas Democrat
Daily Illini
Daily Oklahoman
Daily Republican (Monongahela, PA)
Dallas Morning News
DeBow's Review
Drainage and Farm Journal
Drainage Journal
Duluth News Tribune
Evening Statesman (Walla Walla, WA)
Fargo Daily Republican
Farmer and Mechanic (Raleigh, NC)
Farmer's Register (Petersburg, VA)
Fisher (MN) Bulletin
Forestry and Irrigation (Washington, DC)
Fredonia (KS) Daily Herald
Freeborn County Standard (Albert Lea, MN)
Galveston Daily News
Genesee Farmer
Grand Forks Herald
Gridley (CA) Herald
Idaho Statesman (Boise, ID)
Illinois Farmer
Irrigation Age
Kittson County Enterprise
Manitoba Weekly Free Press
Minneapolis Star Tribune
Minnesota Farmer
Moorhead (MN) Weekly News
Neodesha (KS) Daily Sun
Neodesha (KS) Register
New England Farmer
New Ulm (MN) Review
New-York Daily Tribune
New York Times
North Carolinian
Oakland Tribune
Ohio Cultivator
Olympia (WA) Record
Oregonian (Portland, OR)
Paragould (AR) Daily Soliphone
People's Press (Owatonna, MN)
Philadelphia Inquirer
Pittsfield (MA) Sun
Polk County Journal (Crookston, MN)
Prairie Farmer
Red River Valley News (Glyndon, MN)
St. Louis Commercial Bulletin and Missouri Literary Register
St. Paul Daily Globe
St. Paul Daily News
Selma (CA) Enterprise
Southern Cultivator
State (Columbia, SC)
Sun (Jacksonville, FL)
Sun (New Bern, NC)
Times-Democrat (New Orleans, LA)
Valley Farmer
Warren (MN) Sheaf
Washington Herald
Washington Post
Watchman and Southron (Sumter, SC)
Wilson County Citizen (Fredonia, KS)
Woodland (CA) Daily Democrat

PUBLISHED SOURCES

Antill, Edward. "An Essay on the Cultivation of the Vine, and the Making and Preserving of Wine, Suited to the Different Climates in North-America." *Transactions of the American Philosophical Society* 1 (January 1769): 117–97.

Bennett, Ira E. "Western Affairs at Washington." *Pacific Monthly* 18 (November 1907): 610–20.

Byrd, William II. *Description of the Dismal Swamp and a Proposal to Drain the Swamp.* Edited by Earl Gregg Swem. Metuchen, NJ: Charles F. Heartman, 1922. https://www.loc.gov/item/22022884/.

Caldwell, Charles. "An Essay on the Nature and Sources of the Malaria or Noxious Miasma, from which originate the Family or Diseases, usually known by the denomination of Bilious Diseases; together with the best means of Preventing the Formation of Malaria, removing the Sources, and obviating their Effects on the Human Constitution, when the Cause cannot be removed." *American Journal of the Medical Sciences* 8 (August 1831): 293–341.

———. *An Oration on the Causes of the Difference, in Point of Frequency and Force, Between the Endemic Diseases of the United States of America, and Those of the Countries of Europe, Delivered, by Appointment, to the "Philadelphia Medical Society," on the Fifth Day of February, 1802.* Philadelphia: T. and William Bradford, 1802.

Chalmers, Lionel. *An Account of the Weather and Diseases of South-Carolina.* Vol. 1. London: Edward and Charles Dilly, 1776.

Colden, Cadwallader. "Account of the Climate and Diseases of New-York." *American Medical and Philosophical Register; or, Annals of Medicine, Natural History, Agriculture, and the Arts* 1 (July 1810): 304–10.

———. "Observations on the Fever which prevailed in the City of New-York in 1741 and 2, written in 1743, by the late Hon. Cadwallader Colden. Communicated to Dr. David Hosack by C. D. Colden, Esp." *American Medical and Philosophical Register; or, Annals of Medicine, Natural History, Agriculture, and the Arts* 1 (July 1810): 310–30.

Coxe, Tench. *A Statement of the Arts and Manufactures of the United States of America, for the Year 1810.* Philadelphia: A. Cornman, 1814.

Currie, William. "An Enquiry into the Causes of the Insalubrity of Flat and Marshy Situations; and Directions for Preventing or Correcting the Effects thereof." *Transactions of the American Philosophical Society* 4 (1799): 127–42.

"Drainage and Irrigation Legislation." *Irrigation Age* 22 (April 1907): 166–67.

"Drainage of the Red River Valley." *Drainage and Farm Journal* 8 (August 1886): 236.

Elliott, C. G. *Engineering for Land Drainage: A Manual for the Reclamation of Lands Injured by Water.* 2nd ed. New York: John Wiley & Sons, 1912.

———. *Practical Farm Drainage: A Manual for Farmer and Student.* 2nd ed. New York: John Wiley & Sons, 1908.

———. *Practical Farm Drainage; Why, When and How to Tile Drain.* Indianapolis, IN: J. J. W. Billingsley, 1882.

———. "Questions about Draining." *Prairie Farmer* 58 (January 30, 1886): 66.

———. "Road Improvement." *Prairie Farmer* 58 (March 20, 1886): 180.

"For Reclamation of Lands in Arkansas and Missouri." *Irrigation Age* 22 (November 1906): 18.

"The Growing Interest in Drainage." *Irrigation Age* 22 (March 1907): 136.

Hansbrough, Henry C. "Reclamation of Swamp Lands." *Independent* 62 (February 7, 1907): 320–22.

Hening, William Waller, ed. *The Statutes at Large; Being A Collection of all the Laws of Virginia, From the First Session of the Legislature, in the Year 1619.* Vol. 3. Philadelphia: Thomas Desilver, 1823.

Hill, James J. *Highways of Progress*. New York: Doubleday, Page, 1910.
Jefferson, Thomas. *Notes on the State of Virginia*. Edited by William Peden. Chapel Hill: University of North Carolina Press, 1982. Originally published in 1785.
Klippart, John H. *The Principles and Practice of Land Drainage: A Brief History of Underdraining; A Detailed Examination of its Operation and Advantages: A Description of Various Kinds of Drains, with Practical Directions for their Construction: The Manufacture of Drain-Tile, Etc*. Cincinnati: Robert Clarke, 1861.
Lippincott, J. B. "General Outlook for Reclamation Work in California." *Forestry and Irrigation* 11 (August 1905): 353.
Marsh, George Perkins. *Man and Nature: or, Physical Geography as Modified by Human Action*. Edited by David Lowenthal. Seattle: University of Washington Press, 2003. Originally published in 1864.
Mitchell, Guy Elliott. "To Farm America's Swamps." *American Review of Reviews* 37 (April 1908): 433–49.
———. "Land Reclamation by Drainage." *Forestry and Irrigation* 12 (March 1906): 134–38.
———. "The National Drainage Problem." *Hearst's International* 13 (August 1907): 777–84.
"The National Drainage Association." *Irrigation Age* 22 (February 1907): 110–12.
Newell, Frederick H. "What May Be Accomplished by Reclamation." *Annals of the American Academy of Political and Social Science* 33 (May 1909): 174–79.
"Reclamation Fund Exhausted." *Irrigation Age* 21 (April 1906): 180–81.
"Reclamation Fund Threatened." *Forestry and Irrigation* 12 (March 1906): 112–13.
Red River Valley Drainage in Minnesota. Report of the Drainage Commission to the Drainage Convention, Held at Crookston, Minn. December 8, 1886. Minneapolis, MN: Harrison & Smith, 1887.
Robertson, William. *The History of America*. Vol. 1. London: Thomas White, 1778.
Ruffin, Edmund. *Nature's Management: Writings on Landscape and Reform, 1822–1859*. Edited by Jack Temple Kirby. Athens: University of Georgia Press, 2000.
Rush, Benjamin. "An Enquiry into the Cause of the Increase of Bilious and Intermitting Fevers in Pennsylvania, with Hints for Preventing Them." *Transactions of the American Philosophical Society* 2 (1786): 206–12.
Shaler, Nathaniel Southgate. "The American Swamp Cypress." *Science* 2, no. 23 (July 13, 1883): 38–40.
———. "European Peasants as Immigrants." *Atlantic Monthly* 71 (May 1893): 646–55.
———. "The Floods of the Mississippi Valley." *Atlantic Monthly* 51 (May 1883): 653–60.
———. "Fluviatile Swamps of New England." *American Journal of Science* 3rd ser., 33 (March 1887): 210–21.
———. "The Future of the Negro in the Southern States." *Popular Science Monthly* 57 (June 1900): 151–55.
———. *Nature and Man in America*. New York: Charles Scribner's Sons, 1891.
———. "The Summing Up of the Story." In *The United States of America: A Study of the American Commonwealth, Its Natural Resources, People, Industries, Manufactures, Commerce, and Its Work in Literature, Science, Education, and Self-Government*, edited by Nathaniel Southgate Shaler. New York: D. Appleton, 1897.

———. "The Swamps of the United States." *Science* 7, no. 162 (March 12, 1886): 232–33.
Shaler, Nathaniel Southgate, and Sophia Penn Page Shaler. *The Autobiography of Nathaniel Southgate Shaler with a Supplementary Memoir by His Wife*. Boston: Houghton Mifflin, 1909.
St. Paul, Minneapolis, and Manitoba Railroad Company. *Letters from Golden Latitudes*. St. Paul, MN, 1885.
"Some Account of the Great Dismal Swamp." *Literary Magazine, and American Register* 3 (March 1805): 170–71.
Valentine, Ezra G. *Statement and Statistics Relating to the Bill Entitled "An Act to Appropriate Monies for the Purpose of Opening of Closed Water Courses Leading into the Red River, and its Tributaries, and for Opening Existing Streams in the Red River Valley in the Counties of Wilkin, Clay, Norman, Polk, Marshall and Kittson in this State." Respectfully Submitted: Red River Valley Drainage Commission*. Breckenridge, MN: Gazette Press Print, 1893.
Waring, George E. *Draining for Profit, and Draining for Health*. 2nd ed. New York: Orange Judd, 1884.
Williamson, Hugh. "An Attempt to Account for the Change of Climate, which has been Observed in the Middle Colonies in North-America." *Transactions of the American Philosophical Society* 1 (1771): 272–80.
Wilson, Herbert M. "Reclaiming the Swamp Lands of the United States." *National Geographic* 18 (May 1907): 292–301.
———. "Swamp Lands and Their Reclamation." *Journal of the American Peat Society* 1 (1908): 8–10.
Wright, Thomas. "On the Mode of Most Easily and Effectually Practicable of Drying up the Marshes of the Maritime Parts of North America." *Transactions of the American Philosophical Society* 4 (1799): 243–46.

Secondary Sources

BOOKS AND ARTICLES

Adams, Bluford. "World Conquerors or a Dying People? Racial Theory, Regional Anxiety, and the Brahmin Anglo-Saxonists." *Journal of the Gilded Age and Progressive Era* 8, no. 2 (April 2009): 189–215.
Allen, Garland E. "'Culling the Herd': Eugenics and the Conservation Movement in the United States, 1900–1940." *Journal of the History of Biology* 46, no. 1 (Spring 2013): 31–72.
Allmendinger, David F., Jr. *Ruffin: Family and Reform in the Old South*. New York: Oxford University Press, 1990.
Anderson, Warwick. "Immunities of Empire: Race, Disease, and the New Tropical Medicine, 1900–1920." *Bulletin of the History of Medicine* 70, no. 1 (1996): 94–118.
Andrews, Richard N. L. *Managing the Environment, Managing Ourselves: A History of American Environmental Policy*. 2nd ed. New Haven, CT: Yale University Press, 2006.
Angevine, Robert G. *The Railroad and the State: War, Politics, and Technology in Nineteenth-Century America*. Stanford, CA: Stanford University Press, 2004.

Apel, Thomas A. *Feverish Bodies, Enlightened Minds: Science and the Yellow Fever Controversy in the Early American Republic*. Stanford, CA: Stanford University Press, 2016.

Appel, Peter A. "The Power of Congress 'Without Limitation': The Property Clause and Federal Regulation of Private Property." *Minnesota Law Review* 86 (November 2001): 1–130.

Armitage, David. *The Ideological Origins of the British Empire*. Cambridge, UK: Cambridge University Press, 2000.

Ash, Eric H. *The Draining of the Fens: Projectors, Popular Politics, and State Building in Early Modern England*. Baltimore, MD: Johns Hopkins University, 2017.

Autobee, Robert. "Every Child in a Garden: George H. Maxwell and the American Homecroft Society." *Prologue: The Journal of the National Archives* 28, no. 3 (1996): 195–206.

Baeten, John. "Making Wet Places Drier: Mapping the Evolution of Drainage Technology in the U.S." Network in Canadian History and Environment, June 10, 2020. https://niche-canada.org/2020/07/14/making-wet-places-drier-mapping-the-evolution-of-drainage-technology-in-the-u-s/.

Baker, Lee D. *From Savage to Negro: Anthropology and the Construction of Race, 1896–1954*. Berkeley: University of California Press, 1998.

Baldwin, Peter C. "How Night Air Became Good Air, 1776–1930." *Environmental History* 8, no. 3 (July 2003): 412–29.

Balogh, Brian. *The Associational State: American Governance in the Twentieth Century*. Philadelphia: University of Pennsylvania Press, 2015.

———. *A Government Out of Sight: The Mystery of National Authority in Nineteenth-Century America*. Cambridge, UK: Cambridge University Press, 2009.

Barnett, LeRoy. "Roads, Railroads, and Recreation: Swamplands and the Building of Michigan." *Michigan History* 72 (July–August 1988): 28–34.

Barnett, Lydia. "The Theology of Climate Change: Sin as Agency in the Enlightenment's Anthropocene." *Environmental History* 20, no. 2 (April 2015): 217–37.

Bates, J. Leonard. "Fulfilling American Democracy: The Conservation Movement, 1907 to 1921." *Mississippi Historical Review* 44, no. 1 (June 1957): 29–57.

Baugher, Sherene. "What Is It? Archaeological Evidence of 19th-Century Agricultural Drainage Systems." *Northeast Historical Archaeology* 30–31 (2001): 23–40.

Beauchamp, Keith H. "A History of Drainage and Drainage Methods." In *Farm Drainage in the United States: History, Status, Prospects*, edited by George A. Pavelis, 13–29. Washington, DC: Economic Research Service, Department of Agriculture, 1987.

Bederman, Gail. *Manliness and Civilization: A Cultural History of Gender and Race in the United States, 1880–1917*. Chicago: University of Chicago Press, 1995.

Beier, Lucinda McCray. *Health Culture in the Heartland, 1880–1980: An Oral History*. Urbana: University of Illinois Press, 2008.

Bennett, John W., and Seena B. Kohl. *Settling the Canadian-American West, 1890–1915: Pioneer Adaptation and Community Building*. Lincoln: University of Nebraska Press, 1995.

Bensel, Richard Franklin. *Yankee Leviathan: The Origins of Central State Authority in America, 1859–1877*. New York: Cambridge University Press, 1990.

Berkeley, Edmund, and Dorothy Berkeley. "Man and the Great Dismal." *Virginia Journal of Science* 27 (Fall 1976): 141–71.
Bewell, Alan. *Romanticism and Colonial Disease*. Baltimore: Johns Hopkins University Press, 1999.
Bidwell, Percy W., and John I. Falconer. *History of Agriculture in the Northern United States, 1620–1860*. New York: Peter Smith, 1941.
Biebighauser, Thomas R. *Wetlands Drainage, Restoration, and Repair*. Lexington: University Press of Kentucky, 2007.
Billington, David P., and Donald C. Jackson. *Big Dams of the New Deal Era: A Confluence of Engineering and Politics*. Norman: University of Oklahoma Press, 2017.
Bogue, Margaret Beattie. "The Swamp Land Act and Wet Land Utilization in Illinois, 1850–1890." *Agricultural History* 25, no. 4 (October 1951): 169–80.
Borca, Federico. "*Palus Omni Modo Vitanda*: A Liminal Space in Ancient Roman Culture." *Classical Bulletin* 73, no. 1 (1997): 3–12.
———. "Towns and Marshes in the Ancient World." In *Death and Disease in the Ancient City*, edited by Valerie M. Hope and Eireann Marshall, 74–84. London: Routledge, 2000.
Bosselman, Fred P. "Limitations Inherent in the Title to Wetlands at Common Law." *Stanford Environmental Law Journal* 15 (1996): 247–337.
Bower, Shannon Stunden. "Watersheds: Conceptualizing Manitoba's Drained Landscape, 1895–1950." *Environmental History* 12, no. 4 (October 2007): 796–819.
———. *Wet Prairie: People, Land, and Water in Agricultural Manitoba*. Vancouver: University of British Columbia Press, 2011.
Bowler, Peter J., and Iwan Rhys Morus. *Making Modern Science: A Historical Survey*. Chicago: University of Chicago Press, 2005.
Bradof, Kristine L. "Ditching of Red Lake Peatland During the Homestead Era." In *The Patterned Peatlands of Minnesota*, edited by H. E. Wright, Jr., Barbara A. Coffin, and Norman E. Aaseng, 263–84. Minneapolis: University of Minnesota Press, 1992.
Brechin, Gray. "Conserving the Race: Natural Aristocracies, Eugenics, and the U.S. Conservation Movement." *Antipode* 28, no. 3 (1996): 229–45.
Briggs, Harold E. "Early Bonanza Farming in the Red River Valley of the North." *Agricultural History* 6, no. 1 (January 1932): 26–37.
Brinkley, Douglas. *The Wilderness Warrior: Theodore Roosevelt and the Crusade for America*. New York: HarperCollins, 2009.
Brodhead, Michael J. *David J. Brewer: The Life of a Supreme Court Justice, 1837–1910*. Carbondale: Southern Illinois University Press, 1994.
Brooke, John L. *Climate Change and the Course of Global History*. Cambridge, UK: Cambridge University Press, 2014.
Brooks, Lisa. "'Every Swamp Is a Castle': Navigating Native Spaces in the Connecticut River Valley, Winter 1675–1677 and 2005–2015." *Northeastern Naturalist* 24, no. 7 (March 2017): 45–80.
Brown, Michael. "From Foetid Air to Filth: The Cultural Transformation of British Epidemiological Thought, ca. 1780–1848." *Bulletin of the History of Medicine* 82, no. 3 (Fall 2008): 515–44.
Burton, Ian, and Robert W. Kates, eds. *Readings in Resource Management and Conservation*. Chicago: University of Chicago Press, 1960.

Cadle, Nathaniel. *The Mediating Nation: Late American Realism, Globalization, and the Progressive State*. Chapel Hill: University of North Carolina Press, 2014.

Calhoun, Charles W. "Political Economy in the Gilded Age: The Republican Party's Industrial Policy." *Journal of Policy History* 8, no. 3 (July 1996): 291–309.

Camporesi, Piero. *The Fear of Hell: Images of Damnation and Salvation in Early Modern Europe*. Translated by Lucinda Byatt. University Park: Pennsylvania State University Press, 1991.

Carlson, Anthony E. "Forging Transcontinental Alliances: The Sacramento River Valley in National Drainage and Flood Control Politics, 1900–1917." In *River City and Valley Life: An Environmental History of the Sacramento Region*, edited by Christopher J. Castaneda and Lee M. A. Simpson, 135–57. Pittsburgh, PA: University of Pittsburgh Press, 2013.

———. "The Other Kind of Reclamation: Wetlands Drainage and National Water Policy, 1902–1912." *Agricultural History* 84, no. 4 (Fall 2010): 451–78.

———. "'There May Be Bloodshed': The US Reclamation Service, Localism, and Water Conflicts in the Montana-Alberta Borderlands, 1900–1910." In *Farming across Borders: A Transnational History of the North American West*, edited by Sterling Evans, 395–419. College Station: Texas A&M University Press, 2017.

———. "'Vast Factories of Febrile Poison': Wetlands, Drainage, and the Fate of American Climates, 1750–1850." In *Governing the Environment in the Early Modern World: Theory and Practice*, edited by Sara Miglietti and John Morgan, 153–71. London: Routledge, 2017.

Carlson, Leonard A. *Indians, Bureaucrats, and Land: The Dawes Act and the Decline of Indian Farming*. Westport, CT: Greenwood Press, 1981.

Carney, Judith. *Black Rice: The African Origins of Rice Cultivation in the Americas*. Cambridge, MA: Harvard University Press, 2001.

Carpenter, Daniel P. *The Forging of Bureaucratic Autonomy: Reputations, Networks, and Policy Innovation in Executive Agencies, 1862–1928*. Princeton, NJ: Princeton University Press, 2001.

Cassedy, James H. "Meteorology and Medicine in Colonial America: Beginnings of the Experimental Approach." *Journal of the History of Medicine and Allied Sciences* 24, no. 2 (April 1969): 193–204.

Chapman, Kim Alan, Adelheid Fischer, and Mary Kinsella Ziegenhagen. *Valley of Grass: Tallgrass Prairie and Parkland of the Red River Region*. St. Cloud, MN: North Star Press of St. Cloud, 1998.

Charters, Erica. *Disease, War, and the Imperial State: The Welfare of the British Armed Forces during the Seven Years' War*. Chicago: University of Chicago Press, 2014.

Chinard, Gilbert. "Eighteenth Century Theories on America as a Human Habitat." *Proceedings of the American Philosophical Society* 91, no. 1 (February 1947): 27–57.

Clements, Kendrick A. *Hoover, Conservation, and Consumerism: Engineering the Good Life*. Lawrence: University Press of Kansas, 2000.

Colden, Claudine. *The Fate of the Mammoth: Fossils, Myth, and History*. Translated by William Rodarmor. Chicago: University of Chicago Press, 2002.

Colten, Craig E. *An Unnatural Metropolis: Wresting New Orleans from Nature*. Baton Rouge: Louisiana State University Press, 2005.

Compendium of History and Biography of Central and Northern Minnesota Containing a History of the State of Minnesota. Chicago: Geo. A. Ogle, 1904.
Conklin, Paul K. "The Vision of Elwood Mead." *Agricultural History* 34, no. 2 (April 1960): 88–97.
Crane, Jeff. *The Environment in American History: Nature and the Formation of the United States*. New York: Taylor & Francis, 2015.
Cronon, William. *Changes in the Land: Indians, Colonists, and the Ecology of New England*. New York: Hill and Wang, 1983.
———. *Nature's Metropolis: Chicago and the Great West*. New York: W. W. Norton, 1991.
Crosby, Alfred W. *Ecological Imperialism: The Biological Expansion of Europe, 900–1900*. 2nd ed. Cambridge, UK: Cambridge University Press, 2004.
Cullather, Nick. "Damming Afghanistan: Modernization in a Buffer State." *Journal of American History* 89, no. 2 (September 2002): 512–37.
———. *The Hungry World: America's Cold War Battle against Poverty in Asia*. Cambridge, MA: Harvard University Press, 2010.
Cumbler, John T. *Northeast and Midwest United States: An Environmental History*. Santa Barbara, CA: ABC-CLIO, 2005.
Danbom, David B. *Born in the Country: A History of Rural America*. 2nd ed. Baltimore, MD: Johns Hopkins University Press, 2006.
Daniels, Roger. "Immigration in the Gilded Age: Change or Continuity?" *OAH Magazine of History* 13, no. 4 (Summer 1999): 21–25.
Darby, H. C. *The Changing Fenland*. Cambridge, UK: Cambridge University Press, 1983.
———. "The Draining of the Fens, A. D. 1600–1800." In *An Historical Geography of England Before A.D. 1800: Fourteen Studies*, edited by H. C. Darby, 444–64. Cambridge, UK: Cambridge University Press, 1961.
Darrah, William Culp. *Powell of the Colorado*. Princeton, NJ: Princeton University Press, 1951.
David, Mark B., Laurie E. Drinkwater, and Gregory F. McIsaac. "Sources of Nitrate Yields in the Mississippi River Basin." *Journal of Environmental Quality* 39, no. 5 (September–October 2010): 1657–67.
Davis, Clifford. "The Law of Diffused Surface Water in Eastern Riparian States." *Connecticut Law Review* 6 (Winter 1973–74): 227–45.
Delbourgo, James. *A Most Amazing Scene of Wonders: Electricity and Enlightenment in Early America*. Cambridge, MA: Harvard University Press, 2006.
Detwiler, Justice Brown, ed. *Who's Who in California: A Biographical Directory, 1928–29*. San Francisco: Who's Who Publishing, 1929.
Dillon, Lindsey. "Civilizing Swamps in California: Formations of Race, Nature, and Property in the Nineteenth Century U.S. West." *Society and Space* 40, no. 2 (July 2022): 258–75.
Diouf, Sylviane A. *Slavery's Exiles: The Story of the American Maroons*. New York: New York University Press, 2014.
Di Palma, Vittoria. *Wasteland: A History*. New Haven, CT: Yale University Press, 2014.

Dobbins, Donald V. "Surface Water Drainage." *Notre Dame Law Review* 36, no. 4 (August 1961): 518–26.

Dobson, Mary J. *Contours of Death and Disease in Early Modern England*. Cambridge, UK: Cambridge University Press, 1997.

Dodds, Gordon B. "The Stream-Flow Controversy: A Conservation Turning Point." *Journal of American History* 56, no. 1 (June 1969): 59–69.

Dolson, William F. "Diffused Surface Water and Riparian Rights: Legal Doctrines in Conflict." *Wisconsin Law Review* 58 (1966): 58–120.

Donahue, Brian. *The Great Meadow: Farmers and the Land in Colonial Concord*. New Haven, CT: Yale University Press, 2004.

Dorsey, Leroy G. *Theodore Roosevelt, Conservation, and the 1908 Governors' Conference*. College Station: Texas A&M University Press, 2016.

Douglas, Lake. "'To Improve the Soil and the Mind': Content and Context of Nineteenth-Century Agricultural Literature." *Landscape Journal* 25, no. 1 (2006): 67–79.

Doyle, Martin. *The Source: How Rivers Made America and America Remade its Rivers*. New York: W. W. Norton, 2018.

Drache, Hiram M. *The Day of the Bonanza: A History of Bonanza Farming in the Red River Valley of the North*. Fargo: North Dakota Institute for Regional Studies, 1964.

Drayton, Richard. *Nature's Government: Science, Imperial Britain, and the "Improvement" of the World*. New Haven, CT: Yale University Press, 2000.

Dugatkin, Lee Alan. *Mr. Jefferson and the Giant Moose: Natural History in Early America*. Chicago: University of Chicago Press, 2009.

Dunbar, Robert G. *Forging New Rights in Western Waters*. Lincoln: University of Nebraska Press, 1983.

Dupree, A. Hunter. *Science in the Federal Government: A History of Policies and Activities to 1940*. Cambridge, MA: Harvard University Press, 1957.

Edelson, S. Max. "Clearing Swamps, Harvesting Forests: Trees and the Making of a Plantation Landscape in the Colonial South Carolina Lowcountry." *Agricultural History* 81, no. 3 (Summer 2007): 381–406.

Engelhardt, Carroll. *Gateway to the Northern Plains: Railroads and the Birth of Fargo and Moorhead*. Minneapolis: University of Minnesota Press, 2007.

Epperson, Bruce D. *Draining the Swamp, Southern Style: North Carolina and Florida Wetlands and the Wright Report Scandal, 1896–1926*. Jefferson, NC: McFarland & Company, 2021.

Espinosa, Mariola. "The Question of Racial Immunity to Yellow Fever in History and Historiography." *Social Science History* 38, nos. 3–4 (Fall–Winter 2014): 437–53.

Fagan, Brian. *The Little Ice Age: How Climate Made History, 1300–1850*. New York: Basic Books, 2000.

Feller, Daniel. *The Public Lands in Jacksonian Politics*. Madison: University of Wisconsin Press, 1984.

Fiege, Mark. *Irrigated Eden: The Making of an Agricultural Landscape in the American West*. Seattle: University of Washington Press, 1999.

———. *The Republic of Nature: An Environmental History of the United States*. Seattle: University of Washington Press, 2012.

———. "The Weedy West: Mobile Nature, Boundaries, and Common Space in the Montana Landscape." *Western Historical Quarterly* 36, no. 1 (Spring 2005): 22–47.
Fite, Gilbert C. *The Farmers' Frontier: 1865–1900*. Albuquerque: University of New Mexico Press, 1974.
Fleming, James Rodger. *Historical Perspectives on Climate Change*. New York: Oxford University Press, 1998.
Fox, Stephen. *The American Conservation Movement: John Muir and His Legacy*. Madison: University of Wisconsin Press, 1981.
Fraley, Jill M. "Water, Water, Everywhere: Surface Water Liability." *Michigan Journal of Environmental and Administrative Law* 5, no. 1 (Fall 2015): 73–116.
Frymer, Paul. *Building an American Empire: The Era of Territorial and Political Expansion*. Princeton, NJ: Princeton University Press, 2017.
Gaddie, Ronald Keith, and James L. Regens. *Regulating Wetlands Protection: Environmental Federalism and the States*. Albany: State University of New York Press, 2000.
Garone, Philip. *The Fall and Rise of the Wetlands of California's Great Central Valley*. Berkeley: University of California Press, 2011.
Gates, Paul Wallace. *History of Public Land Law Development*. Washington, DC: Public Land Law Review Commission, 1968.
———. "The Promotion of Agriculture by the Illinois Central Railroad, 1855–1870." *Agricultural History* 5, no. 2 (April 1931): 57–76.
Gerbi, Antonello. *The Dispute of the New World: The History of a Polemic, 1750–1900*. Translated by Jeremy Moyle. Pittsburgh, PA: University of Pittsburgh Press, 1973.
Glacken, Clarence J. *Traces on the Rhodian Shore: Nature and Culture in Western Thought from Ancient Times to the End of the Eighteenth Century*. Berkeley: University of California Press, 1967.
Goetzmann, William H. *Army Exploration in the American West, 1803–1863*. New Haven, CT: Yale University Press, 1959.
Golinski, Jan. "American Climate and the Civilization of Nature." In *Science and Empire in the Atlantic World*, edited by James Delbourgo and Nicholas Dew, 153–74. New York: Routledge, 2008.
———. *British Weather and the Climate of Enlightenment*. Chicago: University of Chicago Press, 2007.
———. "Debating the Atmospheric Constitution: Yellow Fever and the American Climate." *Eighteenth-Century Studies* 49, no. 2 (Winter 2016): 149–65.
———. *Science as Public Culture: Chemistry and Enlightenment in Britain, 1760–1820*. Cambridge, UK: Cambridge University Press, 1992.
Goode, Abby L. *Agrotopias: An American Literary History of Sustainability*. Chapel Hill: University of North Carolina Press, 2022.
Graham, Jennifer S. "The Reasonable Use Rule in Surface Water Law." *Missouri Law Review* 57, no. 1 (Winter 1992): 223–45.
Graham, Otis L., Jr. *Presidents and the American Environment*. Lawrence: University Press of Kansas, 2015.
Gregg, Sara M. "Imagining Opportunity: The 1909 Enlarged Homestead Act and the Promise of the Public Domain." *Western Historical Quarterly* 50, no. 3 (Autumn 2019): 257–79.

Grob, Gerald N. *The Deadly Truth: A History of Disease in America*. Cambridge, MA: Harvard University Press, 2002.

Grove, Richard H. *Green Imperialism: Colonial Expansion, Tropical Island Edens and the Origins of Environmentalism, 1600–1860*. Cambridge, UK: Cambridge University Press, 1995.

Grunwald, Michael. *The Swamp: The Everglades, Florida, and the Politics of Paradise*. New York: Simon & Schuster, 2006.

Hafermehl, Louis N. "To Make the Desert Bloom: The Politics and Promotion of Early Irrigation Schemes in North Dakota." *North Dakota History: Journal of the Northern Plains* 59, no. 3 (Summer 1992): 13–27.

Hager, Willi H. *Hydraulicians in the USA 1800–2000: A Biographical Dictionary of Leaders in Hydraulic Engineering and Fluid Mechanics*. London: Taylor & Francis, 2015.

Hall, Marcus. "The Provincial Nature of George Perkins Marsh." *Environment and History* 10, no. 2 (May 2004): 191–204.

———. "Restoring the Countryside: George Perkins Marsh and the Italian Land Ethic (1861–1882)." *Environment and History* 4, no. 1 (February 1998): 91–103.

Haller, John S., Jr. "Nathaniel Southgate Shaler: A Portrait of Nineteenth-Century Academic Thinking on Race." *Essex Institute Historical Collections* 107 (1971): 173–93.

Hamilton, David E. "Building the Associative State: The Department of Agriculture and American State-Building." *Agricultural History* 64 (Spring 1990): 207–18.

Hankins, Thomas L. *Science and the Enlightenment*. Cambridge, UK: Cambridge University Press, 1985.

Hannaway, Caroline. "Environment and Miasmata." In *Companion Encyclopedia of the History of Medicine*, edited by W. F. Bynum and Roy Porter, 292–308. London: Routledge, 1993.

Hanson, Mark J. "Damming Agricultural Drainage: The Effect of Wetland Preservation and Federal Regulation on Agricultural Drainage in Minnesota." *William Mitchell Law Review* 13, no. 1 (1987): 135–91.

Harrison, Robert. *Congress, Progressive Reform, and the New American State*. Cambridge, UK: Cambridge University Press, 2004.

Harrison, Robert W. *Alluvial Empire: A Study of State and Local Efforts toward Land Development in the Alluvial Valley of the Lower Mississippi River, Including Flood Control, Land Drainage, Land Clearing, Land Forming*. Little Rock, AR: Delta Fund in Cooperation with Economic Research Service, US Department of Agriculture, 1961.

Harrison, Robert W., and Walter M. Kollmorgen. "Land Reclamation in Arkansas under the Swamp Land Grant of 1850." *Arkansas Historical Quarterly* 6, no. 4 (Winter 1947): 369–418.

Hart, John F. "Colonial Land Use Law and Its Significance for Modern Takings Doctrine." *Harvard Law Review* 109, no. 6 (April 1996): 1252–1300.

Hays, Samuel P. *Conservation and the Gospel of Efficiency: The Progressive Conservation Movement, 1890–1920*. Cambridge, MA: Harvard University Press, 1959.

Herget, James E. "Taming the Environment: The Drainage District in Illinois." *Illinois State Historical Society Journal* 71, no. 2 (May 1978): 107–18.

Hewes, Leslie. "The Northern Wet Prairie of the United States: Nature, Sources of Information, and Extent." *Annals of the Association of American Geographers* 41, no. 4 (December 1951): 307–23.

———. "Some Features of Early Woodland and Prairie Settlement in a Central Iowa County." *Annals of the Association of American Geographers* 40, no. 1 (March 1950): 40–57.

Hewes, Leslie, and Phillip E. Frandson. "Occupying the Wet Prairie: The Role of Artificial Drainage in Story County, Iowa." *Annals of the Association of American Geographers* 42, no. 1 (March 1952): 24–50.

Hidy, Ralph, W., Muriel E. Hidy, Roy V. Scott, and Don L. Hofsommer. *The Great Northern Railway: A History*. Cambridge, MA: Harvard Business School, 1988.

Higgs, Robert. *Crisis and Leviathan: Critical Episodes in the Growth of American Government*. New York: Oxford University Press, 1987.

Higham, John. *Strangers in the Land: Patterns of American Nativism, 1860–1925*. New Brunswick, NJ: Rutgers University Press, 1983.

Hixon, Walter. *American Settler Colonialism: A History*. New York: Palgrave Macmillan, 2013.

Hogarth, Rana Asali. "The Myth of Innate Racial Differences between White and Black People's Bodies: Lessons from the 1793 Yellow Fever Epidemics in Philadelphia, Pennsylvania." *American Journal of Public Health* 109, no. 10 (October 2019): 1339–41.

Holcombe, R. I., and William H. Bingham. *Compendium of History and Biography of Polk County, Minnesota*. Minneapolis: W. H. Bingham and Co., 1916.

Horsman, Reginald. *Race and Manifest Destiny: The Origins of American Racial Anglo-Saxonism*. Cambridge, MA: Harvard University Press, 1981.

Howe, Daniel Walker. *What Hath God Wrought: The Transformation of America, 1815–1848*. New York: Oxford University Press, 2007.

Hoxie, Frederick E. *A Final Promise: The Campaign to Assimilate the Indians, 1880–1920*. Lincoln: University of Nebraska Press, 1984.

Humphreys, Margaret. *Malaria: Poverty, Race, and Public Health in the United States*. Baltimore: Johns Hopkins University Press, 2001.

Hundley, Norris, Jr. *Dividing the Waters: A Century of Controversy between the United States and Mexico*. Berkeley: University of California Press, 1966.

———. *Water and the West: The Colorado River Compact and the Politics of Water in the American West*. Berkeley: University of California Press, 1975.

Hurt, R. Douglas. *American Agriculture: A Brief History*. Rev. ed. West Lafayette, IN: Purdue University Press, 2002.

Ihde, Aaron J. *The Development of Modern Chemistry*. New York: Harper & Row, 1964.

Imlay, Samuel J., and Eric D. Carter. "Drainage on the Grand Prairie: The Birth of a Hydraulic Society on the Midwestern Frontier." *Journal of Historical Geography* 38, no. 2 (April 2012): 109–22.

Irmscher, Christoph. *Louis Agassiz: Creator of American Science*. New York: Houghton Mifflin, 2013.

Isenberg, Andrew C., ed. *The Oxford Handbook of Environmental History*. New York: Oxford University Press, 2014.

Isenberg, Nancy. *White Trash: The 400-Year Untold History of Class in America.* New York: Viking, 2016.
Jackson, Donald C. "The Engineer as Lobbyist: John R. Freeman and the Hetch Hetchy Dam (1910–13)." *Environmental History* 21, no. 2 (April 2016): 288–314.
———. "Engineering in the Progressive Era: A New Look at Frederick Haynes Newell and the U.S. Reclamation Service." *Technology and Culture* 34, no. 3 (July 1993): 539–74.
Jacobson, Matthew Frye. *Barbarian Virtues: The United States Encounters Foreign Peoples at Home and Abroad, 1876–1917.* New York: Hill and Wang, 2000.
Jacoby, Karl. *Crimes against Nature: Squatters, Poachers, Thieves, and the Hidden History of American Conservation.* Berkeley: University of California Press, 2001.
Jankovic, Vladimir. "Climates as Commodities: Jean Pierre Purry and the Modelling of the Best Climate on Earth." *Studies in History and Philosophy of Modern Science Part B: Studies in History and Philosophy of Modern Physics* 41, no. 3 (September 2010): 201–7.
Jarcho, Saul. "Cadwallader Colden as a Student of Infectious Disease." *Bulletin of the History of Medicine* 29, no. 2 (March–April 1955): 99–115.
Johnson, Benjamin Heber. "Environment: Nature, Conservation, and the Progressive State." In *A Companion to the Gilded Age and Progressive Era,* edited by Christopher McKnight Nichols and Nancy C. Unger, 71–83. West Sussex, UK: Wiley Blackwell, 2017.
———. *Escaping the Dark, Gray City: Fear and Hope in Progressive-Era Conservation.* New Haven, CT: Yale University Press, 2017.
Johnson, W. Carter, Brett Werner, Glenn R. Guntenspergen, et al. "Prairie Wetland Complexes as Landscape Functional Units in a Changing Climate." *BioScience* 60, no. 2 (February 2010): 128–40.
Judd, Richard W. *The Untilled Garden: Natural History and the Spirit of Conservation in America, 1740–1840.* Cambridge, UK: Cambridge University Press, 2009.
Kaatz, Martin R. "The Black Swamp: A Study in Historical Geography." *Annals of the Association of American Geographers* 45, no. 1 (March 1955): 1–35.
Kane, Lucile M. *The Waterfall That Built a City: The Falls of St. Anthony in Minneapolis.* St. Paul: Minnesota Historical Society, 1966.
Kantor, Isaac. "Ethnic Cleansing and America's Creation of National Parks." *Public Land & Resources Law Review* 28 (June 2007): 41–64.
Keller, Morton. *America's Three Regimes: A New Political History.* New York: Oxford University Press, 2007.
Kelley, Robert. *Battling the Inland Sea: Floods, Public Policy, and the Sacramento Valley.* Berkeley: University of California Press, 1989.
———. "The Context and the Process: How They Have Changed Over Time." In *Water Resources Administration in the United States: Policy, Practice, and Emerging Issues,* edited by Martin Reuss, 10–22. East Lansing: American Water Resources Association and Michigan State University Press, 1993.
Kinyon, Stanley V., and Robert C. McClure. "Interferences with Surface Waters." *Minnesota Law Review* 24, no. 7 (June 1940): 891–939.
Kirby, Jack Temple. "Introduction." In *Nature's Management: Writings on Landscape and Reform, 1822–1859,* edited by Jack Temple Kirby, xi–xxxi. Athens: University of Georgia Press, 2000.

Kirsch, Scott. "The Allison Commission and the National Map: Towards a Republic of Knowledge in Late Nineteenth-Century America." *Journal of Historical Geography* 36, no. 1 (January 2010): 29–42.

Kline, Benjamin. *First Along the River: A Brief History of the U.S. Environmental Movement*. 5th ed. Lanham, MD: Rowman & Littlefield, 2022.

Kluger, James R. *Turning On Water with a Shovel: The Career of Elwood Mead*. Albuquerque: University of New Mexico Press, 1992.

Koch, Alexander, Chris Brierley, Mark M. Maslin, and Simon L. Lewis. "Earth System Impacts of the European Arrival and Great Dying in the Americas after 1492." *Quaternary Science Reviews* 207 (2019): 13–36.

Koelsch, William A. "The Legendary 'Rediscovery' of George Perkins Marsh." *Geographical Review* 102, no. 4 (October 2012): 510–24.

Koppes, Clayton R. "Efficiency/Equity/Esthetics: Towards a Reinterpretation of American Conservation." *Environmental Review* 11, no. 2 (Summer 1987): 127–46.

Krakoff, Sarah. "Settler Colonialism and Reclamation: Where American Indian Law and Natural Resources Law Meet." *Colorado Natural Resources, Energy and Environmental Law Review* 24, no. 2 (2013): 261–86.

Kramer, Paul A. "Empires, Exceptions, and Anglo-Saxons: Race and Rule between the British and United States Empires, 1880–1910." *Journal of American History* 88, no. 4 (March 2002): 1315–53.

Krueger, Lillian. "Motherhood on the Wisconsin Frontier (II)." *Wisconsin Magazine of History* 29, no. 3 (March 1946): 333–46.

Kupperman, Karen Ordahl. "Fear of Hot Climates in the Anglo-American Colonial Experience." *William and Mary Quarterly*, 3rd ser., 41, no. 2 (April 1984): 213–40.

———. "The Puzzle of the American Climate in the Early Colonial Period." *American Historical Review* 87, no. 5 (December 1982): 1262–89.

Lang, M. W., J. C. Ingebritsen, and R. K. Griffin. *Status and Trends of Wetlands in the Conterminous United States 2009–2019: Report to Congress*. Washington, DC: US Department of the Interior, Fish and Wildlife Service, 2024. https://www.fws.gov/sites/default/files/documents/2024-03/wetlands-status-and-trends-2009-2019-signed.pdf.

Langston, Nancy. *Where Land and Water Meet: A Western Landscape Transformed*. Seattle: University of Washington Press, 2003.

Lansing, Michael J. "From Wheat to Wheaties: Minneapolis, the Great Plains, and the Transformation of American Food." In *The Greater Plains: Rethinking a Region's Environmental Histories*, edited by Brian Frehner and Kathleen A. Brosnan, 232–34. Lincoln: University of Nebraska Press, 2021.

LeMay, Michael C. *Guarding the Gates: Immigration and National Security*. Westport, CT: Praeger Security International, 2006.

Leonard, Thomas C. *Illiberal Reformers: Race, Eugenics, and American Economics in the Progressive Era*. Princeton, NJ: Princeton University Press, 2016.

Lepore, Jill. *The Name of War: King Philip's War and the Origins of American Identity*. New York: Vintage, 1999.

Leshy, John D. *Our Common Ground: A History of America's Public Lands*. New Haven, CT: Yale University Press, 2021.

Levere, Trevor H. "Measuring Gases and Measuring Goodness." In *Instruments and Experimentation in the History of Chemistry*, edited by Frederic L. Holmes and Trevor H. Levere, 105–35. Cambridge, MA: MIT Press, 2000.

Lewis, William M., Jr. *Wetlands Explained: Wetland Science, Policy, and Politics in America.* New York: Oxford University Press, 2001.

Lilly, J. Paul. "A History of Swamp Land Development in North Carolina." In *Pocosin Wetlands: An Integrated Analysis of Coastal Plain Freshwater Bogs in North Carolina*, edited by Curtis Richardson, Mary L. Matthews, and Stephen A. Anderson, 20–39. Stroudsburg, PA: Hutchinson Ross, 1981.

Lindemann, Mary. *Medicine and Society in Early Modern Europe.* Cambridge, UK: Cambridge University Press, 1999.

Littlefield, Douglas R. *Conflict on the Rio Grande: Water and the Law, 1879–1939.* Norman: University of Oklahoma Press, 2008.

Livingstone, David N. "A Geologist by Profession, a Geographer by Inclination: Nathaniel Southgate Shaler and Geography at Harvard." In *Science at Harvard University: Historical Perspectives*, edited by Clark A. Elliott and Margaret Rossiter, 146–66. Bethlehem, PA: Lehigh University Press, 1992.

———. *Nathaniel Southgate Shaler and the Culture of American Science.* Tuscaloosa: University of Alabama Press, 1987.

———. "Nature and Man in America: Nathaniel Southgate Shaler and the Conservation of Natural Resources." *Transactions of the Institute of British Geographers* 5, no. 3 (1980): 369–82.

———. "Science and Society: Nathaniel S. Shaler and Racial Ideology." *Transactions of the Institute of British Geographers* 9, no. 2 (1984): 181–210.

Lockley, Tim, and David Doddington. "Maroon and Slave Communities in South Carolina before 1865." *South Carolina Historical Magazine* 113, no. 2 (April 2012): 125–45.

Lovett, Laura L. *Conceiving the Future: Pronatalism, Reproduction, and the Family in the United States, 1890–1938.* Chapel Hill: University of North Carolina Press, 2007.

———. "Land Reclamation as Family Reclamation: The Family Ideal in George Maxwell's Reclamation and Resettlement Campaigns, 1897–1933." *Social Politics: International Studies in Gender, State & Society* 7, no. 1 (Spring 2000): 80–100.

———. "'Rooted in the Soil': Family Ideals, Land Reclamation, and Irrigation Resettlement as Welfare in the United States, 1897–1933." In *Families of a New World: Gender, Politics, and State Development in a Global Context*, edited by Lynne Haney and Lisa Pollard, 85–98. New York: Routledge, 2003.

Lowenthal, David. *George Perkins Marsh: Prophet of Conservation.* Seattle: University of Washington Press, 2000.

Lubetkin, M. John. *Jay Cooke's Gamble: The Northern Pacific Railroad, the Sioux, and the Panic of 1873.* Norman: University of Oklahoma Press, 2006.

Lurie, Edward. *Louis Agassiz: A Life in Science.* Baltimore: Johns Hopkins University Press, 1988.

M., H. H. "Surface Water: The Rights of Abutting Property Owners." *Virginia Law Review* 15 (January 1929): 288–93.

Maass, Arthur, and Raymond L. Anderson. *. . . and the Desert Shall Rejoice: Conflict, Growth, and Justice in Arid Environments.* Cambridge, MA: MIT Press, 1978.

Macleod, David I. "Food Prices, Politics, and Policy in the Progressive Era." *Journal of the Gilded Age and Progressive Era* 8, no. 3 (July 2009): 365–406.

Malone, Michael P. *James J. Hill: Empire Builder of the Northwest*. Norman: University of Oklahoma Press, 1996.

Manganiello, Christopher J. *Southern Water, Southern Power: How the Politics of Cheap Energy and Water Scarcity Shaped a Region*. Chapel Hill: University of North Carolina Press, 2015.

Manning, Thomas G. *Government in Science: The U.S. Geological Survey, 1867–1894*. Lexington: University Press of Kentucky, 1967.

Marti, Donald B. "Agricultural Journalism and the Diffusion of Knowledge: The First Half-Century in America." *Agricultural History* 54, no. 1 (January 1980): 28–37.

Martin, Albro. *James J. Hill and the Opening of the Northwest*. Oxford: Oxford University Press, 1976.

McCally, David. *The Everglades: An Environmental History*. Gainesville: University Press of Florida, 1999.

McCorvie, Mary R., and Christopher L. Lant. "Drainage District Formation and the Loss of Midwestern Wetlands, 1850–1930." *Agricultural History* 67 (Fall 1993): 13–39.

McKenna, Owen P., Samuel Kucia, David Mushet, Michael Anteau, and Mark Wiltermuth. "Synergistic Interaction of Climate and Land-Use Drivers Alter the Function of North American, Prairie-Pothole Wetlands." *Sustainability* 11, no. 23 (2019): 1–20.

McMurry, Sally. "Who Read the Agricultural Journals? Evidence from Chenango County, New York, 1839–1865." *Agricultural History* 63, no. 4 (Autumn 1989): 1–18.

McNally, Robert Aquinas. *Cast Out of Eden: The Untold Story of John Muir, Indigenous Peoples, and the American Wilderness*. Lincoln: University of Nebraska Press, 2024.

McNeill, J. R. *Mosquito Empires: Ecology and War in the Greater Caribbean, 1620–1914*. Cambridge, UK: Cambridge University Press, 2010.

———. *Something New Under the Sun: An Environmental History of the Twentieth-Century World*. New York: W. W. Norton, 2000.

McWilliams, James E. *American Pests: The Losing War on Insects from Colonial Times to DDT*. New York: Columbia University Press, 2008.

Meindl, Christopher F., Derek H. Alderman, and Peter Waylen. "On the Importance of Environmental Claims-Making: The Role of James O. Wright in Promoting the Drainage of Florida's Everglades in the Early Twentieth Century." *Annals of the Association of American Geographers* 92, no. 4 (December 2002): 682–701.

Meinig, D. W. *The Shaping of America: A Geographical Perspective on 500 Years of History*. Vol. 2, *Continental America, 1800–1867*. New Haven, CT: Yale University Press, 1993.

Melosi, Martin. *Water in North American Environmental History*. New York: Routledge, 2022.

Menand, Louis. "Morton, Agassiz, and the Origins of Scientific Racism in the United States." *Journal of Blacks in Higher Education* 34 (Winter 2001–2): 110–13.

Mercer, Lloyd J. *Railroads and Land Grant Policy: A Study in Government Intervention*. New York: Academic Press, 1982.

Merchant, Carolyn. *Ecological Revolutions: Nature, Gender, and Science in New England*. Chapel Hill: University of North Carolina Press, 1989.

Meyer, Melissa L. "The Red Lake Ojibwe." In *The Patterned Peatlands of Minnesota*, edited by H. E. Wright, Jr., Barbara A. Coffin, and Norman E. Aaseng, 251–61. Minneapolis: University of Minnesota Press, 1992.

———. "'We Can Not Get a Living as We Used to': Dispossession and the White Earth Anishinaabeg, 1889–1920." *American Historical Review* 96, no. 2 (April 1991): 368–94.

———. *The White Earth Tragedy: Ethnicity and Dispossession at a Minnesota Anishinaabe Reservation*. Lincoln: University of Nebraska Press, 1994.

Meyer, William B. *Americans and Their Weather*. Oxford: Oxford University Press, 2000.

———. "From Past to Present: A Historical Perspective on Wetlands." In *Wetlands*, edited by Sharon L. Spray and Karen L. McGlothlin, 84–100. Lanham, MD: Rowman & Littlefield, 2004.

———. *Human Impact on the Earth*. Cambridge, UK: Cambridge University Press, 1996.

———. *The Progressive Environmental Prometheans: Left-Wing Heralds of a "Good Anthropocene."* Basingstoke, UK: Palgrave Macmillan, 2016.

———. "When Dismal Swamps Became Priceless Wetlands." *American Heritage* 45, no. 3 (May–June 1994): 108–16.

Miller, Charles A. *Jefferson and Nature: An Interpretation*. Baltimore: Johns Hopkins University Press, 1988.

Miller, David C. *Dark Eden: The Swamp in Nineteenth-Century American Culture*. Cambridge, UK: Cambridge University Press, 1989.

Miller, Ralph N. "American Nationalism as a Theory of Nature." *William and Mary Quarterly* 12, no. 1 (January 1955): 74–95.

Mitsch, William J., and James G. Gosselink. *Wetlands*. 4th ed. Hoboken, NJ: John Wiley & Sons, 2007.

Monaco, C. S. *The Second Seminole War and the Limits of American Aggression*. Baltimore: Johns Hopkins University Press, 2019.

Morris, Christopher. *The Big Muddy: An Environmental History of the Mississippi and Its Peoples, from Hernando de Soto to Hurricane Katrina*. New York: Oxford University Press, 2012.

Morris, J. Brent. *Dismal Freedom: A History of the Maroons of the Great Dismal Swamp*. Chapel Hill: University of North Carolina Press, 2022.

Mulry, Kate Luce. *An Empire Transformed: Remolding Bodies and Landscapes in the Restoration Atlantic*. New York: New York University Press, 2021.

Murray, Stanley N. "Railroads and the Agricultural Development of the Red River Valley of the North, 1870–1890." *Agricultural History* 31, no. 4 (October 1957): 57–66.

———. *The Valley Comes of Age: A History of Agriculture in the Valley of the Red River of the North, 1812–1920*. Fargo: North Dakota Institute for Regional Studies, 1967.

Nagle, John Copeland. "From Swamp Drainage to Wetlands Regulation to Ecological Nuisances to Environmental Ethics." *Case Western Reserve Law Review* 58, no. 3 (2008): 787–812.

Nahlik, A. M., and M. S. Fennessy. "Carbon Storage in US Wetlands." *Nature Communications* 7 (December 2016): 1–9.

Nash, Linda. *Inescapable Ecologies: A History of Environment, Disease, and Knowledge*. Berkeley: University of California Press, 2006.

Navakas, Michele Currie. *Liquid Landscape: Geography and Settlement at the Edge of Early America*. Philadelphia: University of Pennsylvania Press, 2017.

Nelson, John William. *Muddy Ground: Native Peoples, Chicago's Portage, and the Transformation of a Continent*. Chapel Hill: University of North Carolina Press, 2023.

Nelson, Megan Kate. "Hidden Away in the Woods and Swamps: Slavery, Fugitive Slaves, and Swamplands in the Southeastern Borderlands, 1739–1845." In *"We Shall Independent Be": African American Place Making and the Struggle to Claim Space in the United States*, edited by Angel David Nieves and Leslie M. Alexander, 251–72. Boulder: University Press of Colorado, 2008.

———. "The Landscape of Disease: Swamps and Medical Discourse in the American Southeast, 1800–1880." *Mississippi Quarterly* 55, no. 4 (Fall 2002): 535–67.

———. *Saving Yellowstone: Exploration and Preservation in Reconstruction America*. New York: Scribner, 2022.

———. *Trembling Earth: A Cultural History of the Okefenokee Swamp*. Athens: University of Georgia Press, 2009.

Nevius, Marcus P. *City of Refuge: Slavery and Petit Marronage in the Great Dismal Swamp, 1763–1856*. Athens: University of Georgia Press, 2020.

———. "New Histories of Marronage in the Anglo Atlantic World and Early North America." *History Compass* 18, no. 5 (May 2020): 1–14.

Numbers, Ronald L. *The Creationists: From Scientific Creationism to Intellectual Design*. Expanded ed. Cambridge, MA: Harvard University Press, 2006.

Nygren, Joshua. *The State of Conservation: Rural America and the Conservation-Industrial Complex since 1920*. Chapel Hill: University of North Carolina Press, 2025.

O'Neill, Karen M. *Rivers by Design: State Power and the Origins of U.S. Flood Control*. Durham, NC: Duke University Press, 2006.

Opie, John. *Nature's Nation: An Environmental History of the United States*. Fort Worth, TX: Harcourt Brace, 1998.

Orsi, Richard J. *Sunset Limited: The Southern Pacific Railroad and the Development of the American West, 1850–1930*. Berkeley: University of California Press, 2005.

Palmer, Ben[jamin]. "Swamp Land Drainage with Special Reference to Minnesota." *Bulletin of the University of Minnesota* 5. University of Minnesota, Studies in the Social Sciences, March 1915.

Parker, Robert. *Miasma: Pollution and Purification in Early Greek Religion*. Oxford: Clarendon Press, 1983.

Pastore, Christopher L. *Between Land and Sea: The Atlantic Coast and the Transformation of New England*. Cambridge, MA: Harvard University Press, 2014.

Patterson, Gordon. *The Mosquito Crusades: A History of the American Anti-Mosquito Movement from the Reed Commission to the First Earth Day*. New Brunswick, NJ: Rutgers University Press, 2009.

———. *The Mosquito Wars: A History of Mosquito Control in Florida*. Tallahassee: University Press of Florida, 2004.

———. "The Trials and Tribulations of Amos Quito: The Creation of the Florida Anti-Mosquito Association." In *Paradise Lost? The Environmental History of Florida*, edited by Jack E. Davis and Raymond Arsenault, 160–76. Gainesville: University Press of Florida, 2005.

Patterson, K. David. "Yellow Fever Epidemics and Mortality in the United States, 1693–1905." *Social Science and Medicine* 34, no. 8 (April 1992): 855–65.

Pawley, Emily. *The Nature of the Future: Agriculture, Science, and Capitalism in the Antebellum North.* Chicago: University of Chicago Press, 2020.

Pelling, Margaret. "Contagion/Germ Theory/Specificity." In *Companion Encyclopedia of the History of Medicine*, edited by W. F. Bynum and Roy Porter, 309–34. London: Routledge, 1993.

Pemble, Richard H. *The Natural History of the Red River Valley Region before European Settlement.* Moorhead: Minnesota State University, 2005.

Penick, James, Jr. "The Progressives and the Environment: Three Themes from the First Conservation Movement." In *The Progressive Era*, edited by Lewis L. Gould, 115–31. Syracuse, NY: Syracuse University Press, 1974.

Peterson, Richard H. "The Failure to Reclaim: California State Swamp Land Policy and the Sacramento Valley, 1850–1866." *Southern California Quarterly* 56, no. 1 (Spring 1974): 45–60.

Phillips, Sarah T. "Antebellum Reform, Republican Ideology, and Sectional Tension." *Agricultural History* 74, no. 4 (Autumn 2000): 799–822.

———. *This Land, This Nation: Conservation, Rural America, and the New Deal.* New York: Cambridge, UK: Cambridge University Press, 2007.

Pickstone, John V. "Dearth, Dirt and Fever Epidemics: Rewriting the History of British 'Public Health,' 1780–1850." In *Epidemics and Ideas: Essays on the Historical Perception of Pestilence*, edited by Terence Ranger and Paul Slack, 125–48. Cambridge, UK: Cambridge University Press, 1992.

Pielou, E. C. *After the Ice Age: The Return of Life to Glaciated North America.* Chicago: University of Chicago Press, 1991.

Pisani, Donald J. "Beyond the Hundredth Meridian: Nationalizing the History of Water in the United States." *Environmental History* 5, no. 4 (October 2000): 466–82.

———. "A Conservation Myth: The Troubled Childhood of the Multiple-Use Idea." *Agricultural History* 76, no. 2 (Spring 2002): 154–71.

———. "Federalism and the American West, 1900–1940." In *Frontier and Region: Essays in Honor of Martin Ridge*, edited by Robert C. Ritchie and Paul Andrew Hutton, 83–108. San Marino, CA: Huntington Library Press, 1997.

———. *From the Family Farm to Agribusiness: The Irrigation Crusade in California and the West, 1850–1931.* Berkeley: University of California Press, 1984.

———. "George Maxwell, the Railroads, and American Land Policy, 1899–1904." *Pacific Historical Review* 63, no. 2 (May 1994): 177–202.

———. "The Irrigation District and the Federal Relationship: Neglected Aspects of Water History in the Twentieth Century." In *The Twentieth-Century West: Historical Interpretations*, edited by Gerald D. Nash and Richard W. Etulain, 257–92. Albuquerque: University of New Mexico Press, 1989.

———. "The Many Faces of Conservation: Natural Resources and the American State, 1900–1940." In *Taking Stock: American Government in the Twentieth Century*, edited by Morton Keller and R. Shep Melnick, 123–55. Cambridge, UK: Cambridge University Press, 1999.

———. *To Reclaim A Divided West: Water, Law, and Public Policy, 1848–1902*. Albuquerque: University of New Mexico Press, 1992.

———. "Reclamation and Social Engineering in the Progressive Era." *Agricultural History* 57, no. 1 (January 1983): 46–63.

———. "State vs. Nation: Federal Reclamation and Water Rights in the Progressive Era." *Pacific Historical Review* 51, no. 3 (August 1982): 265-82.

———. "A Tale of Two Commissioners: Frederick Newell and Floyd Dominy." In *The Bureau of Reclamation: History Essays from the Centennial Symposium*, edited by Brit Storey, 637–50. Denver: Bureau of Reclamation, 2008.

———. *Water and American Government: The Reclamation Bureau, National Water Policy, and the West, 1902–1935*. Berkeley: University of California Press, 2002.

———. "Water Planning in the Progressive Era: The Inland Waterways Commission Reconsidered." *Journal of Policy History* 18, no. 4 (Fall 2006): 389–418.

Porter, Roy. *The Greatest Benefit to Mankind: A Medical History of Humanity*. New York: W. W. Norton and Company, 1997.

Powell, Miles A. *Vanishing America: Species Extinction, Racial Peril, and the Origins of Conservation*. Cambridge, MA: Harvard University Press, 2016.

Prince, Hugh. "A Marshland Chronicle, 1830–1960: From Artificial Drainage to Outdoor Recreation in Central Wisconsin." *Journal of Historical Geography* 21, no. 1 (January 1995): 3–22.

———. *Wetlands of the American Midwest: A Historical Geography of Changing Attitudes*. Chicago: University of Chicago Press, 1997.

Purcell, Aaron D. *Arthur Morgan: A Progressive Vision for American Reform*. Knoxville: University of Tennessee Press, 2014.

———. "Plumb Lines, Politics, and Projections: The Florida Everglades and the Wright Report Controversy." *Florida Historical Quarterly* 80, no. 2 (Fall 2001): 161–97.

Rabbitt, Mary C. *A Brief History of the U.S. Geological Survey*. Washington, DC: US Department of the Interior, 1986. https://pubs.usgs.gov/book/1986/rabbitt-brief-history/report.pdf.

Rae, John B. "The Great Northern's Land Grant." *Journal of Economic History* 12, no. 2 (Spring 1952): 140–45.

Reese, Ronald. "Under the Weather: Climate and Disease, 1700–1900." *History Today* 46, no. 1 (January 1996): 35–41.

Reiger, John F. *American Sportsmen and the Origins of Conservation*. 3rd ed. Corvallis: Oregon State University Press, 2001.

Reisner, Marc. *Cadillac Desert: The American West and its Disappearing Water*. New York: Viking, 1986.

Reuss, Martin. *The Corps of Engineers and Water Resources in the Progressive Era (1890–1920)*. Kansas City, MO: American Public Works Association, 2009.

———. *Designing the Bayous: The Control of Water in the Atchafalaya Basin, 1800–1995*. College Station: Texas A&M University Press, 2004.

Richardson, Elmo. *The Politics of Conservation: Crusades and Controversies, 1897–1913*. Berkeley: University of California Press, 1962.

Rigby, Kate. *Topographies of the Sacred: The Poetics of Place in European Romanticism.* Charlottesville: University of Virginia Press, 2004.

Righter, Robert W. *The Battle Over Hetch Hetchy: America's Most Controversial Dam and the Birth of Modern Environmentalism.* New York: Oxford University Press, 2005.

Riley, James C. *The Eighteenth-Century Campaign to Avoid Disease.* London: Macmillan, 1987.

Robinson, Elwyn B. *History of North Dakota.* Lincoln: University of Nebraska Press, 1966.

Rodgers, Daniel T. *Atlantic Crossings: Social Politics in a Progressive Age.* Cambridge, MA: Harvard University Press, 1998.

Roger, Philippe. *The American Enemy: A Story of French Anti-Americanism.* Translated by Sharon Bowman. Chicago: University of Chicago Press, 2005.

Rome, Adam. "What Really Matters in History? Environmental Perspectives on Modern America." *Environmental History* 7, no. 2 (April 2002): 303–18.

Rook, Robert E. "An American in Palestine: Elwood Mead and Zionist Water Resource Planning, 1923–1936." *Arab Studies Quarterly* 22, no. 1 (Winter 2000): 71–89.

Ron, Ariel. *Grassroots Leviathan: Agricultural Reform and the Rural North in the Slaveholding Republic.* Baltimore: Johns Hopkins University Press, 2020.

Rosenberg, Gabriel N. "No Scrubs: Livestock Breeding, Eugenics, and the State in the Early Twentieth-Century United States." *Journal of American History* 107, no. 2 (September 2020): 362–87.

Ross, John F. *The Promise of the Grand Canyon: John Wesley Powell's Perilous Journey and His Vision for the American West.* New York: Penguin, 2019.

Rowley, William D. *The Bureau of Reclamation: Origins and Growth to 1945.* Denver: Bureau of Reclamation, 2006.

———. "Introduction." *Agricultural History* 76, no. 2 (Spring 2002): 137–41.

———. *Reclaiming the Arid West: The Career of Francis Newlands.* Bloomington: University of Indiana Press, 1996.

Royster, Charles. *The Fabulous History of the Dismal Swamp Company: A Story of George Washington's Times.* New York: Vintage Books, 2000.

Rozum, Molly P. *Grasslands Grown: Creating Place on the U.S. Northern Plains and Canadian Prairies.* Lincoln: University of Nebraska Press, 2021.

Rumer, Tom. *Unearthing the Land: The Story of Ohio's Scioto Marsh.* Akron, OH: University of Akron Press, 1999.

Rusnock, Andrea. "Hippocrates, Bacon, and Medical Meteorology at the Royal Society, 1700–1750." In *Reinventing Hippocrates*, edited by David Cantor, 136–53. Aldershot, Hampshire, UK: Ashgate Press, 2002.

Saikku, Mikko. *This Delta, This Land: An Environmental History of the Yazoo-Mississippi Floodplain.* Athens: University of Georgia Press, 2005.

Sanders, Elizabeth. *Roots of Reform: Farmers, Workers, and the American State, 1877–1917.* Chicago: University of Chicago Press, 1999.

Sayers, Daniel O., P. Brendan Burke, and Aaron M. Henry. "The Political Economy of Exile in the Great Dismal Swamp." *International Journal of Historical Archaeology* 11, no. 1 (March 2007): 60–97.

Scheiber, Harry N. "American Federalism and the Diffusion of Power: Historical and Contemporary Perspectives." *University of Toledo Law Review* 9 (Summer 1978): 619–80.

———. "Federalism and the Constitution: The Original Understanding." In *American Law and the Constitutional Order: Historical Perspectives*, edited by Lawrence M. Friedman and Harry N. Scheiber, 85–98. Cambridge, MA: Harvard University Press, 1988.

Schmid, James A. "Wetlands as Conserved Landscapes in the United States." In *Cultural Encounters with the Environment: Enduring and Evolving Geographic Themes*, edited by Alexander B. Murphy, Douglas L. Johnson, and Viola Haarmann, 133–55. Lanham, MD: Rowman & Littlefield, 2000.

Schulman, Bruce J. "Governing Nature, Nurturing Government: Resource Management and the Development of the American State, 1900–1912." *Journal of Policy History* 17, no. 4 (October 2005): 375–403.

Scott, James C. *Against the Grain: A Deep History of the Earliest States*. New Haven, CT: Yale University Press, 2017.

———. *Seeing Like a State: How Certain Schemes to Improve the Human Condition Have Failed*. New Haven, CT: Yale University Press, 1998.

Scott, Roy V. "Land Use and American Railroads in the Twentieth Century." *Agricultural History* 53, no. 4 (October 1979): 683–703.

Severson, Kieth E., and Carolyn Hull Sieg. *The Nature of Eastern North Dakota: Pre-1880 Historical Ecology*. Fargo: North Dakota Institute for Regional Studies, 2006.

Shallat, Todd. *Structures in the Stream: Water, Science, and the Rise of the U.S. Corps of Engineers*. Austin: University of Texas Press, 1994.

Sherow, James Earl. *Watering the Valley: Development along the High Plains Arkansas River, 1870–1950*. Lawrence: University Press of Kansas, 1990.

Skowronek, Stephen. *Building a New American State: The Expansion of National Administrative Capacities, 1877–1920*. New York: Cambridge University Press, 1982.

Sleeper-Smith, Susan. *Indigenous Prosperity and American Conquest: Indian Women of the Ohio River Valley, 1690–1792*. Chapel Hill: University of North Carolina Press, 2018.

Smith, Kimberly K. *The Conservation Constitution: The Conservation Movement and Constitutional Change, 1870–1930*. Lawrence: University Press of Kansas, 2019.

Spence, Mark David. *Dispossessing the Wilderness: Indian Removal and the Making of the National Parks*. New York: Oxford University Press, 1999.

Spiro, Jonathan Peter. *Defending the Master Race: Conservation, Eugenics, and the Legacy of Madison Grant*. Burlington: University of Vermont Press, 2009.

Stegner, Wallace. *Beyond the Hundredth Meridian: John Wesley Powell and the Second Opening of the West*. Boston: Houghton Mifflin, 1954.

Steinberg, Theodore L. *Down to Earth: Nature's Role in American History*. 4th ed. New York: Oxford University Press, 2018.

———. *Slide Mountain: or the Folly of Owning Nature*. Berkeley: University of California Press, 1995.

Stern, Alexandra Minna. *Eugenic Nation: Faults and Frontiers of Better Breeding in Modern America*. 2nd ed. Oakland: University of California Press, 2016.

Stewart, Mart. *"What Nature Suffers to Groe": Life, Labor, and Landscape on the Georgia Coast, 1680–1920*. Athens: University of Georgia Press, 1996.

Stine, Jeffrey K. *America's Forested Wetlands: From Wasteland to Valued Resource*. Durham, NC: Forest History Society, 2008.

Stoll, Steven. *Larding the Lean Earth: Soil and Society in Nineteenth-Century America.* New York: Hill and Wang, 2002.

Strausberg, Stephen F. "Indiana and the Swamp Lands Act: A Study in State Administration." *Indiana Magazine of History* 73, no. 3 (September 1977): 191–203.

Strom, Claire. *Profiting from the Plains: The Great Northern Railway and Corporate Development of the American West.* Seattle: University of Washington Press, 2003.

Suazo, Matthew E. "Translating Dumont de Montigny's Frog: Or, Locating the *Mémoires Historiques Sur La Louisiane* in an Ecology of American Degeneracy." In *Swamp Souths: Literary and Cultural Ecologies,* edited by Kirstin L. Squint, Eric Gary Anderson, Taylor Hagood, and Anthony Wilson, 111–25. Baton Rouge: Louisiana State University Press, 2020.

Sutter, Paul S. "'The First Mountain to Be Removed': Yellow Fever Control and the Construction of the Panama Canal." *Environmental History* 21, no. 2 (April 2016): 250–59.

———. *Let Us Now Praise Famous Gullies: Providence Canyon and the Soils of the South.* Athens: University of Georgia Press, 2015.

———. "Nature's Agents or Agents of Empire? Entomological Workers and Environmental Change during the Construction of the Panama Canal." *Isis: A Journal of the History of Science* 98, no. 4 (2007): 724–54.

———. "The World with Us: The State of American Environmental History." *Journal of American History* 100, no. 1 (June 2013): 94–119.

Swain, Donald C. *Federal Conservation Policy, 1921–1933.* Berkeley: University of California Press, 1963.

Tanji, K. K., and C. G. Keyes, Jr. "Water Quality Aspects of Irrigation and Drainage: Past History and Future Challenges for Civil Engineers." In *Perspectives in Civil Engineering: Commemorating the 150th Anniversary of the American Society of Civil Engineers,* edited by Jeffrey S. Russell, 101–9. Reston, VA: American Society of Civil Engineers, 2003.

Taylor, Dorceta E. *The Rise of the American Conservation Movement: Power, Privilege, and Environmental Protection.* Durham, NC: Duke University Press, 2016.

Taylor, Ryan W. *Federalism of Wetlands.* London: Routledge, 2014.

Teale, Chelsea. "Agricultural Wetland Use and Management in the Dutch-Settled Northeast, 1620 to 1800." *New York History* 98, no. 2 (Spring 2017): 177–204.

Teisch, Jessica B. *Engineering Nature: Water, Development, and the Global Spread of American Environmental Expertise.* Chapel Hill: University of North Carolina Press, 2011.

Tester, John R. *Minnesota's Natural Heritage: An Ecological Perspective.* Minneapolis: University of Minnesota Press, 1995.

Thompson, John. *Wetlands Drainage, River Modification, and Sectoral Conflict in the Lower Illinois Valley, 1890–1930.* Carbondale: Southern Illinois University Press, 2002.

Thompson, Kenneth. "Historic Flooding in the Sacramento Valley." *Pacific Historical Review* 29, no. 4 (November 1960): 349–60.

Thornton, Russell. *American Indian Holocaust and Survival: A Population History Since 1492.* Norman: University of Oklahoma Press, 1987.

Tiner, Ralph W. *In Search of Swampland: A Wetland Sourcebook and Field Guide.* 2nd ed. New Brunswick, NJ: Rutgers University Press, 2005.

———. *Wetland Indicators: A Guide to Wetland Formation, Identification, Delineation, Classification, and Mapping.* 2nd ed. Boca Raton, FL: CRC Press, 2017.
Tomes, Nancy. *The Gospel of Germs: Men, Women, and the Microbe in American Life.* Cambridge, MA: Harvard University Press, 1998.
Trachtenberg, Alan. *The Incorporation of America: Culture and Society in the Gilded Age.* New York: Hill and Wang, 1982.
Tuck, Eve, and K. Wayne Yang. "Decolonization Is Not a Metaphor." *Decolonization: Indigeneity, Education & Society* 1, no. 1 (2012): 1–40.
Tuffnell, Stephen. *Made in Britain: Nation and Emigration in Nineteenth-Century America.* Berkeley: University of California Press, 2020.
Tyrrell, Ian. *Crisis of the Wasteful Nation: Empire and Conservation in Theodore Roosevelt's America.* Chicago: University of Chicago Press, 2015.
Urban, Michael A. "An Uninhabited Waste: Transforming the Grand Prairie in Nineteenth Century Illinois, USA." *Journal of Historical Geography* 31, no. 4 (October 2005): 647–65.
Valenčius, Conevery Bolton. *The Health of the Country: How American Settlers Understood Themselves and the Their Land.* New York: Basic Books, 2002.
Van Hise, Charles Richard. *The Conservation of Natural Resources in the United States.* New York: Macmillan, 1921.
Veenendaal, Augustus J., Jr. *The Saint Paul and Pacific Railroad: An Empire in the Making, 1862–1879.* DeKalb: Northern Illinois University Press, 1999.
Veracini, Lorenzo. "Introducing: Settler Colonial Studies." *Settler Colonial Studies* 1 (January 2011): 1–12.
———. *Settler Colonialism: A Theoretical Overview.* New York: Palgrave Macmillan, 2010.
Vileisis, Ann. *Discovering the Unknown Landscape: A History of America's Wetlands.* Washington, DC: Island Press, 1997.
Vogel, Brant. "The Letter from Dublin: Climate Change, Colonialism, and the Royal Society in the Seventeenth Century." *Osiris* 26, no. 1 (2011): 111–28.
Warren, Louis S. *The Hunter's Game: Poachers and Conservationists in Twentieth-Century America.* New Haven, CT: Yale University Press, 1997.
Wear, Andrew. "Making Sense of Health and the Environment in Early Modern England." In *Medicine in Society: Historical Essays,* edited by Andrew Wear, 119–48. Cambridge, UK: Cambridge University Press, 1992.
———. "Place, Health, and Disease: The *Air, Waters, Places* Tradition in Early Modern England and North America." *Journal of Medieval and Early Modern Studies* 38, no. 3 (Fall 2008): 443–65.
Weaver, Marion M. *History of Tile Drainage (In America Prior to 1900).* Waterloo, NY: M. M. Weaver, 1964.
Weidenhammer, Erich. "Patronage and Enlightened Medicine in the Eighteenth-Century British Military: The Rise and Fall of Dr. John Pringle, 1707–1782." *Social History of Medicine* 29, no. 1 (February 2016): 21–43.
Wellock, Thomas R. *Preserving the Nation: The Conservation and Environmental Movements: 1870–2000.* Wheeling, IL: Harlan Davidson, 2007.
White, Richard. *"It's Your Misfortune and None of My Own": A New History of the American West.* Norman: University of Oklahoma Press, 1991.

———. *Railroaded: The Transcontinentals and the Making of Modern America*. New York: W. W. Norton, 2011.

———. *The Republic for Which it Stands: The United States during Reconstruction and the Gilded Age, 1865–1896*. Oxford: University of Oxford Press, 2017.

White, Sam. "Unpuzzling American Climate: New World Experience and the Foundations of a New Science." *Isis* 106, no. 3 (September 2015): 544–66.

White, Sam, Kenneth M. Sylvester, and Richard Tucker. "North American Climate History." In *Cultural Dynamics of Climate Change and the Environment in North America*, edited by Bernd Sommer, 109–36. Leiden: Brill, 2015.

White, W. Thomas. "A Gilded Age Businessman in Politics: James J. Hill, the Northwest, and the American Presidency, 1884–1912." *Pacific Historical Review* 57, no. 4 (November 1988): 439–56.

Whyte, Kyle. "Settler Colonialism, Ecology, and Environmental Injustice." *Environment and Society: Advances in Research* 9 (September 2018): 125–44.

Wilbur, Ray Lyman, and William Atherton Du Puy. *Conservation in the Department of the Interior*. Washington, DC: Government Printing Office, 1931.

Wilhelm, Peter W. "Draining the Black Swamp: Henry and Wood Counties, Ohio, 1870–1920." *Northwest Ohio Quarterly* 56, no. 3 (Summer 1984): 79–95.

Williams, Michael. "Agricultural Impacts in Temperate Wetlands." In *Wetlands: A Threatened Landscape*, edited by Michael Williams, 181–216. Oxford: Basil Blackwell, 1990.

Willoughby, Christopher D. "'His Native, Hot Country': Racial Science and Environment in American Medical Thought." *Journal of the History of Medicine and Allied Sciences* 72, no. 3 (July 2017): 328–51.

Wilson, Anthony. *Shadow and Shelter: The Swamp in Southern Culture*. Jackson: University Press of Mississippi, 2006.

———. *Swamp: Nature and Culture*. London: Reaktion, 2018.

Wilson, Robert M. *Seeking Refuge: Birds and Landscapes of the Pacific Flyway*. Seattle: University of Washington Press, 2010.

Wittfogel, Karl. *Oriental Despotism: A Comparative Study of Total Power*. New Haven, CT: Yale University Press, 1957.

Wolfe, Patrick. "Settler Colonialism and the Elimination of the Native." *Journal of Genocide Research* 8, no. 4 (December 2006): 387–409.

Wood, Gordon S. *Empire of Liberty: A History of the Early Republic, 1789–1815*. Oxford: Oxford University Press, 2009.

———. "Environmental Hazards, Eighteenth-Century Style." In *Old World, New World: America and Europe in the Age of Jefferson*, edited by Leonard J. Sadosky, Peter Nicolaisen, Peter S. Onuf, and Andrew O'Shaughnessy, 15–31. Charlottesville: University of Virginia Press, 2010.

Worster, Donald. *A River Running West: The Life of John Wesley Powell*. Oxford: Oxford University Press, 2001.

———. *Rivers of Empire: Water, Aridity, and the Growth of the American West*. New York: Pantheon Books, 1985.

———. *Shrinking the Earth: The Rise and Decline of Natural Abundance*. New York: Oxford University Press, 2016.

———. *An Unsettled Country: Changing Landscapes of the American West*. Albuquerque: University of New Mexico Press, 1994.

Wrobel, David M. *The End of American Exceptionalism: Frontier Anxiety from the Old West to the New Deal*. Lawrence: University Press of Kansas, 1993.

Wyant, William. *Westward in Eden: Public Lands and the Conservation Movement*. Berkeley: University of California Press, 1982.

Yochelson, Ellis L. *Charles Doolittle Walcott, Paleontologist*. Kent, OH: Kent University Press, 1998.

Zabilka, Ivan L. "Nathaniel Southgate Shaler and the Kentucky Geological Survey." *Register of the Kentucky Historical Society* 80, no. 4 (Autumn 1982): 408–31.

Zavodnyik, Peter. *The Rise of the Federal Colossus: The Growth of Federal Power from Lincoln to F.D.R.* Santa Barbara, CA: Praeger, 2011.

Zilberstein, Anya. *A Temperate Empire: Making Climate Change in Early America*. Oxford: Oxford University Press, 2016.

Zuidervaart, Huib J. "An Eighteenth-Century Medical-Meteorological Society in the Netherlands: An Investigation of Early Organization, Instrumentation and Quantification. Part 2." *British Journal for the History of Science* 39, no. 1 (March 2006): 49–66.

Zwiers, Maarten. "Swamp Manifesto: Manifestations of the Swamp." *Soapbox: Journal for Cultural Analysis* 5 (2024): 21–39.

UNPUBLISHED THESES AND DISSERTATIONS

Allen, Davis. "A Deep History of Shallow Waters: Enclosing the Wetland Commons in the Era of Improvement." PhD diss., Case Western Reserve University, 2022.

Almquist, Alton Wilhelm. "Farm Drainage in the Red River Valley of Minnesota." Master's thesis, University of Minnesota, 1955.

Berg, Walter Louis. "Nathaniel Southgate Shaler: A Critical Study of an Earth Scientist." PhD diss., University of Washington, 1957.

Eisenstadt, Peter R. "The Weather and Weather Forecasting in Colonial America." PhD diss., New York University, 1990.

Otto, Joseph W. "Plumbing the Prairies: Water Management in the Agricultural Midwest, 1850–1920." PhD diss., University of Oklahoma, 2023.

Poe, Cynthia R. "Reconstructing the Levees: The Politics of Flooding in Nineteenth-Century Louisiana." PhD diss., University of Wisconsin, 2006.

Soffar, Allan J. "Differing Views on the Gospel of Efficiency: Conservation Controversies between Agriculture and Interior, 1898–1938." PhD diss., Texas Tech University, 1974.

Stalcup, Samuel R. "Public Interest, Private Lands: Soil Conservation in the United States, 1890–1940." PhD diss., University of Oklahoma, 2014.

Windsor, Roger Andrew. "Artificial Drainage of East Central Illinois, 1820–1920." PhD diss., University of Illinois, 1975.

Index

African Americans. *See* Black people
Agassiz, Louis, 56, 97–100, 207n12
Alabama, 32, 51, 143, 212n22, 214n42
American Farmer (journal), 39, 42
American Forestry Association, 127
American Philosophical Society, 25, 28
American Review of Reviews (magazine), 141
American Revolution, 94
American West, 9–10, 105, 193n25, 199n39
Anglo-Saxons, 11, 95, 97, 114–20
Anishinaabe lands, 12, 15, 122–25, 129, 132, 145, 148–50, 182
Anishinaabe people, 32, 122–23, 146, 182–83, 189
Antill, Edward, 24
Antiquities Act (1906), 124
Appalachian Mountains, 38, 50, 106. *See also* trans-Appalachia
Arkansas, 38, 51, 137, 139, 144, 168–71, 176, 212n22, 214n42
Arkansas River, 138–39
Atchafalaya basin, 15
Atlantic Coast Division, 96, 105, 115. *See also* US Geological Survey (USGS)
Atlantic Monthly (magazine), 115–16

Bacon, Augustus O., 144–45
Baker, A. H., 84–86
Balogh, Brian, 11, 197n6
Baltimore, MD, 142
Bay-mway-way-be-nais (Anishinaabe leader), 182, 184, 186
Beacon Society (Boston), 112
Benedict Drainage District (Kansas), 152–53
Bernard, A. G., 136, 170

Bible, 25–26, 98; Genesis, 99. *See also* Noah's flood
Bien, Morris, 140, 148, 215n55
Black Oak Swamp (Indiana), 26
Black people, 30, 32, 117; bodies of and racism against, 5, 98, 116–17, 119; communities of, 5; history and knowledge of, 13, 97, 108, 114, 120, 185; resistance of, 33. *See also* race
Board of Agriculture (Oklahoma Territory), 135
Board of State Drainage Commissioners (Minnesota), 90
Boen, Peter (Peder O.), 54–55, 71
bogs, 4, 8, 17, 23, 37–38, 46, 112, 192n11. *See also* wetlands
Bonanza farms, 60–61, 69–70, 74, 80–81
Boston Cultivator (newspaper), 42
Brahmins, 115, 118
Brevik, Ole O., 66, 202n29
Brewer, David J., 139–40, 142
Britain, 100, 102, 109, 115. *See also* England
Broward, Napoleon Bonaparte, 142, 144
Brown, Frederick P., 89
Buffalo River (Minnesota), 54–55, 73–74
Buffon, Comte de (Georges-Louis Leclerc), 18, 22–23, 25
Bureau of Indian Affairs, 183; Commissioner of, 182
Bureau of Reclamation. *See* Reclamation Service
Byrd, William, II, 4

Caldwell, Charles, 28–30
California, 51, 101, 137, 157–58; Indigenous peoples of, 31; politicians and attorneys of, 92, 126, 134–36, 169, 172

California Water and Forest Association, 157
Canada, 56, 96, 126; Manitoba, 10, 57, 187, 199n39; Nova Scotia, 22; Ontario, 57, 62
canals, 65–66, 70, 112, 127, 151, 158, 163
Cannon, Joseph G., 134
Cannon, Thomas L., 135
Caribbean, 27
Carter, Thomas H., 126
Casey, Thomas Lincoln, 87
Central Valley, 31, 184
Charlotte Observer (newspaper), 142
Chesapeake, 22
Chicago River, 32
Chicago's muddy portage, 31–32, 184
Childs, E. D., 79
Chippewa, 122. *See also* Lake Superior Chippewa
Christians, 17; Catholic, 95, 117–18; Protestants, 95
Civil War: antebellum, 37, 41, 44, 62, 187; during and post-war, 44, 58–59, 99–100, 105, 126, 155, 186
Clarke, James, 144
Clean Water Act (1972), 186
climate: changes, 25–26; crisis, 187; manipulation of, 19, 24, 44, 48; unhealthy, 14, 17–18, 21–23, 27, 29, 33, 38, 40
climatology, 26, 33
Cobb, Collier, 109, 111
Colden, Cadwallader, 24
Colorado, 105, 128, 138–39, 144
Colorado River, 106, 156
Columbia Magazine, 25
Columbia River, 136, 156
Columella, Lucius Junius Moderatus, 20
committees (congressional), 86–87, 126, 151, 153, 157, 161, 170; Committee on Agriculture and Forestry, 134; Committee on Geological Survey, 136; Committee on Resolutions, 142; House Committee on Agricultural Expenditures, 179; House Committee on Agriculture, 171; House Committee on Irrigation of Arid Lands, 128, 133; House Committee on Public Lands, 133–34; Senate Committee on Public Lands, 143–44
common enemy doctrine, 72
Comstock, Solomon G., 74
Conger, A. B., 47
Congress, 7; acts of, 93, 127, 149; agency creation, 9, 11, 127, 157–58; land cessions and transfers, 12, 52, 87, 122, 156, 178, 182; policies of, 8, 14, 34; reports and surveys, 4, 103, 105, 109–10, 134, 146, 155–56, 166; subsidies, 53, 58–59, 91. *See also* Swamp Land Acts (1849, 1850, and 1860)
conservationism, 11, 145
conservation movement, 120, 123, 146, 180, 184; federal, goals of, 12–13, 124–25, 150; federal, initiatives, 78, 114, 132–33, 161; rise of, 11, 15, 97–98, 102, 112; studies of, 9
constitutionality: state, 84–86, 164; US, 135, 139, 141–42, 144
Cooke, Jay, 59
Cornell University, 187
Country Gentleman (magazine), 60
courts, 50, 56, 70, 72, 164. *See also specific courts*
Cowles, Henry, 41
Coxe, Tench, 24, 30, 185
Crooks, A. D., 151
Crookston, MN, 1–3, 7, 11, 14, 61, 76, 81, 94, 164; conventions, 1–3, 6, 12, 78–79, 82, 92, 123, 160
Crookston Times (newspaper), 79
crop rotation, 36, 52
Cuba, 96
cultivation of land, 30, 32, 43, 73, 83; boundary of, 6–7, 36; failures related to, 23–24, 172; impact of, 143; issues with, 4, 17, 62, 64, 174; potential of, 96, 111–12, 161–63
Cultivator (journal), 39–42, 45–48
Currie, William, 27–28

Dakota Territory, 70, 77. *See also* North Dakota
dams, 9, 54, 103–4, 112, 128, 139, 189
Darwin, Charles, 98–100
Day, Frank Arah, 87
DeBow's Review (magazine), 49
Deep Fork Drainage Association (Oklahoma), 135
deforestation, 18, 22–25, 27, 103
Democrats, 128, 134–35, 144, 171; Southern, 125, 129–30, 143. *See also specific politicians*
Department of the Interior, 9, 105, 131, 141–42, 156–57, 169, 214n41; Secretary of, 128, 138, 142, 144. *See also* General Land Office (GLO)
Depression of 1893, 126
Des Plaines River, 32
Devil's Lake (Wisconsin), 77
disease, 33, 39, 46–47, 51, 156, 169, 184–86; rural, 4, 33. *See also* malaria; miasma; yellow fever
dispossession, 7, 13, 93–94, 120; Indigenous, 3, 11–12, 96, 123, 131, 149, 179, 183, 185. *See also* reclamation; settler colonialism
District of Columbia. *See* Washington, DC
Donaldson, H. M., 89
Douglas, Stephen, 58
drainage: considerations, 2–3, 16, 47, 56, 66, 69–70, 79, 82, 162; failures of, 67, 73; funding for, 128, 130, 176; impact of, 8, 14, 16, 43, 52, 97, 101, 123–24, 154, 176, 181–82, 187, 207n18; need for, 6, 21–22, 24, 29–30, 35–36, 104; opposition to, 81, 182–83; problems related to, 63, 74, 84, 170; proposals for, 70, 133–34, 138, 141–43, 164, 167; purpose of, 11–13, 45, 132; relationship to conservation movement, 11–13, 123–26, 161; responsibility for, 134, 154–55, 170. *See also* drainage districts; drainage systems; drainage tiles; drains and underdrains; dredging and dredges; improvements; Indigenous: lands; Midwest; reclamation; Swamp Land Acts (1849, 1850, and 1860)

Drainage and Farm Journal, 84, 160
drainage districts: formation of, 11, 14, 37, 50–53, 80, 85–86, 90, 151–52, 163–64; problems related to, 63, 72, 122, 131, 189–90; responsibilities, 8, 14, 16, 55, 83, 131, 143–44, 172, 190; support for, 15, 79, 135, 153, 180, 183
Drainage Journal, 79, 160
drainage systems, 57, 132, 164–65, 172, 187. *See also* drains and underdrains
drainage tiles, 2, 7, 43–44, 53, 122, 165. *See also* drains and underdrains; Johnston, John
drains and underdrains, 28, 37, 44, 51, 69, 79–80, 154, 162, 165–66, 183
dredging and dredges, 7–8, 33, 155, 160
drought, 42
Duffy, Mark, 188

easements, 66
ecology, 3, 13, 15, 53, 104, 114, 141, 181, 183, 186
ecosystems, 32, 75, 101, 165, 169, 178, 183, 186–87; non–North American, 196n36; premodern definition of, 195n12
Edmundson, W. G., 48
Ellet, Charles, Jr., 103, 208n23
Elliott, C. G. (Charles Gleason), 1–2, 6, 14–15, 79–84, 86–87, 151–54, 158–67, 169–73, 176, 178–81
engineers, 8, 16, 75, 82, 123, 159; civil, 74, 157; federal, 136, 150, 152, 160–63, 165–66, 185; ODI, 15, 125, 153–54, 168–69, 178–79; USDA, 135, 142, 174; railroad, 55, 65–66, 73, 75
England, 7, 39–40, 44, 60, 100–101, 103, 118; London, 22. *See also* Great Britain
Enlightenment, 17–18
enslaved people, 5, 8, 26, 33, 36, 184. *See also* Black people
Escher, Johannes, 98
estuaries, 4, 188. *See also* wetlands
eudiometers, 17, 21
eugenics, 3, 11, 93, 97–98, 114–15, 118–20, 162, 183; eugenicists, 100, 105

Europe, 26, 29, 102, 109, 119; eastern, 95, 105, 114, 117–18, 126; intellectuals in, 23; land management in, 26, 107, 109, 134, 172, 185; northern, 60, 116, 118; peasants in, 116–17; physicians and scientists in, 4, 17–20, 22; settlers and colonization of, 22, 25, 113; southern, 95, 114
Evening Statesman (newspaper), 130
Everglades. *See* Florida

Fall River (Kansas), 151
Fanning, J. T., 84
Farmers' Bulletin, 160
Farmers' Register (periodical), 35, 38
farms, diversification of, 35, 41, 52, 92. *See also* bonanza farms; wetlands: conversion of
federalism: and drainage, 3, 10, 142, 148; environmental, 52; legacy of, 189–90; problems related to, 15, 83, 125, 145; violations of, 133–34, 141
Ferguson, Adam, 23
Fiege, Mark, 49
Fisher, Walter L., 145
Fisher Bulletin, 66, 70
Fiske, John, 118
Flint, Frank P., 134–38, 143–44, 169, 172, 176
Florida, 26, 51, 174, 176, 178, 212n22, 214n42; Everglades, 142, 168, 178–79; politicians of, 142, 144, 178–79, 215n53
Fontana, Felice, 21
Forestry and Irrigation (journal), 131–32
Forsyth, Andrew N., 71
Foster, John, 43
France, 72, 103
Frantz, Frank, 135
Fredonia Daily Herald, 152, 155
French naturalists, 18, 22, 99
Frymer, Paul, 12
Fulton, Charles William, 134

Gaa-Miskwaabikaang, 188
Garfield, James A., 142–44
General Land Office (GLO), 6, 141, 143. *See also* Department of the Interior

Genesee Farmer (journal), 39, 42, 46
geology, 96–101, 103–7, 109–10, 113, 115, 159; geologists, 10, 56, 93, 100, 109
George II (king), 4
Georgia, 27, 134, 162, 165; politicians of, 144, 164–65
Georgia Medical Society, 30
Germany, 60, 118
Gilded Age, 1, 10, 61, 64, 89, 95, 97, 100, 187
Gilfillan, C. J., 73
Gougé-Powless, Carolyn L.C., 188
Grand Canyon, 106
Grand Forks Herald (newspaper), 66
Gray, Asa, 99
grazing, 39, 124, 130, 156. *See also* livestock
Great Britain, 100, 102, 109, 115. *See also* England
Great Dismal Swamp, 4–5, 32, 109–11, 124, 130–31, 134
Great Northern Railway (GN), 1, 14, 55, 77, 88–89, 92, 126, 130
Great Plains, 4, 94
Greece, 19, 21
Gronna, Asle, 130
Grunsky, C. E., 131, 137
Guinea Coast (West Africa), 117
Gulf of Mexico, 187
Gunnison Project (Colorado), 128

Hamilton, Alexander, 24
Hamilton, David E., 154
Hansbrough, Henry C., 129–31
Harbaugh, Springer, 70, 81–82
Hardman, Lamartine Griffin, 164–65
Harvard University, 10, 93–94, 98–101, 109. *See also* Lawrence Scientific School
Hatch Act (1887), 156
Hawaii, 96
Hayden, Ferdinand V., 105
Hayes, Rutherford B., 60, 105
Hays, Samuel P., 9
Hearst's International (magazine), 141
Hetch Hetchy Valley, 146
Hill, James J., 1, 7, 55–56, 61–67, 69, 74–80, 88–89, 92, 126, 130

Hill, Lewis W., 130
Hippocrates of Cos, 19
Hippocratic tradition, 17–20. *See also* neo-Hippocratic tradition and new Hippocratism
Hippocratic Corpus, 19
Histoire naturelle (Leclerc), 22
History of America (Robertson), 23
Hitchcock, Ethan A., 128, 137
Hogenson, Hogen M., 67, 69–75
Holcomb, Silas A., 126
Holland, 7, 60, 84, 104
Homestead Act (1862), 61, 122
Hubbard, Lucius F., 85
Hudson Bay, 56
Humphreys, Margaret, 219n55
Hungary, 115
Huxley, Thomas, 100
hydrology, 56, 110, 141, 182; hydrologists, 15

Iceland, 60
Idaho Statesman (newspaper), 130
Illinois, 38, 124, 160; drainage programs in, 51, 55, 79–80, 86, 190; ecosystems, 57, 172, 187; politicians of, 58, 134, 158
Illinois Central Railroad, 58
Illinois Farmer (journal), 38
Illinois River, 158
Illinois Wesleyan College, 158
illness. *See* disease; malaria; miasma; yellow fever
immigration, 95, 97, 100, 114–16, 118, 120, 123, 132; new immigration
Immigration Restriction League (Boston), 118
imperialism, 11, 21, 115, 119
improvements: agricultural, 14, 24, 36, 41, 43–44, 52; health, 16, 40; navigation, 103; practices, 35, 157; problems related to, 149, 151; responsibility for, 84–85, 146, 174, 176; rural, 13, 66. *See also* canals; drainage
Indiana, 26, 51, 55, 57, 79–80, 124, 144, 172, 187
Indigenous communities, 12, 22, 31, 108, 184, 189, 192n11

Indigenous history and knowledge, 13, 36, 97, 108, 114, 120, 185
Indigenous lands: cession of, 122, 124, 126, 129, 131, 140, 145; connections to, 146, 150; claims to, 3, 6, 34, 36. *See also* dispossession: Indigenous; wetlands: Indigenous relationships with
Indigenous people, 5, 8, 24, 26, 30, 32–33. *See also specific groups*
infrastructure, 6–7, 9, 90, 96, 184–85
Inland Waterways Commission, 214n41
Irish people, 26, 44
Iron Mountain Railroad, 171
irrigation: impact of, 10, 108, 114, 124, 134, 139, 145, 158; practice of, 9, 12, 63–64, 106, 140–41; programs and projects, 9, 92, 120, 123, 126–30, 132–33, 135–46, 156–58, 180, 184; structures, 9, 103, 128, 137; studies and surveys of, 10, 109–10, 157
Irrigation Age (journal), 131
Isenberg, Nancy, 26
Isthmian Canal Commission, 137
Italian physicians, 21, 39

Jefferson, Thomas, 18, 25, 30, 146
Jeffersonianism, 133, 162
Johns Hopkins University, 142
Johnston, John, 41, 43–44, 47
Journal of the American Peat Society, 141

Kankakee Marsh, 124
Kansas, 54, 57, 95, 139, 153; politicians of, 151, 171
Kansas Supreme Court, 71
Kentucky, 98–100, 103–4; politicians of, 101, 105
Kentucky Geological Survey, 97, 117
Kerr, T., 40
King, Clarence, 105
King, D. E., 171
King Philip's War, 5
Konkow, 31

Lake Agassiz, 56–57, 82
Lake Erie, 31

Lake Michigan, 46
Lake Superior, 188
Lake Superior Chippewa: Lac du Flambeau Band of, 189; Lac Vieux Desert Band of, 189; Red Cliff Band of, 188
Lake Traverse, 87
Lamarck, Jean-Baptiste, 99
Lambert, Lewis, 46–47
land, conversion of, 3, 45, 52–53, 72, 108, 114; speculation, 58, 91, 129, 154, 178. *See also* dispossession: Indigenous; reclamation; wetlands
Land Ordinance (1785), 6
Landriani, Marsilio, 21
Lane, Alfred C., 109
Latimer, Asbury, 130, 134–35, 142–44
Latin America, 116
Lawrence Scientific School, 115. *See also* Harvard University
lawsuits, 56, 71, 73–74. *See also* US Supreme Court
Leclerc, Georges-Louis, 18, 22–23, 25
Lepore, Jill, 192n11
Leslie, Preston H., 101
Lester, H. W., 46
levees, 47, 51, 62, 104, 126, 151–53, 158, 160, 174, 208n23; boards, 50, 53
Lever, Asbury F., 134
Literary Magazine, and American Register, 5
Little Ice Age, 22
Little River (Oklahoma), 135
livestock, 36–37, 39–40, 46, 128, 156
localism, 58, 123, 145, 150, 189–90
Lockhart, Charles, 70
Lommeland, Andrew, 54–57, 61, 67, 69–70, 74–75
Los Angeles County Superior Court, 136
Louisiana, 51, 168, 170–71, 174, 176, 212n22, 214n42. *See also* New Orleans, LA
Lyell, Charles, 100

Macon, Robert B., 171
Maine, 27, 108; politicians of, 134
malaria, 4–5, 18, 28, 33, 35, 38–40, 104, 169, 172, 180. *See also* disease

Man and Nature (Marsh), 97, 101–3
Manvel, Allen, 73–74
maroon communities, 32. *See also* Black people
Marsh, George Perkins, 97, 113, 120
marshes. *See* wetlands
Massachusetts, 106, 136; politicians of, 128
Massachusetts Institute of Technology, 127
Massachusetts Supreme Court, 72
Maxwell, George H., 92, 120, 126–27, 130–31, 133, 157, 162, 170
McCormick, Cyrus, 44, 61, 76
McEathron, W. J., 153
McGill, A. R., 86
McKinley, William, 127
McNeill, J. R., 27
Mead, Elwood, 15, 120, 156–58, 162–63, 167, 170–71
Merritt, E. B., 182
miasma, 4, 6, 17–21, 24, 26–29, 36–40, 46. *See also* disease
Michigan, 50–51, 55, 92, 172, 189
Michigan Geological Survey, 109
Middle Rivers, 82
Midwest: drainage in, 78, 82, 91, 93, 96, 158; farms, 52; organizations in, 35, 51, 80, 152; railroads in, 58; reclamation funds in, 130, 133; tile drainage in, 44; wet prairies in, 23, 34, 37–38, 50, 55–56
Milk River Project (Montana), 128
Miller, N. D., 89
Milton, John, 17
Minneapolis Star Tribune (newspaper), 66, 185
Minnesota, 53, 60, 130, 160, 174, 187; Becker County, 74; Beltrami Swamp, 76–77, 89; Breckenridge, 58, 62; Cass County, 130; Clay County, 54–55, 67, 71, 74, 80, 89; Clay County board of commissioners, 86; drainage, 91, 138, 182–83, 185; Duluth, 69, 88; Fisher, 76, 93; Grant County, 88; Keystone, 70, 81; Kittson County, 66, 79–82, 86, 88, 90; laws and doctrines, 72, 75, 80, 83–85, 150; Marshall County, 79–80, 86, 88, 90; Martin County, 87;

Minneapolis, 56, 59, 61–62, 84; Moland Township, 54–55, 67, 70–71, 74; Moorhead, 70–71; Morken Township, 54–55, 70–71; Norman County, 79–80, 86, 88, 90; Olmstead County, 67; politicians of, 12, 122; Polk County, 64–66, 69–70, 76–77, 79–80, 86, 88, 90; public lands, 129; railroads, 14, 58–59; reclaimed lands, 87, 90, 122, 124, 126, 145, 148; Red Lake County, 90; St. Anthony, 58–59; statehood, 51, 56; St. Cloud, 167; Stillwater, 58; St. Paul, 56, 58, 62, 70–71, 87–88; Thief River Falls, 146; Traverse County, 88; Wilkin County, 88
Minnesota Drainage League, 136
Minnesota Engineers and Surveyors Society, 167
Minnesota Supreme Court, 73–74, 123
Mississippi, 51, 168, 170, 176, 212n22, 214n42; politicians of, 145
Mississippi River, 29, 36, 62, 96, 103, 111, 124, 126, 139, 187
Mississippi River Valley, 8, 38–39, 48, 106
Missouri, 51, 101, 168, 172; politicians of, 7, 135
Missouri Pacific Railroad, 171
Missouri River, 156
Mitchell, Guy Elliot, 131–32, 141
Miwok people, 31
Mondell, Frank W., 133
Montana, 77, 92, 105, 126, 128
Moorhead Weekly News, 65, 67
Morgan, Arthur E., 167–69, 178–79
Mosness, O., 70–72
Mud River (Minnesota), 146
multipurpose conservation, 137–38. *See also* conservation movement
Murray, A. J., 38

Nah-gon-nway-we-dung (Anishinaabe leader), 182, 184, 186
Narragansett Basin, 106
National Academy of Sciences, 60, 105
National Drainage Association (NDA), 142, 170

national drainage congress, 135–36
National Geographic Society, 127
National Irrigation Association (NIA), 92–93, 127
National Irrigation Congress, 12
national parks, 8, 13, 124. *See also specific parks*
Native Americans. *See* Indigenous people
native-borns (Euro-Americans), 11, 95, 97, 114–15, 118–20. *See also* Anglo-Saxons
nativism and nativists, 118
Nebraska, 54, 57, 95, 105; politicians of, 126
Nelson, Knute, 88–89
neo-Hippocratic tradition and new Hippocratism, 19, 21–23, 27, 29, 33. *See also* Hippocratic tradition
neo-Lamarkism, 95, 99–100, 117
Nevada, 128; politicians of, 127, 156
New Deal, 154, 165
Newell, Frederick H., 12, 15, 120, 124–25, 127–33, 138, 143–44, 150, 156–57, 162
New England, 14, 53, 103, 106–8, 111–13, 115, 118–19
New England Farmer (journal), 39, 52–53
New Era Grading Company, 70
New Hampshire, 26
New Jersey, 24, 174
Newlands, Francis G., 127–28, 156
New Mexico, 8, 176
New Orleans, LA, 27, 49
New York, 43–44, 108; New York City, 24, 26; politicians of, 127; State Fair, 47
New York Times, 179
New-York Tribune, 60
Nichols, D. A. A., 39
Noah's flood, 4, 14, 25, 104
North American Review (magazine), 25, 115
North Canadian River, Deep Fork Branch of (Oklahoma), 135
North Carolina, 5, 32, 110, 124, 134, 168, 179; politicians of, 130, 143
North Dakota, 10, 163, 176, 187; boundaries of, 14, 56, 58–60; ditches, 66; drainage, 131, 136, 138; Fargo, 61, 69; Grand Forks, 61, 70, 76, 130; land sales in, 129–30; politicians of, 129; railroads, 77

Northern Pacific Railway (NP), 59–61, 69, 92, 126
Northwest Experiment Farm (University of Minnesota), 164
Norwegian, 53, 67, 76
Notes on the State of Virginia (Jefferson), 25

Oberlin Agricultural and Horticultural Society, 41
Oberlin College, 158
Office of Drainage Investigations (ODI), 11, 15, 125, 151–56, 165, 167–72, 174, 176, 178–81. *See also* US Department of Agriculture (USDA)
Office of Irrigation and Drainage Investigations (OIDI), 158, 161–67, 178. *See also* US Department of Agriculture (USDA)
Office of Irrigation Investigations (OII), 156–58, 167. *See also* US Department of Agriculture (USDA)
Ohio, 49, 51, 55, 57, 101, 172, 187; Cincinnati, 98; Guernsey County, 43; Toledo, 46–47
Ohio Cultivator (journal), 40–43, 46, 48–49
Ohio Geological Survey, 127
Ohio River, 98, 103–4, 208n23, 212n22, 214n42
Ohio River Valley, 31–32, 184
Oklahoma, 135, 139; Oklahoma City, 138, 142, 169–70
Oklahoma City Chamber of Commerce, 135
Oklahoma City drainage congress, 138, 167, 169
Old Northwest, 8, 36
Oneida Nation, 188–89
On the Origin of Species (Darwin), 99
Oregon, 51; Malheur Basin, 163; politicians of, 134
Orsi, Richard, 63
Ostrem, Arne, 71
Otter Trail River (Minnesota), 83, 89

Panic of 1873, 59
Pardee, George, 157
Patwin, 31
Pauw, Cornelius de, 23
Pawley, Emily, 40

Pedersen, Jens, 71
Penrose, Spencer, 109
Philadelphia, PA, 26–27
Philippines, 119
Philosophical Transactions (journal), 20, 25
Pillsbury, C. A., 59
Pinchot, Gifford, 12–13, 127, 184
Pisani, Donald J., 63, 128, 214n41
Poland, 115
Polk County Board of Commissioners (Minnesota), 77
Pollio, Marcus Vitruvius, 20
Powell, John Wesley, 60, 97, 105–6, 109–10, 114, 120, 127
Power, James B., 59–60, 69–70
Prairie Farmer (newspaper), 40, 46
Prentice, Ed (Anishinaabe leader), 182, 184, 186
Prince, Hugh, 81, 91
Progressive Era. *See also* conservation movement
Prunty, Frank, 152
Puget Sound, 59

race, 5, 11, 15, 33, 95, 97–100, 105, 114–20, 132. *See also* Anglo-Saxon; Black people; eugenics; Indigenous people; scientific racism
railroad. *See specific companies*
Rainey, Henry T., 158
Recherches philosophiques sur les Américains (De Pauw), 23
reclamation, 10, 37; definition of, 12; federal, 125, 127–31, 134, 137–40, 146; land, 78, 120, 150, 157, 176, 185–86; movement for, 8, 135–36, 151; purpose and benefit of, 13, 114, 118; responsibility for, 80, 91–92; of swamps and wetlands, 51, 93, 108, 120, 123, 129, 131, 141, 143, 145, 163, 184. *See also* dispossession; drainage; drainage districts; Indigenous lands; irrigation; Reclamation Service; *and specific states*
Reclamation Act (1902), 93, 127–29, 137–39, 144

Reclamation Service, 12, 15, 133, 149–50; authority of, 125, 130; creation of, 123, 126–27, 156–57; leaders of, 124, 127; problems with, 131–32, 139–40, 144–46, 153–54, 169–71, 180; projects, 128–29, 135–37, 141–43, 163, 176
Red Lake, Upper (Minnesota), 183
Red Lake Reservation, 122, 182, 184
Red Lake River (Minnesota), 77, 83, 87
Red River, 56–58, 60, 73, 77, 80, 82, 87; channelization of, 83–84, 86, 93
Red River Board of Audit, 2, 85, 88–89
Red River of the North, valley of. *See* Red River Valley
Red River Valley, 14–15, 54, 56–59, 64, 66, 69, 76–78, 80, 94, 163–64; bonanzas, 60; drainage of, 2–4, 55, 82–83, 87–88, 96, 130, 160; "Red River Boom," 57, 60, 62, 65, 74–75
Red River Valley News, 65
Reis, George C., 64–65
religion. *See* Christians
Republicans, 125, 127–28, 133–35, 143–44. *See also specific politicians*
reservation. *See* Indigenous lands; *and specific reservations*
Revolutionary War. *See* American Revolution
Rhode Island, 8
Richards, W. A., 141
River and Harbors Act (1890), 87
rivers. *See specific rivers*
Robertson, F. M., 152
Robertson, William, 23, 164
Rocky Mountains, 174
Roosevelt, Franklin Delano, 167
Roosevelt, Theodore: administration, 15, 78, 125, 184; land reclamation, 149; Reclamation Service, 127, 131, 214n41; support for conservation, 12, 123, 128, 130, 155
Ross, Ronald, 161
Royal Agricultural College, 39
Royal Society, 20
Ruffin, Edmund, 5–6, 35–37, 40–41, 48–49

rural communities and settlers, 7, 40, 96–97, 119–20, 143, 170
rural improvement, 7, 13, 112, 124, 132, 144–46, 154, 176, 190
rural nostalgia, 132–33, 162; press, 36, 39, 44–45, 47–48, 52–53
rural settlement and landscape, 3, 10, 15, 35, 95, 113, 125–26, 163, 171, 185
Rush, Benjamin, 27–28
Russia, 115

Sacramento River (California), 136, 156
Sacramento River Valley, 31, 136–37
Salt River Project (Arizona), 128
Sampson, Bernard, 66, 69
Sanders, Edgar, 46
Sand Hill River (Minnesota), 76–77, 81–82, 89, 93
San Joaquin River, 156
San Joaquin Valley, 32, 137
San Louis Valley, 168
Santa Fe Railroad Company, 92, 126
Savannah City Council, 30
Science (magazine), 94–96, 108, 111, 115
scientific racism, 3, 11, 114, 118. *See also* eugenics; race
Scotland, 43
Scott, Charles Frederick, 71
Scott, James C., 196n36
Scott, John, 7–8
Scribner's Magazine, 115
Seabrook, Whitemarsh B., 49
Sells, Cato, 183
Seminole Wars, 5, 192n11
Seneca County Agricultural Society, 44
Seneca Lake (New York), 43
settler colonialism, 3, 13, 60, 94, 97, 149, 186. *See also* imperialism
Shaler, Nathaniel Southgate: background, 98–101; and George Perkins Marsh, 101–2; investigations at USGS, 14, 93, 96, 106–10, 114, 120–21, 134, 143, 162, 173–74; issues related to eugenics, race, and immigration, 95, 97, 100, 114–20; publications by, 94, 96; scientific

Shaler, Nathaniel Southgate (*continued*):
beliefs and legacy, 97, 103–5, 111–14, 120, 174, 183, 185
Singleton, S. M., 152
Small, John Humphrey, 130
Smithsonian Institution, 105; Bureau of Ethnology, 106
Smythe, William Ellsworth, 120
Snake River, 70, 82
social Darwinism, 162. *See also* eugenics
Sorsby, N. T., 40
South America, 28
South Carolina, 32, 173; James Island, 172; politicians of, 49, 128, 130, 134, 142, 144
South Dakota, 95, 126, 187
Southern Pacific Railroad, 92, 126
Steenerson, Christopher, 77–79, 93
Steenerson, Elias, 76–78, 81, 93
Steenerson, Halvor, 141, 145–46; and Anishinaabe land cessions, 122–25, 129; and Crookston drainage convention, 1–2, 78; and drainage legislation, 131–33, 138, 148–49, 182; leader of national drainage movement, 1–2, 12, 15
Stewart, John T., 164
St. Francis Drainage District (Arkansas), 172
St. Francis River basin (Arkansas), 168, 171–72
Stiles, Ezra, 25
St. Paul, Minneapolis and Manitoba Railway (SPM&M), 55–57, 62, 65–67, 69–71, 73–83, 86–92. *See also* Great Northern Railway (GN)
St. Paul and Pacific Railroad, 58–62. *See also* Great Northern Railway (GN)
St. Paul Daily Globe, 86
Strabo, 20
surface water: institutional responses, 81, 83, 151; laws and legal issues related to, 8, 71–73, 75–76, 80, 123; problems resulting from, 29, 40, 46–49, 64–65, 67, 70, 164; removal or diversion of, 7, 28, 36–37, 41, 43–44, 50, 53, 66, 84; reports on, 107, 111. *See also* common enemy doctrine; drainage; swamps

swamp-disease nexus, 18, 27–28, 39
Swamp Land Acts (1849, 1850, and 1860): authorization of, 8, 51, 84; impact of, 14, 37, 55–56, 140, 176, 178; problems related to, 63, 75, 85, 88, 96, 125, 142; purpose, 10; responsibility, 91, 137, 141
swamps. *See under* drainage; reclamation: of swamps; wetlands
Sweetwater Project (Wyoming), 128
Switzerland, 60
Syverson, Ole, 71

Taft, William Howard, 180
Tamarac River (Minnesota), 82
Teele, R. P., 142
Teller, Henry M., 144
Tennessee Valley Authority, 167
Texas, 10, 123
Thoburn, J. B., 135
Timber Culture Act (1873), 61
Tirrell, Charles Q., 128, 130
topography: features, 1, 19, 29, 47, 57; surveys, 2, 14–15, 81, 87, 106, 135, 152–53, 160
trans-Appalachia, 38. *See also* Appalachian Mountains
tribes (Native), 123, 183, 188–89, 210n1. *See also* Indigenous people; *specific tribal groups*
Trinity Mountains, 136
Truckee-Carson Project (Nevada), 128
True, A. C., 166
Turner, Frederick Jackson, 126, 206n3
Tyndall, John, 100

Union Pacific Railroad, 59, 92
University of California–Berkeley, 157, 162
University of Heidelberg, 109
University of Illinois, 79, 159–60, 187
University of Minnesota's Northwest Experiment Farm, 164
University of North Carolina, 109
urbanization, 16, 18, 47, 123, 132–33, 141, 146
US Army Corps of Engineers, 87, 125, 142

US Army Corps of Topographical Engineers, 105
US Circuit Court (District of Minnesota), 71
US Department of Agriculture (USDA), 8, 15, 124–25, 140, 155–56, 164–66, 171–72, 174; leaders of, 6, 79, 142–43, 154, 160, 170, 179; drainage programs, 130–31, 134–36, 138, 145, 157, 159, 180; leaders of, 6, 79, 142–43, 154, 160, 170, 179. *See also* Office of Drainage Investigations (ODI); Office of Irrigation and Drainage Investigations (OIDI); Office of Irrigation Investigations (OII)
US Fish and Wildlife Service, 186
US Forest Service, 12, 184
US Geological Survey (USGS), 8, 109–10, 123, 131–32, 140–41, 156; investigations of, 10, 52, 119, 134–35, 146; leaders of, 14, 60, 93–94, 105–6, 120–21, 123, 127, 129, 145; proposals of, 11; reports of, 107–8, 110–11, 114–15, 137, 162, 173. *See also* Atlantic Coast Division
US Patent Office, 8, 155
US Supreme Court, 180, 186; *Kansas v. Colorado* (1907), 138, 140, 143–44; *United States v. Rio Grande Irrigation Company* (1899), 139

Valentine, Ezra G., 86–90
Van Breda, Jacob, 28
Varro, Marcus Terentius, 19
Verdigris River (Kansas), 151
Vermillion River (Illinois), 159
Virginia, 5, 32, 35, 111, 115, 134. *See also* Great Dismal Swamp
Volstead, Andrew, 125, 148–49, 182
Volstead Act (1908), 149–50

Walcott, Charles D., 123–24, 129, 131, 136, 145
Washburn, William, 59
Washington (state), 126, 130, 139, 162, 184
Washington, DC, 9, 86, 93, 123, 127, 133, 150, 153, 165, 180, 182

Washington Post, 130, 140
Washita River (Oklahoma), 135, 171
Webster, Noah, 27
Western Rural (journal), 39
West Point, 98
wetlands: categorization and surveys of, 10, 97, 107, 114, 134, 142, 172–74; conversion and drainage of, 15–16, 34, 45, 49, 52, 56, 114, 126, 129; destruction and cession of, 37, 55, 87, 180; historical views of, 17, 72, 113; history of 8, 91; impact of, 29, 48, 57, 84, 93, 137, 161–62; Indigenous relationships with, 14, 30–33, 184–85, 188–90; resources and functions of, 18, 21, 40, 101, 103–4, 110, 186–87. *See also* drainage; reclamation: of swamps; reclamation: of wetlands
wet prairies, 8, 15, 57, 61–62, 64, 69, 75, 130, 172, 185–87. *See also* Midwest: wet prairies in; wetlands
Wheeler, George, 105
White Earth Reservation, 122
whiteness. *See also* eugenics; race
Whitney, Milton, 165–66
Whyte, Kyle, 13
Williams, John Sharp, 145
Williamson, Hugh, 24–25
Wilson, James, 166, 172, 178–79
Wilson County Citizen (newspaper), 153
Wisconsin, 51, 55–57, 92, 101, 174, 187–89
Wittfogel, Karl, 193n23
Woolverton, E., 39
Works, John D., 157–68
World War I, 3, 9, 52, 145, 185
Worster, Donald, 103, 193n23
Wright, J. O., 171, 174, 176, 178–79, 181
Wukchumni Yokuts, 32
Wyoming, 105, 126, 128; Greybull Valley, 162; leaders of, 133, 156

Yale University, 25
Yarnell, D. L., 153
yellow fever, 4–5, 26–28, 30. *See also* disease
Yosemite National Park, 146

www.ingramcontent.com/pod-product-compliance
Lightning Source LLC
Chambersburg PA
CBHW030734250426
43671CB00035B/357